METHODS IN CELL BIOLOGY

VOLUME XI

Yeast Cells

Contributors to This Volume

T. Biliński

M. Brendel

B. L. A. Carter

Michael S. Esposito

Rochelle E. Esposito

W. W. Fäth

A. Fiechter

G. R. Fink

Seymour Fogel

W. Gajewski

Helen Greer

James E. Haber

Harlyn O. Halvorson

D. C. Hawthorne

Friedrich Kopp

S.-C. Kuo

F. Lacroute

W. Laskowski

Barbara Shaffer Littlewood

J. Litwińska

J. M. Mitchison

Juan-R. Mor

R. K. Mortimer

P. Munz

John R. Pringle

David N. Radin

C. F. Robinow

Elissa P. Sena

Fred Sherman

Juliet Welch

Reed B. Wickner

S. Yamamoto

J. Żuk

Methods in Cell Biology

Edited by

DAVID M. PRESCOTT

DEPARTMENT OF MOLECULAR, CELLULAR AND
DEVELOPMENTAL BIOLOGY
UNIVERSITY OF COLORADO
BOULDER, COLORADO

VOLUME XI

Yeast Cells

1975

ACADEMIC PRESS • New York San Francisco London
A Subsidiary of Harcourt Brace Jovanovich, Publishers

ACADEMIC PRESS, INC.
111 Fifth Avenue, New York, New York 10003

United Kingdom Edition published by
ACADEMIC PRESS, INC. (LONDON) LTD.
24/28 Oval Road, London NW1

LIBRARY OF CONGRESS CATALOG CARD NUMBER: 64-14220

ISBN 0-12-564111-7

PRINTED IN THE UNITED STATES OF AMERICA

CONTENTS

LIST OF CONTRIBUTORS

Numbers in parentheses indicate the pages on which the authors' contributions begin.

T. BILIŃSKI, Institute of Biochemistry and Biophysics, Polish Academy of Sciences, Warsaw, Poland (89)

M. BRENDEL, Fachbereich Biologie der J. W. Goethe-Universität, Frankfurt am Main, Germany (287)

B. L. A. CARTER, Department of Zoology, University of Edinburgh, Edinburgh, Scotland (201)

MICHAEL S. ESPOSITO, Erman Biology Center, Department of Biology, University of Chicago, Chicago, Illinois (303)

ROCHELLE E. ESPOSITO, Erman Biology Center, Department of Biology, University of Chicago, Chicago, Illinois (303)

W. W. FÄTH, Fachbereich Biologie der J. W. Goethe-Universität, Frankfurt am Main, Germany (287)

A. FIECHTER, Institute of Microbiology, Swiss Federal Institute of Technology, Zürich, Switzerland (97)

G. R. FINK, Department of Genetics, Development and Physiology, Cornell University, Ithaca, New York (247)

SEYMOUR FOGEL, Genetics Department, University of California, Berkeley, California (71)

W. GAJEWSKI, Institute of Biochemistry and Biophysics, Polish Academy of Sciences, Warsaw, Poland (89)

HELEN GREER, Department of Genetics, Development and Physiology, Cornell University, Ithaca, New York (247)

JAMES E. HABER, Department of Biology and Rosenstiel Basic Medical Sciences Research Center, Brandeis University, Waltham, Massachusetts (45)

HARLYN O. HALVORSON, Department of Biology and Rosenstiel Basic Medical Sciences Research Center, Brandeis University, Waltham, Massachusetts (45)

D. C. HAWTHORNE, Department of Genetics, University of Washington, Seattle, Washington (221)

FRIEDRICH KOPP, Institute for Cell Biology, Swiss Federal Institute of Technology, Zürich, Switzerland (23)

S.-C. KUO,[1] Waksman Institute of Microbiology, Rutgers University, The State University of New Jersey, New Brunswick, New Jersey (169)

F. LACROUTE, Institut de Biologie Moléculaire et Cellulaire, Strasbourg, France (235)

W. LASKOWSKI, Zentralinstitut 5 der Freien Universität, Berlin, Germany (287)

BARBARA SHAFFER LITTLEWOOD, Department of Physiological Chemistry, School of Medicine, University of Wisconsin, Madison, Wisconsin (273)

J. LITWIŃSKA, Institute of Biochemistry and Biophysics, Polish Academy of Sciences, Warsaw, Poland (89)

J. M. MITCHISON, Department of Zoology, University of Edinburgh, Edinburgh, Scotland (201)

JUAN-R. MOR, Institute of Microbiology, Swiss Federal Institute of Technology, Zürich, Switzerland (131)

R. K. MORTIMER, Division of Medical Physics, Donner Laboratory, University of California, Berkeley, California (221)

P. MUNZ, Institute of General Microbiology, University of Bern, Bern, Switzerland (185)

[1] *Present address:* Department of Biological Research, Lederle Laboratories, Pearl River, New York.

JOHN R. PRINGLE, Institute of Microbiology, Swiss Federal Institute of Technology, Zürich, Switzerland (131)

DAVID N. RADIN, Genetics Department, University of California, Berkeley, California (71)

C. F. ROBINOW, Department of Bacteriology and Immunology, University of Western Ontario, London, Ontario, Canada (1)

ELISSA P. SENA, Genetics Department, University of California, Berkeley, California (71)

FRED SHERMAN, Department of Radiation Biology and Biophysics, University of Rochester School of Medicine and Dentistry, Rochester, New York (189)

JULIET WELCH, Genetics Department, University of California, Berkeley, California (71)

REED B. WICKNER, Laboratory of Biochemical Pharmacology, National Institute of Arthritis, Metabolism, and Digestive Diseases, National Institutes of Health, Bethesda, Maryland (295)

S. YAMAMOTO,[2] Waksman Institute of Microbiology, Rutgers University, The State University of New Jersey, New Brunswick, New Jersey (169)

J. ŻUK, Institute of Biochemistry and Biophysics, Polish Academy of Sciences, Warsaw, Poland (89)

[2] *Permanent address:* Department of Agricultural and Biological Chemistry, Kochi University, Nankoku, Kochi, Japan.

PREFACE

In the absence of firsthand personal instruction, researchers are often reluctant to adopt new techniques. This hesitancy probably stems chiefly from the fact that descriptions in the literature do not contain sufficient detail concerning methodology; in addition, the information given may not be sufficient to estimate the difficulties or practicality of the technique or to judge whether the method can actually provide a suitable solution to the problem under consideration. The presentations in this volume are designed to overcome these drawbacks. They are comprehensive to the extent that they may serve not only as a practical introduction to experimental procedures but also to provide, to some extent, an evaluation of the limitations, potentialities, and current applications of the methods. Only those theoretical considerations needed for proper use of the method are included. Special emphasis has been placed on inclusion of much reference material in order to guide readers to early and current pertinent literature.

Volume XI of *Methods in Cell Biology* continues with the presentation of techniques and methods in cell research that have not been published or have been published in sources that are not readily available. Much of the information on experimental techniques in modern cell biology is scattered in a fragmentary fashion throughout the research literature. In addition, the general practice of condensing to the most abbreviated form materials and methods sections of journal articles has led to descriptions that are frequently inadequate guides to techniques.

This volume is devoted to research methods that have been developed with yeast cells. Yeast cells are becoming increasingly useful in the analyses of problems in cell biology, and the aim here is to bring together into one compilation complete and detailed treatments of various research procedures used with this cell type. Originally, we had planned to cover yeast methods in a single volume, but the subject has proven to be too large. Coverage of yeast methods will therefore continue in Volume XII of this series.

DAVID M. PRESCOTT

Chapter 1

The Preparation of Yeasts
for Light Microscopy

C. F. ROBINOW[1]

Department of Bacteriology and Immunology,
University of Western Ontario,
London, Ontario, Canada

How yeasts are prepared for light microscopy depends on what the investigator wants to know about them but, regardless of the nature of the question asked, a sample of yeasts is morphologically most revealing when it is uniform and when the yeasts are immobilized in one plane of focus, either in a medium suitable for growth or flat on glass, well-preserved, and selectively stained. The preparations should be sufficiently well illuminated to permit examination and photography at magnifications at which important details are easily visible. This chapter describes ways in which these conditions may be met.

[1]Supported by a grant from the Medical Research Council.

1

I.   Light Microscopy of Living yeasts

A.   Ordinary Microscopy of Living Yeasts

1.   SAMPLING OF ISOLATED COLONIES OF CONFLUENT GROWTH ON AGAR MEDIA OR OF CELLS GROWN IN LIQUIDS

a. Hanging Agar Block.   A wire loop, about 1 mm in diameter, is used to spread a small volume of the cells to be sampled across the surface of the agar surrounding the colony, or over fresh agar medium in another dish. The latter course must be followed with samples of confluent growth and with the deposit from a centrifuged pilot sample from a liquid culture.

A small slab of agar covered with yeasts is cut out and lifted from the culture dish with a fine scalpel, or a flexible blade fashioned from flattened resistance wire, and is placed, cells down, against a cover slip which is next, agar down, placed over a hole cut out of a plastic slide having a thickness slightly greater than the depth of the agar medium. A *small* drop of water applied with the tip of a finger to the edge of the cover slip provides sufficiently firm adhesion of slip to slide to permit examination with an oil-immersion objective. No other seal is required. Immobilized between cover slip and agar, in the same plane of focus and well separated from each other, yeasts are less refractile, hence reveal more of their interior than if they are clumped together and adrift in the relatively large volume of water of the usual wet mount. Admittedly, even wet mounts are useful if they are suitably illuminated, as proved by the handsome views of *Nadsonia* cells obtained by Havelkova (1973) using Nomarski optics.

b. Slide Cultures (Plate I, Fig. 1).   Hanging blocks merely provide general information on what one is dealing with. A thinner assembly is required for purposes of photography. Many forms have been described. A simply and quickly prepared reliable slide culture is one derived from a technique used by Fortner (1930). A microslide is immersed in 20 ml of molten agar medium in a petri dish, lifted out with forceps, held horizontally, wiped clean on one side, and placed in an empty dish. A second coat may be applied about 1 minute later, but a single one usually suffices. Yeasts are placed on the agar along a few crisscrossing streaks made with a thin, flexible glass fiber obtained by heating the middle of a length of glass rod to the melting point over a blue bunsen flame and pulling it out rapidly from both ends. Several such fibers may be stored in a wide test tube. Yeasts do not stick well to dry glass. To avoid disappointments inoculate in the following manner. A loopful of the starter culture, which may be an overnight slant or a young isolated colony, is spread on fresh agar over an area of about 1 × 4 cm. A slab is cut out of the inoculated field and placed close to the dish holding the freshly prepared thin film of agar. A loopful of water is added to the cells

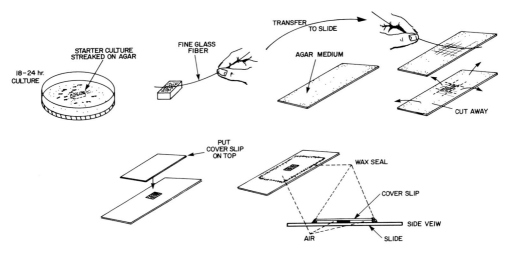

PLATE I. FIG. 1. Steps in the preparation of thin agar slide cultures. See text for explanation of details.

on the cutout and mixed with them. The glass fiber readily picks up cells from this slurry. Cross-streaking of the agar-coated slide achieves good separation, important for time-lapse photography of growing and dividing cells. The agar on the slide is next trimmed down to a narrow rectangle measuring about 2 × 4 mm with the help of a scalpel, flattened wire, or cataract knife. These knives, which have many uses in microculture and cytological work are expensive but, unlike the surgeon, the cytologist can continue using them long after they have lost their original superlative sharpness. The cropped agar is covered with a no. 1 or no. 0 glass slip. Wax from a small birthday candle, the wick serving as brush, provides an adequate seal. Finished cultures are stored in petri dishes together with a wad of wet paper tissue to guard against desiccation. For studies of growth and multiplication, one selects isolated cells near the edge of the agar. A more elaborate kind of thin-culture chamber suitable for time-lapse photography at high magnification (Coles and Kendrick, 1968) has been used successfully by Poon and Day (1974) in a study of mitosis in the yeastlike sporidial phase of *Ustilago violacea*.

Apart from vacuoles and, in large cells, a faint image of the nucleus (Plate II, Fig. 2), there is little to be seen in living yeasts growing between glass and agar. Fortner cultures are useful in studying asci. The spores in asci of *Saccharomyces* are more easily counted immobilized between agar and a cover slip than when they are suspended in water. In such cultures the germination of ascospores may be followed at high magnification (Plate II, Figs. 8 and 9). Watching this process over several hours conveys better

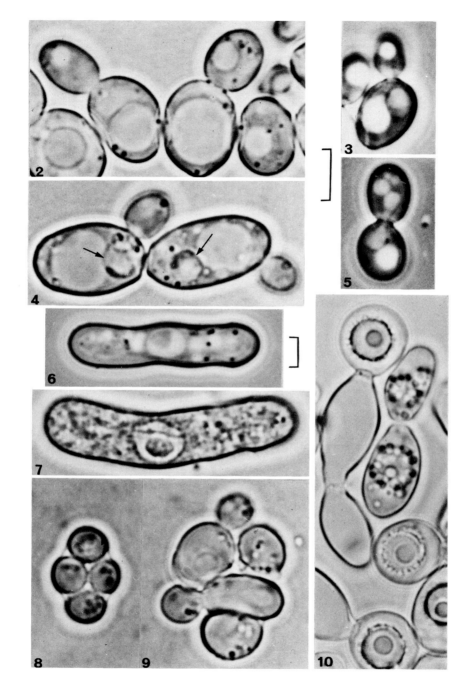

than many a published diagram the great increase in volume that accompanies spore germination.

c. Cytological Variants of Fortner Slide Cultures. Uses of Acetic Acid and Formalin. Lifelike habit pictures may be obtained from specimens prepared as just described but with 2.5% Formalin (40% solution of formaldehyde) in the agar. Fixation enhances the normally weak contrasts between the cytoplasm, vacuole, and cell wall (Plate II, Fig. 10). Moreover, a demonstration slide of this kind does not change during the course of a teaching session.

The nucleus of baker's yeast is visible in the bright-field microscope as a sharply drawn clear vesicle with a narrow peripheral crescent of dense (i.e., shrunken) nucleolar material when yeasts are examined in Fortner slides with 0.5–1.0% acetic acid in the agar (Plate II, Fig. 4), (Robinow and Marak, 1966, and earlier authors). The visibility of the nuclei of living, unstained, unfixed fission yeasts is equally enhanced by this rather crude method (Plate II, Fig. 7).

B.   Phase-Contrast Microscopy of Living Yeasts

As Barer *et al.* (1953), Mueller (1956), Mason and Powelson (1956), and Girbardt (1960) showed for animal cells, yeasts, bacteria, and fungal

---

PLATE II.   Slide cultures. All except Fig. 10 are of *S. cerevisiae.* All but Figs. 3, 5, and 6 enlarged ×2700. Scales denote 5 μm.

FIG. 2.   Living cells of Hawthorne's tetraploid no. 1376 in slide culture of agar medium. Ordinary microscopy. One sees lipid granules and vacuoles. In cell left of center, the nucleus is faintly visible between the vacuole and bud.

FIG. 3.   Phase-contrast microscopy. Lallemand brand of baking yeast growing in 21% gelatin medium. Nucleus between bright vacuole and bud. Its lower pole is occupied by a relatively dense *nucleolar* crescent. ×1650.

FIG. 4.   Same strain and preparation as in Fig. 2, with 0.5% acetic acid in the agar. Nuclei (arrows) with dense nucleolar crescents are now visible in both cells. Note cell wall in focus around buds.

FIG. 5.   Same material and same technique as in Fig. 3; Nucleus, a pale "bow-tie" shape, about to be divided between mother cell below and daughter. ×1650.

FIG. 6.   Living cell of *Schizosaccharomyces japonicus* var. *versatilis* in 21% gelatin medium. One sees lipid droplets, vacuoles, and nucleus with large nucleolus. ×1650.

FIG. 7.   *Schizosaccharomyces japonicus* var. *versatilis* on 0.5% acetic agar. The nucleus contains a nucleolus and a spindle. For electron micrographs of this organelle in another fission yeast, *S. pombe*, see McCully and Robinow (1971).

FIG. 8.   Living ascus of Lallemand yeast on agar medium. Note tautly stretched wall.

FIG. 9.   The ascus of Fig. 8 photographed again 7 hours later. Germling cells of several neighbouring asci had fused in pairs.

FIG. 10.   *Nadsonia elongate* on agar containing 5% Formalin.

hyphae, respectively, rewarding phase-contrast microscopy requires close attention to the level of refractility of the culture medium. Phase-contrast microscopy of yeasts in water or agar media is unrewarding, except as an aid in counting or when checking states of growth (Wiemken *et al.*, 1970). But vacuoles, nuclei and, in some species, even the mitochondria stand out clearly from the cytoplasm of yeasts growing in thin films of nutrient media containing 21–25% gelatin.

A microculture for phase-contrast microscopy at high magnification in which yeasts may be grown in a thin film of gelatin medium surrounded by an ample sealed volume of air has been described by Robinow and Marak (1966). It has consistently given good results and is set up as follows. A small loopful of molten 21–25% gelatin medium containing 0.5–1% Difco yeast extract and 2% glucose is solidified by placing it in the center of a slide which rests on the lid of a petri dish filled to the brim with ice-cold water. The tiny mound of congealed medium is inoculated with the help of a fine, flexible glass fiber charged with cells from young growth on agar, first stirred into a slurry with a little water (see Section I,A,b). A cover glass is placed over the inoculated gelatin, slightly flattening it at the point of contact. The slide with the gelatin and its attached cover slip is next pressed against the lid of a second dish filled to overflowing with warm water (about 60 °C). As soon as the gelatin begins to spread beneath the cover glass, the slide is returned to the cold dish to arrest further expansion. The assembly is sealed with wax from a small birthday candle. One looks for cells close to the edge of the much flattened drop. Contrasts are optimal about 1 hour after the start of the culture (Plate II, Figs. 3, 5, and 6).

Closely spaced phase-contrast sequences of the growth and division of cells of budding yeasts have been published elsewhere (Robinow and Marak, 1966; Matile *et al.*, 1969). A third published time-lapse phase-contrast division sequence, that of the fission yeast *Schizosaccharomyces japonicus* var. *versatilis*, is notable for its clear view of the elongating spindle inside the nucleus (Robinow and Bakerspigel, 1965).

During mitosis, as is now well known, the nucleolar and the chromatinic (chromosomal) components of yeast nuclei are divided together within the intact nuclear envelope and distributed to daughter nuclei by the elongation and constriction of the mother nucleus. *Saccharomyces cerevisiae* is not a species in which this process is easily followed in life, even with phase-contrast microscopy. In this yeast the low density of the chromatinic portion of dividing nuclei stands out clearly enough from the cytoplasm (Plate II, Fig. 5), but the behavior of the associated nucleolar material is less obvious. The nucleolar component remains distinct, however, in dividing nuclei of fission yeasts and in those of the large budding yeast *Wickerhamia fluorescens*. The use of polyvinylpyrrolidone, instead of gelatin, to increase the refractility of the growth medium (Schaechter *et al.*, 1962) seems worth exploring, since it might permit the phase-contrast microscopy of yeasts growing at temperatures above the melting point of gelatin.

C. Direct Staining of Unfixed Yeasts

1. CELL WALL, BUD SCARS, AND BIRTH SCARS

Fluorescence microscopy of budding yeasts stained with primulin (Streiblová and Beran, 1963) has proved of great value in studies of the behavior of the cell wall during growth and division of budding, as well as of fission yeasts. Excellent illustrations and detailed discussions of what can be learned by this technique can be found in reviews by Beran (1968) Streiblová *et al.* (1966), and Streiblová and Wolf (1972). Calcofluor, a fluorescent brightener used to detect areas of wall differentiation in fungi (Gull and Trinci, 1974), recently also proved its worth in a study of cell division in the fission yeast *Schizosaccharomyces pombe* (Johnson *et al.*, 1974).

2. MITOCHONDRIA

A variety of histochemical methods for staining the mitochondria of both living and fixed *Saccharomyces* has been elaborated by Reiss (1969). There is no reason to doubt the validity of his detailed protocols, but he has been poorly served by the printers of his illustrations. Mitochondria are best studied in electron micrographs of sections but, to disgress briefly from the subject of this section, the mitochondria of fission yeasts, small as they are, are still just visible in the phase-contrast microscope, and long, snaky mitochondria are readily seen during life in the relatively large budding aseosporous yeast *W. fluorescens* (Plate III, Fig. 15), where they are also sometimes preserved by Helly fixation and can then be made visible with acid fuchsin (Plate III, Fig. 16a and b).

3. LIPID

Droplets of lipid in yeast cells are readily colored by 0.5% Sudan black B in ethylene glycol, prepared a day in advance and filtered before use. Yeasts are either directly stirred into the Sudan solution or first picked up on eggwhite-coated cover slips which are then mounted over a drop of the colorant. Growing, budding yeasts, as is well known, contain varying numbers of small droplets of lipid (Plate III, Fig. 12). More numerous and larger droplets are found in yeasts undergoing meiosis (Plate III, Figs. 13 and 14). At this stage of the life cycle, shape, size, and arrangement of the droplets of lipid combine to endow them with a fortuitous resemblance to chromosomes (of higher eukaryotes). However, the evidence from many

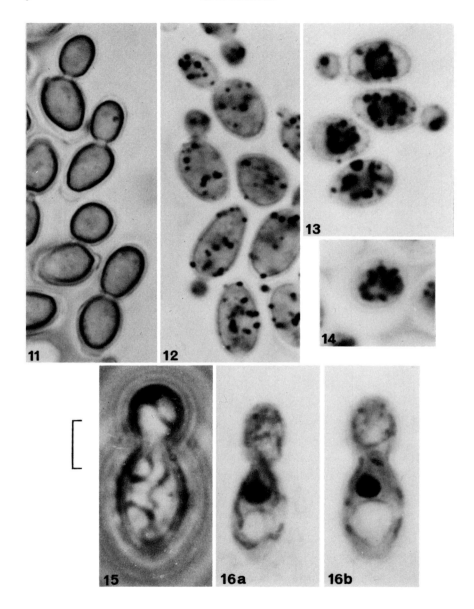

independently made mutually consistent photomicrographs of yeast nuclei in meiosis specifically stained for chromatin (see Section D, 2, b, ii) disposes of this notion, as does also the successive staining of the same group of cells for chromatin and lipid (C. F. Robinow, unpublished observation).

II.   Fixed and Stained Preparations

Microscopy and photomicrography of stained specimens requires that the yeasts be made to *adhere to cover slips*, not only for ease of handling but also in order that the cells may be brought closer to the objective lens of the microscope than is feasible with cells attached to slides (unless the cells are first dried on the slide and then covered with immersion oil, a procedure not here contemplated).

Although attachment to glass is *followed* by fixation, it is convenient to list fixatives first and then consider strategies for making yeasts adhere to cover slips.

A.   Fixatives

1. *Modified Helly's. Stock solution.* Mercuric chloride, 5 gm; potassium dichromate, 3 gm; distilled water, 100 ml. Immediately before use add Formalin to make 5%. Fix for 10–20 minutes, transfer, and wash with and store in 70% ethanol. If fixed preparations are not to be processed within a day or two, it is advisable to keep them in the cold in the preservative of Newcomer (1953).

PLATE III.   *Saccharomyces cerevisiae.* All except Fig. 14 magnified ×2700. Scale denotes 5 μm.

FIG. 11.   Intact cells fixed with osmium tetroxide vapor, postfixed with saturated solution of mercuric chloride in water. The surface of the protoplast faintly tinted with Victoria blue 4R. Curvature of the cells exaggerates depth of stain. From Robinow and Murray (1953).

FIG. 12.   Growing cells of Lallemand baking yeast photographed in a solution of Sudan black B in ethylene glycol. Note that some of the droplets of lipid seem to lie within the cell wall.

FIGS. 13 and 14.   Same yeast as in Fig. 12 at meiosis. Lipid colored with Sudan black B. Cell wall and surface of protoplast sharply outlined. Note deceptive resemblance to chromosomes, especially in Fig. 14. Fig. 14: ×3600.

FIG. 15.   Mitochondria in living cell of *W. fluorescens* growing in a gelatin medium.

FIG. 16.   Mitochondria seen at two levels of focus of a Helly-fixed cell of *Wickerhamia* stained with acid fuchsin. The short, cigar-shaped object some distance above the deeply stained nucleolus in 16b is a short intranuclear *spindle*. For another view of this organelle see Fig. 21.

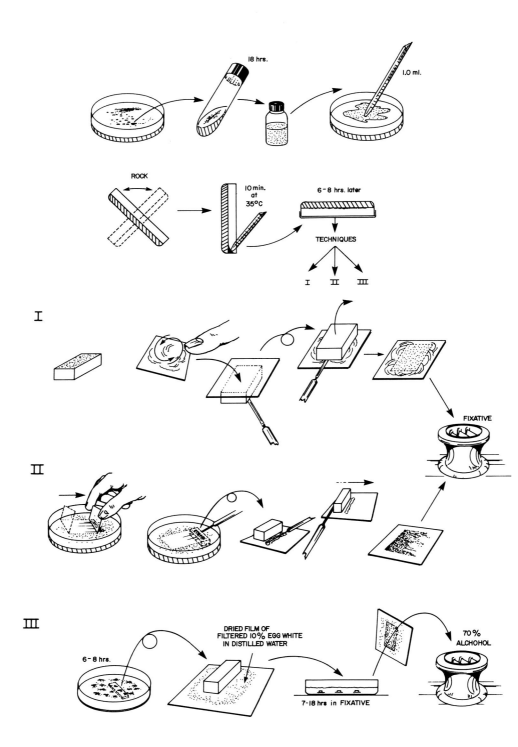

18 hrs.

1.0 ml.

ROCK

10 min.
at
35°C

6-8 hrs. later

TECHNIQUES

I   II   III

I

FIXATIVE

II

III

6-8 hrs.

DRIED FILM OF
FILTERED 10% EGG WHITE
IN DISTILLED WATER

7-18 hrs. in FIXATIVE

70%
ALCHOHOL

2. *Schaudinn's. Half strength.* Absolute ethanol, 1 part; saturated solution of mercuric chloride, 2 parts; water, 3 parts. Mix, and add glacial acetic acid to make 2%. Fix for 10 minutes. Transfer to, wash with, and store in 70% ethanol.

3. *AFA*. Ethanol (95%), 50 ml; glacial acetic acid, 5 ml; Formalin, 5 ml; water, 40 ml. Fixation of impressions on eggwhite (see Section B,1) for 45 minutes has proved satisfactory, but the transfer of adequate numbers of yeasts to glass by fixation with this reagent *through* the growth medium may take overnight. An AFA mixture with more ethanol and less acetic acid (70% ethanol, 90 ml; acetic acid, 3 ml; Formalin, 5 ml) is recommended by Lindegren *et al.* (1956). It has provided instructive preparations of *Saccharomyces* in meiosis.

B. Techniques for Sampling Yeasts Growing on Agar Media

Different species differ considerably in the ease with which they can be detached from the surface of an agar medium and made to adhere to a glass cover slip (see Plate IV, top). One or the other of the following strategies will resolve most of the difficulties likely to be encountered.

1. METHOD I (Plate IV)

This method is suitable for budding growth of *Saccharomyces*. To ensure uniformity of states of growth, it is advisable to start a petri dish culture with cells from an 18- to 24-hour slant culture previously seeded with cells from a single, well-isolated colony. Growth from the slant is dispersed in a few milliliters of tap water (sterility need not be observed in experiments lasting but a few hours) to give a "gray" suspension of about 10^8 cells/per ml. Agar medium in a standard petri dish (9 cm in diameter) is flooded with 1 ml of this suspension. The medium ought to be clear, to allow scanning of the growth under the microscope. A volume of 15–20 ml per dish should not be exceeded. The inoculated dish is tilted in various directions to ensure even distribution of the liquid inoculum and is then held upright. Excess inoculum is removed with the pipette that delivered it and with pieces of blotting paper. Following 5- to 10-minute drying (and further draining) of the upright open dish in a 35 °C incubator, the closed dish is incubated for 6–8 hours or held at bench temperature until a lawn of contiguous but not overlapping microcolonies has developed. Oblong slabs measuring 6×12 mm

PLATE IV.   FIG. 17.   Three ways of transferring agar-grown yeasts to cover glasses. See text for explanation of details.

are cut, lifted out of the petri dish, and placed, cells up, on the bench or a glass plate. A cover slip bearing a thin, moist film of fresh eggwhite is next lowered onto the growth. A scalpel slipped underneath the agar help to turn the assembly upside down. With the cover slip held firmly against the bench, the slab of agar is flicked off by one swift upward movement of the tip of a flattened wire or the aforementioned cataract knife which has been inserted between agar and cover slip. The wet film of cells left behind on the cover glass is instantly plunged into fixative standing ready in Columbia staining jars (Bowen, 1929). These are Coplin-style, slotted vessels of about 12-ml capacity designed to hold four cover slips (eight, back to back). Indispensable in cytological work with bacteria, yeasts, and mycelial fungi, they are manufactured by the Arthur H. Thomas Co., Philadelphia, Pa.

2. METHOD II (Plate IV)

Yeasts that tend to adhere to the agar and leave too few cells behind on the cover slip when impressions are attempted may be dealt with in the following way. A dish of agar medium is inoculated as described above, but the cells of the single-layer, nearby confluent growth are not *directly* transferred to an eggwhite-coated glass slip. Instead, using the edge of a cover slip, some of the growth is first pushed off the surface of the agar and piled up into a shallow ridge. This, together with the underlying agar, is then cut out and placed, cells *down*, on a fresh cover slip near its left edge (assuming you are right-handed). A thin streak of eggwhite is drawn along the cutout with the help of a small wire loop. The agar is then pushed across the streak of eggwhite and over the rest of the coverslip with a single, swift movement of the side of the blade of a small scalpel. The thin, mixed film of eggwhite and yeasts left on the glass is instantly immersed in fixative. (*Note*: Time for adequate growth of *Saccharomyces*, fixation, extraction, staining, and photography may be found in a single working day of usual length if cultures are started a day ahead, late in the afternoon, and are kept overnight at 15 °C. Return of the cultures to room temperature or to 35 °C early the next morning provides, 3–5 hours later, a lawn of budding and dividing cells dense enough to give useful imprints on albumen-coated cover slips.)

For *fission yeasts* which are both relatively slow growing and hard to detach from agar, a variant of the last-described technique is recommended. A streak of a heavy inoculum taken from a young, growing culture is laid across a peripheral segment of agar medium in a petri dish and incubated for about 15 hours. It is then spread over the remaining agar surface with the help of a bent glass rod moistened with a few drops of water. A single layer of closely spaced growing cells is thus obtained, which after several hours of further incubation is harvested by the "scrape–cut out–push over egg-white" procedure described in the preceding paragraph.

3.   METHOD III (Plate IV)

This method involves direct transfer to glass by fixation through the agar. Simple but time-consuming, this procedure, when used with AFA provides more instructive views of the chromosomes of the nuclei of growing, budding yeasts than other methods in common use. Thinly poured agar medium (10 ml per standard petri dish) is relatively sparsely inoculated in the manner of method I. Narrow strips are cut from the agar as soon as a lawn of closely spaced, small, single-layer microcolonies has been obtained. The cutouts, thin flexible mats of agar, are placed, cells *down*, on cover slips bearing a thin, *dried* film of 10% albumen obtained by adding fresh eggwhite to distilled water in a measuring cylinder and filtering the supernatant after the insoluble globulins have sunk to the bottom. The small volume of filtrate carried by a wire loop of 1 mm diameter suffices for a film covering the greater part of the surface of a 22-mm square cover slip. The albumen-coated glass slips carrying agar cutouts are next immersed in AFA fixative in a petri dish. It is important to prepare the fixative at least 1 hour before it is needed, because when first prepared the mixture becomes about 5° warmer than the temperature of the room. When used immediately, it causes air bubbles to arise in the agar, which tend to float the cutouts off their cover slips prematurely. Fixation for six hours or longer ensures the complete transfer of the growth from agar to glass. At the end of this time the agar is flicked off with the light touch of a scalpel; the cover slips, now bearing a single layer of well-spaced, fixed yeasts, are washed with and stored in 70% ethanol or, for longer periods, in Newcomer's (1953) preservative.

C.   Handling of Yeasts Growing in Liquid Media

Cells cultured in liquids can be transferred to glass by either one of two procedures. The first, which has given excellent results in the hands of Pontefract and Miller (1962), is described by its authors as follows: "A loopful of cell suspension (centrifugation recommended, CR.) was first mixed on a grease-free coverglass with a drop of Haupt's adhesive prepared by Johannsen's formula (1940) except that 3% formalin was added as preservative. Several minutes were allowed for the cells to settle to the bottom of the drop and the cover glass was then lowered carefully into a petri dish containing either a modified Carnoy fixative comprising three parts absolute ethanol and one part glacial acetic acid or Helly's fixative prepared by the formula of Gurr (1957). The cells were fixed for 10 min. with the former of for 20 min. with the latter." Haupt's adhesive is prepared as follows: 1 gm of pure, finely divided gelatin is dissolved in 100 ml of distilled water at 30°C; 2 gm of phenol crystals and 15 ml of glycerin are added.

Alternatively, fixation of cells growing in liquids, may be carried out in the test tube. The yeasts are spun down, takin up in fixative, and washed by centrifugation in three changes of 70% ethanol. The sample is spun down once more, and the deposit, now usually badly clumped, is taken up in 95% ethanol and broken up as much as possible by working it up and down in a Pasteur pipette. Single drops of suspension, concentrated by sedimentation or by one more round of centrifugation, are dropped from a height of about 3 inches onto the center of a cover slip thinly smeared with fresh eggwhite. The ethanol, spreading rapidly outward from the point of impact, entraps the yeasts in a coagulating film of albumen, thus attaching them firmly to the cover slip. This procedure is a simplified version of a reliable one recommended by Mackinnon and Hawes (1961) for the fixation of protozoa, in which collodion rather than eggwhite serves as adhesive.

The instructions given in Section II,B are concerned with preparations of *intact* yeast cells. Flattened *protoplasts* have recently been recommended for nuclear studies of yeasts by Fischer *et al.* (1973) and Egel and Pentzos-Daponte (1974). They seem most useful for counting meiotic chromosomes (see Section II,D,2,ii).

D.   Staining of Fixed Preparations

Chemically fixed, suitably stained preparations of yeasts can provide information on the plasma membrane, the mitochondria (as mentioned in Section II,C,2), the structure and behavior of the nucleus in growing, dividing cells, and the chromosomes of nuclei in meiosis. This list does of course not exhaust the range of visible cell components, but it includes what is dealt with here.

1.   PLASMA MEMBRANE

The both water- and lipid-soluble dye Victoria blue 4R in 0.05% solution in water stains a thin layer at the surface of the protoplast, just inside the cell wall (Plate III, Fig. 11) if yeasts fixed by osmium tetroxide vapors or by Helly solution are postfixed for 5 minutes with a saturated solution of mercuric chloride. The latter treatment apparently serves to block entry of the dye into the cytoplasm of cells that were normal and intact at the time of fixation. This procedure, first published in 1953 by Robinow and Murray, has recently been found useful in following the initiation of and recovery from cell surface damage sustained by cells of *Wickerhamia* grown for several hours at the unsuitably high temperature of 37 °C (J. W. McDonald, unpublished). A sensitive indicator of a cell's soundness, staining with Victoria blue may conceivably have some uses in experimental work on yeast cell life cycles.

2. STAINING THE NUCLEUS

a. General Remarks.   The size and shape of the yeast nucleus and of the parts within, nucleolus and spindle, are well preserved by Helly's fixative. However, the components of the nucleus can, at present, not all be demonstrated at the same time by a single staining procedure. Two complementary views of yeast nuclei can be produced—both instructive and both incomplete. Lifelike images of yeast nuclei with nucleolus and internal spindle standing out sharply against a faintly and evenly tinted nucleoplasm are provided by brief staining with acid solutions of acid fuchsin. Conversely, after acid hydrolysis and staining with Giemsa, one obtains less naturallooking preparations in which only the chromatinic moiety of the nucleus is properly stained (more or less solidly in some species and in the form of neat chromosomes in others) and in which spindles are lacking and the nuclei enclose nucleoli so faintly stained as to be easily overlooked.

b. Details of Staining Techniques. i. Acid fuchsin.   This technique should be used after Helly or Schaudinn fixation; it fails after AFA fixation. It is suitable for cells from growing cultures of budding and fission yeasts, but is not suitable for *Saccharomyces* in meiosis. Cover slips bearing fixed yeasts are removed from 70% ethanol, rinsed several times with 1% glacial acetic acid in water, and immersed in acid fuchsin in acid solution in a Columbia staining jar. The required concentration of the dye and length of staining vary with the species. They are found by staining pilot specimens and examining them in a small, shallow dish in 1% acetic acid with a *water-immersion objective* of 40 or 70 × power (made by Carl Zeiss as well as E. Leitz). We recommend 1 minute and 40 seconds to 2 full minutes in acid fuchsin 1:40,000 in 1% acetic acid for budding cells of baker's yeast, and 4 minutes for fission yeasts. The stained preparation is rinsed with and mounted over a drop of 1% acetic acid. Excess mountant is removed by applying pieces of blotting paper to the edges of the cover slip. When the latter has settled on the slide and is no longer floating on it, the preparation is gently flattened by hand pressure transmitted through a stack of about a dozen sheets of blotting paper. It is next sealed with wax from a small birthday candle or, better, with Gurr's Glyceel, obtainable from Searle Scientific Services, High Wycombe, Bucks, England, or their North American representatives.

One point needs to be strongly made: Instructive views of acid fuchsin–stained nuclei are seen only in preparations that are optimally illuminated (Plate III, Fig. 16b; Plate V, Figs. 19 and 21). What one is looking for are clear-cut differences in light absorption. Such differences do exist between the various structures in well-stained specimens, but they are more subtle than the "noisy" differences in refractility between cell wall, cytoplasm,

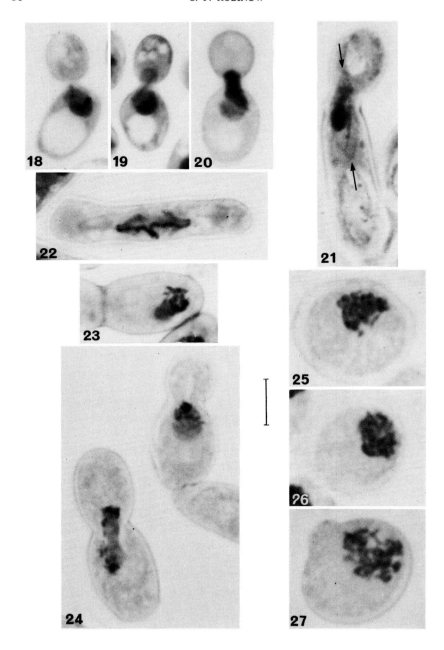

granular inclusions, and vacuole, which are all that is perceived in a wet mount of stained cells examined in the dim light cast by a substage condenser wrongly placed and improperly illuminated. Gentle flattening (*not* squashing) of the stained yeasts helps considerably, and a green light filter greatly enhances the visibility of the delicately stained nuclear structures, but these measures will be of no avail if care is not taken to obtain proper illumination.

    ii. Giemsa.   A reliable schedule is as follows. Impressions on eggwhite on cover slips, fixed by Helly's, half-strength Schaudinn's, or AFA fixative and stored in 70% ethanol are extracted for 1 hour with 1% NaCl at 60°C. This step, recommended by Ganesan and Swaminathan (1958), following Opie and Lavin (1946), enhances the contrast between cytoplasm and chromatin in the finished preparation. The films are next treated for 8–10 minutes with 1 N HCl at the same temperature, rinsed with tap water and

---

PLATE V.   All figures ×2700.   Scale denotes 5 μm.

FIG. 18.   Red Star baking yeast. Helly fixation, acid fuchsin. The nucleus lies between the vacuole and the bud. At its base is the cup-shaped nucleolus. A short spindle is seen within (in reality, traversing) the lightly stained nucleoplasm.

FIG. 19.   Same brand of yeast as in Fig. 18. Helly fixation, acid fuchsin. Nucleus fixed in the process of entering the bud. The nucleolus is still in the form of a thick crescent at the base of the nucleus. The intranuclear spindle is long and straight and extends between short transverse bars, poorly resolved images of the spindle pole bodies seen in electron micrographs.

FIG. 20.   Same brand of yeast as in Fig. 18 in a state of nuclear division comparable to that shown in Fig. 19. Helly fixation, HCl, Giemsa. The faintly stained *nucleolus* (lower pole of nucleus) contains specks of chromatin, but the deeply stained nucleoplasm cannot be resolved into the chromosomes it is known to contain. AFA fixative holds promise of more instructive preparations. (See also Fischer *et al.*, 1973.)

FIG. 21.   *Wickerhamia.* Helly fixation, acid fuchsin. The nucleus (between the arrows) has been fixed in the process of entering the bud. It contains an apparently softened, flowing nucleolus, and a short, straight spindle terminating in distinct knobs representing optically unresolved spindle pole bodies.

FIG. 22.   Anaphase of mitosis in the fission yeast *Schiz. japonicus* var. *versatilis.* Helly fixation, HCl, Giemsa.

FIG. 23.   Chromosomes beside the nucleolus in one of a pair of young sister cells of *Wickerhamia.* AFA fixation through the agar (method III). HCl, Giemsa, squashed after staining.

FIG. 24.   Chromosomes and nucleolus in budding cells of *Wickerhamia.* Prepared in the same way as in Fig. 23. Among the chromosomes of the nucleus at top right is seen the deeply stained "chromatin attachment point" of Robinow (1972).

FIGS. 25 and 26.   Chromosomes at meiosis I in Lallemand baking yeast. Schaudinn fixation, HCl, Giemsa.

FIG. 27.   Chromosomes at meiosis in tetraploid no. 1376. Helly fixation, HCl, Giemsa.

Giemsa buffer [$M/15$ phosphates (pH 6.8), available in tablet form] and left for 1 or 2 hours in Columbia jars with 10 ml of buffer and 18 drops of Gurr's improved Giemsa solution R 66 or other reliable product. Stained films are examined in buffer with a water-immersion lens. Overstaining of cells or of albumen in the background can be corrected by extraction with distilled water to 30 ml of which in a petri dish *lid* one has added one or two loopfuls of acetic acid. The stained film, held in forceps, is repeatedly moved about in the acidulated water for a few seconds at a time, transferred back to buffer in the *bottom* half of a petri dish (sticking to conventions helps avoiding mistakes), and checked for desired contrasts in buffer in a small, shallow dish under the water-immersion lens. The finished specimen is mounted over 1 drop of buffer containing a small amount of Giemsa solution (2 drops/12 ml). If habit preparations are desired in which normal proportions are preserved and the yeasts retain their rounded, three-dimensional shapes, one gently blots the preparation through several layers of filter paper. If it is chromosomes that are looked for, the stained film will have to be vigorously flattened through pressure on the cover slip. There are several ways of doing this. We have found the device designed by Miller and Colaiace (1970) (Plate VI, Fig. 28) well suited for the purpose. The usable area of squashed preparations tends to be somewhat reduced by air-filled spaces encroaching on the remaining film of buffer, but enough

PLATE VI. FIG. 28.   Pressure device copied from Miller and Colaiace (1970). The white card has the dimensions of an ordinary microslide.

of it can be saved by a good seal applied instantly. What is seen in Giemsa preparations varies with the species and in some instances with the method of fixation. Neat chromosomes are always be found in dividing nuclei of *Lipomyces* (Robinow, 1961), *Schiz. pombe* (McCully and Robinow, 1971), and in *Schiz. japonicus* var. *versatilis* (Plate V, Fig. 22). But in baker's yeast and the cytologically similar apiculate sporing yeast *Wickerhamia*, the appearance of the nucleus varies markedly with fixation. HCl–Giemsa preparations of Helly- or Schaudinn-fixed cells of these yeasts, with their brightly stained blobs of chromatin, look neat at first sight but disappoint through lack of detail (Plate V, Fig. 20). The chromatin appears as a more-or-less solidly packed mass of barely resolvable granules, and the nonchromatinic matrix of the nucleolus is often hardly visible even when the small chromosomes that extend into it from the main mass of chromatin are adequately stained as they are in Figs. 1, 4, and 5 of Ganesan and Swaminathan (1958). If such preparations are used as the sole source of knowledge of yeast nuclei, erroneous notions are bound to arise. The nucleolus of *Saccharomyces* becomes visible if it is stained long enough (Plate V, Fig. 20). The nucleus, fixed with AFA and vigorously squashed after staining, can often be resolved into minute, chromosomelike rodlets, visible but, so far, hard to photograph. However, in this way informative preparations, not unlike those of spread protoplasts of *Saccharomyces carlsbergensis* described by Fischer *et al.* (1973), have been obtained of intact cells of *Wickerhamia* fixed directly through the agar (Plate V, Figs. 23 and 24) by Robinow (1972).

iii. Notes on Giemsa staining. It is customary to use 1 N HCl as the hydrolyzing agent before Giemsa (Piekarski, 1937), but Lindegren and his collaborators (Lindegren *et al.*, 1956; McClary *et al.*, 1956; Miller *et al.*, 1963) obtained equally good results with 10% perchloric acid allowed to act at 4°C overnight. It is not known why acid hydrolysis should be required to endow yeast chromatin with an affinity for the Giemsa stain, which it normally appears to lack. However, before accepting hydrolysis as an indispensable step, one would want to know whether yeast nuclei can be stained by the Giemsa method of Piéchaud (1954), by which bacterial "nucleoids" are brightly colored *without* having first been hydrolyzed.

If Giemsa stains nuclei too solidly, then acetoorcein should be tried as an alternative. The stain is prepared by dissolving 1 part of synthetic orcein (Fisher Scientific Company, Fair Lawn, New Jersey) in 100 parts of 60% acetic acid in water and filtering before use. Extraction of fixed cells with 1% NaCl can be dispensed with. They are hydrolyzed in the usual way for about 8 minutes, rinsed with water, soaked for a few minutes in 60% acetic acid, and stained with acetoorcein for 20–30 minutes before being mounted over a drop of the same stain. Preparations often look their best after an interval of 1 or 2 days.

3. MEIOSIS

The chromosomes of *S. cerevisiae* and of related spore-forming yeasts fixed early in meiosis are more distinct than mitotic chromosomes of growing, budding cells. Cells sporulating in liquid media can be fixed after attachment to cover slips by the procedure of Pontefract and Miller (1962) (see Section II,C). For growth from crowded agar plates the "scrape–cut out–push over eggwhite" method (Plate IV) is recommended. The choice of fixative does not seem important. Helly's, Schaudinn's, and AFA all give similar results. Acid hydrolysis followed by Giemsa is the staining procedure of choice. The chromosomes of meiosis I are tightly crammed together inside the intact nucleus. Their arrangement can be loosened to some extent by strong pressure on the cover slip. Many, mutually consistent photographs of constellations of meiotic chromosomes prepared in these or similar ways are to be found in the literature (Lindegren *et al.*, 1956; McClary *et al.*, 1956; Pontefract and Miller, 1962; Miller *et al.*, 1963; Robinow and Bakerspigel, 1965), and further and representative examples are provided by Plate V, Figs. 25–27. In such preparations the chromosomes can be seen but are not easily counted. A better chance, at least in fission yeasts, for doing this is offered by *flattened protoplasts* prepared by the method of Egel and Pentzos-Daponte (1974). Spread protoplasts prepared in a similar manner have also been used in work on *S. carlsbergensis* (Fischer *et al.*, 1973). The counts obtained by the former group are in good agreement with counts of mitotic chromosomes of the same species made by MCully and Robinow (1971), but the work on *Saccharomyces* is more difficult to interpret, partly because of the apparent absence of well-preserved chromatinic constellations corresponding to the patterns of anaphase I and II familiar from preparations of flattened, *intact* cells (Pontefract and Miller, 1962; Matile *et al.*, 1969). There is little doubt, however, that spread protoplasts prepared either in the manner of Fischer *et al.* (1973) or in some other way will prove most useful for counting meiotic chromosomes of *S. cerevisiae*.

III. Summary and Cautionary Remarks

Two species, *S. cerevisiae* and *Schiz. pombe*, have been studied more frequently and more thoroughly than other yeasts, because they provide rewarding material for biochemical, genetic, and general cell biology studies. But the numbers of yeasts are large, and these organisms are of interest also in contexts other than those listed above. To take but one aspect of their biology: Different species differ markedly in *nuclear* structure and behavior. In fact, it is not unlikely that yeasts will eventually provide useful information on chromosome fine structure and on the dynamics of (certain forms of) mitosis. Meanwhile, the field has to be charted by light microscopy. The cell wall, plasma membrane, mitochondria, vacuole, nucleus, spindle, and chromosomes of yeasts are suitable subjects for light microscopy, supplementing and sometimes initiating the more subtle morphological analysis that is the province of the electron microscope. However, even the best of preparations is wasted unless it is examined under a microscope sufficiently well illuminated to make important detail clearly and comfortably visible at high magnification. The principles of light microscopy are the same for yeasts as for other objects, and their explanation is beyond the scope of this chapter. They have not changed since Shillaber wrote his admirable *Photomicrography in Theory and Practice* (1944). Only a few pointers are offered. Those who have need of morphology will find it convenient to have two microscopes available—one for "kitchen" work such as the scanning of cultures on agar, checking the results of staining, and looking for rewarding sights later to be photographed on a more exactly assembled instrument. The "kitchen" microscope need not be an expensive one, but it must be illuminated with a lamp equipped with imaging lens and a "field stop" diaphragm, permitting an approximation to Koehler illumination. The use of a 15 × ocular in all work in place of the customary 10 × one is strongly recommended. It provides *comfortable* magnification, provided the microscope is well illuminated. Photography demands bright, deliberately and rationally adjusted light, and the maximum of resolution compatible with good visibility (contrast) of detail.

REFERENCES

Barer, R., Ross, R. F. A., and Tkczyk, S. (1953). *Nature (London)* **171**, 720.

Beran, K. (1968). *Advan. Microbiol. Physiol.* **2**, 143.

Bowen, R. H. (1929). *Stain Technol.* **4**, 57.

Cole, G. T., and Kendrick, W. B. (1968). *Mycologia* **60**, 340.

Egel, R., and Pentzos-Daponte, A. (1974). *Arch. Mikrobiol.* **95**, 319.

Fischer, P., Weingartner, B., and Wintersberger, V. (1973). *Exp. Cell Res.* **79**, 452.

Fortner, J. (1930). *Zentralbl. Bakteriol., Parasitenk. Infektienskr., Abt. 1: Orig.* **115**, 96.

Ganesan, A. T., and Swaminathan, M. S. (1958). *Stain Technol.* **33**, 115.

Girbardt, M. (1960). *Ber. Deut. Bot. Ges.* **73**, 227.

Gull, K., and Trinci, A. P. J. (1974). *Arch. Mikrobiol.* **96**, 53.

Gurr, G. T. (1957). "Biological Staining Methods," G. T. Gurr Ltd., London.

Havelková, M. (1973). *Arch. Mikrobiol.* **90**, 77.

Johannsen, D. A. (1940). "Plant Microtechnique," McGraw-Hill, New York.

Johnson, B. F., Yoo, Y. B., and Calleja, G. B. (1974). *In* "Cell Cycle Controls" (G. M. Padilla, I. L. Cameron, and A. M. Zimmerman, eds.), p. 153. Academic Press, New York.

Lindegren, C. C., Williams, M. A., and McClary, D. O. (1956). *Antonie van Leeuwenhoek, J. Microbiol. Serol.* **22**, 1.

McClary, D. O., Williams, M. A., Lindegren, C. C., and Ogur, M. (1956). *J. Bacteriol.* **73**, 360.

McCully, E. K., and Robinow, C. F. (1971). *J. Cell Sci.* **9**, 475.
MacKinnon, D. L., and Hawes, R. S. J. (1961). "An Introduction to the Study of Protozoa," Oxford Univ. Press, London and New York.
Mason, D. J., and Powelson, D. M. (1956). *J. Bacteriol.* **71**, 474.
Matile, P., Moor, H., and Robinow, C. F. (1969). *In* "The Yeasts" (A. H. Rose and J. S. Harrison, eds.), Vol. 1, p. 219. Academic Press, New York.
Miller, G. R., McClary, D. O., and Bowers, W. D., Jr. (1963). *J. Bacteriol.* **85**, 725.
Miller, M. W., and Colaiace, J. D. (1970). *Stain Technol.* **45**, 81.
Mueller, R. (1956). *Naturwissenschaften* **43**, 428.
Newcomer, E. H. (1953). *Science* **118**, 61.
Opie, E. L., and Lavin, G. I. (1946). *J. Exp. Med.* **84**, 107.
Piechaud, M. (1954). *Ann. Inst. Pasteur, Paris* **86**, 787.
Piekarski, G. (1937). *Arch. Mikrobiol.* **8**, 428.
Pontefract, R. D., and Miller, J. J. (1962). *Can. J. Microbiol.* **8**, 573.
Poon, N. H., and Day, A ,W. (1974). *Can. J. Microbiol.* **20**, 739.
Reiss, J. (1969). *Stain Technol.* **44**, 5.
Robinow, C. F. (1961). *J. Biophys. Biochem. Cytol.* **9**, 879.
Robinow, C. F. (1972). *In* "Fermentation Technology Today" (G. Terui, ed.), p. 833. Soc. Ferment. Technol., Kyoto, Japan.
Robinow, C. F., and Bakerspigel, A. (1965). *In* "The Fungi" (G. C. Ainsworth and A. E. Sussman, eds.), Vol. 1, pp. 119–142. Academic Press, New York.
Robinow, C. F., and Marak, J. (1966). *J. Cell Biol.* **29**, 129.
Robinow, C. F., and Murray, R. G. E. (1953). *Exp. Cell. Res.* **4**, 390.
Schaechter, M., Williamson, J. P., Hood, J. R., and Koch, A. L. (1962). *J. Gen Microbiol.* **29**, 421.
Shillaber, C. P. (1944). "Photomicrography in Theory and Practice." Wiley, New York.
Streiblová, E., and Beran, K. (1963). *Exp. Cell. Res.* **30**, 603.
Streiblová, E., and Wolf, A. (1972). *Z. Allg. Mikrobiol.* **12**, 673.
Streiblová, E., Malek, I., and Beran, K. (1966). *J. Bacteriol.* **91**, 428.
Wiemken, A., Matile, P., and Moor, H. (1970). *Arch. Mikrobiol.* **70**, 89.

Chapter 2

Electron Microscopy of Yeasts

FRIEDRICH KOPP

Institute for Cell Biology,
Swiss Federal Institute of Technology,
Zürich, Switzerland

I. General Remarks

One of the main problems encountered in the electron microscopy of biological material is the interpretation of the ultrastructures revealed. This is not surprising if one considers the complicated molecular organization and the complex chemistry of the objects. Usually, painstaking collection of information is necessary until enough evidence is gained to prove a certain surmise. By means of the electron microscope, we advance into spaces of minute dimensions and therefore comprehend only tiny parts of the reality.

An electron micrograph of biological material normally has the character of a "snapshot" of a random sample. This should be taken into account

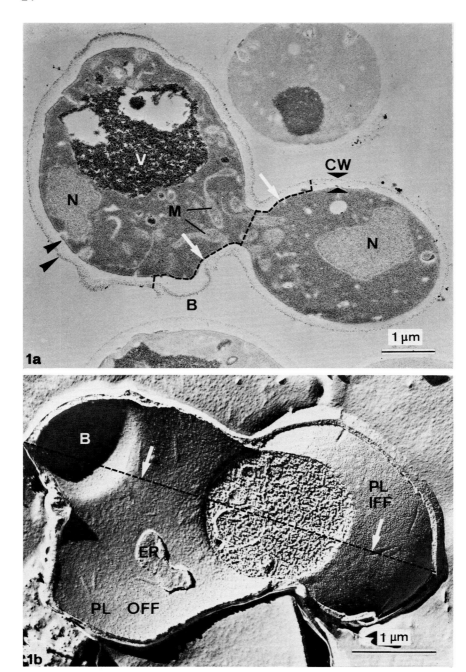

when the phenomena discovered are interpreted. It is a great help and saves much time if the object is prepared systematically before the specific work-up for electron microscopy takes place. Increasing the number of objects that show a certain state, for example, by synchronization of cells or by cleaning of membrane preparations by means of a gradient system, can be useful.

To obtain an image of any material, electrons have to be scattered in the object. Light atoms such as nitrogen, carbon, and oxygen give only weak scattering, and the image produced shows no contrast. It is therefore necessary to introduce heavy atoms in some way or other.

A. Ultrathin-Sectioning

In the case of ultrathin-sectioning, heavy atoms are introduced into the object during chemical fixation and staining. Afterward the material has to be dehydrated, embedded in resin, and cut with an ultramicrotome. With or without poststaining, which again introduces heavy atoms, the sections of the object are examined directly in the electron microscope. The advantage of this method is that not only information about the ultrastructure but also about the distribution of active chemical groups can be gained. Figure 1a, for example, shows the distribution of silver-reducing groups after oxidation with periodic acid in the budding yeast *Saccharomyces cerevisiae*. The appearance of structural artefacts due to chemical and mechanical stress during the course of the different preparative steps is an obvious disadvantage which can partially be controlled by comparing different procedures.

B. Freeze-Etching

See Steere (1957), Moor (1969, 1971), Bullivant (1973), and Mühlethaler (1973). Physical stabilization of biological objects is achieved by rapid

---

FIG. 1. (a) Ultrathin section of a budding yeast, *S. cerevisiae*, stained with silver–hexamethylenetetramine after oxidation with periodic acid. A stratum at the outside of the cell wall takes the stain, and a thin, intense line of silver grains on the border of the protoplast delimits the cell wall. Note the strong reaction in the vacuole. The fracture plane of the cell in (b) is shown by a dotted line. ×14,000. (b) The connection between mother cell and daughter cell frozen in the living state [cf. (a)]. In the mother cell a bud scar (B) is visible, and the outer fracture face of the plasmalemma is revealed (OFF). The fracture crosses the connection between the cells through the plasma and passes along the plasmalemma in the daughter cell, where the inner fracture face (IFF) is revealed. CW, cell wall; ER, endoplasmic reticulum; M, mitochondria; N, nucleus; PL, plasmalemma; V, vacuole. ×22,000.

freezing (Moor, 1964, 1973a). In the frozen state the objects are broken under vacuum and with (freeze-etching) or without (freeze-fracturing) sublimation of some of the ice, the split samples are replicated by evaporation of about 2 nm of a heavy metal (normally platinum). The metal replica is than reinforced by a layer of evaporated carbon, about 20 nm thick, which is translucent for electrons. After thawing and cleaning, the replica can be studied in the electron microscope. The micrographs obtained by this means show the objects in a condition very close to the living state. Yeast cells have been proved to survive freezing under the conditions used in this technique (Moor and Mühlethaler, 1963). Shadow-casting at an oblique angle produces a three-dimensional impression (Fig. 2). (In this article all micrographs have been photographically reversed, so that the evaporated platinum appears white and the shadows of the objects appear black. To simplify visual interpretation all the micrographs are arranged in such a way that the shadow-casting is directed from top left to bottom right.)

In ultrathin-sectioning membranes are shown mainly in profile, whereas freeze-fracturing and freeze-etching allow us to visualize different face views of membranes (Mühlethaler, 1971). A combination of both methods, thin-sectioning and freeze-etching, can provide a more detailed insight into the relation of structure and chemistry, as has been demonstrated with the plasmalemma of yeast by Kopp (1972, 1973).

The resolution in ultrathin-sectioning is limited, among other reasons, by the fact that the contrast arises through superposition of the effects from many heavy atoms distributed over the whole thickness of the thin section of about 50 nm. In freeze-etch replicas, however, lateral resolution is given by the size of the crystallites of the heavy metal atoms and their interactions with the substrate (Moor, 1971, 1973b).

II.   Fixation and Staining

A.   Aldehyde Fixation

To obtain a good representation of yeast cells in ultrathin-sectioning is not without problems. The comparatively thick cell wall probably prevents the rapid penetration of fixatives and stains. Many micrographs of yeasts have been published in which the membranes, especially the plasmalemma,

---

FIG. 2.   Electron micrograph of a freeze-etch replica of a cross-fractured yeast cell, *Schiz. pombe*. The cell was frozen in the living state. The different organelles have a well-rounded shape, and the plasmalemma adheres closely to the cell wall. L, lipid droplets; G, Golgi apparatus. ×26,000.

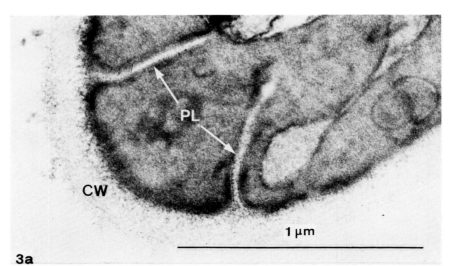

3a

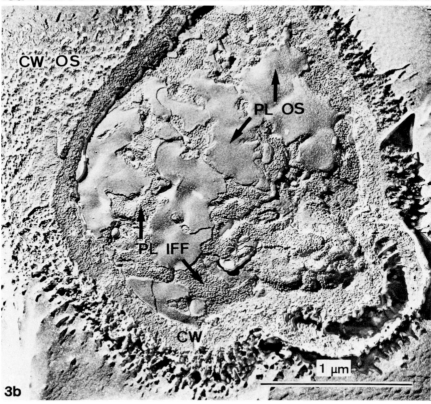

3b

look deeply pleated. However, micrographs of freeze-fractured yeast rapidly frozen in the living state show all organelles to be well-rounded, and the plasmalemma lying close to the inner surface of the cell wall (Fig. 2).

It is very important to pay attention to the tonicity of the fixation medium used. The osmotic pressure of the medium should be adapted to the material under investigation. Checking with an osmometer is useful. Figure 3a shows a thin section of a cell of *Schizosaccharomyces pombe* that was fixed with a hypertonic glutaraldehyde–formaldehyde medium and then stained with potassium permanganate, uranyl acetate, and lead citrate. In all the cells of this sample, a great number of artificial folds is found in the plasmalemma. Figure 3b shows a freeze-etched cell of *S. cerevisiae* that was fixed in a hypertonic aldehyde medium. The protoplast is revealed by fracturing away part of the cell wall. It displays a complicated convoluted surface with smooth areas and areas covered by particles. A possible interpretation of the structures is discussed later.

If the cells are fixed isotonically, micrographs of freeze-etched material do not look very different from micrographs of cells frozen in the living state (Figs. 1b and 4). Figure 4 shows *S. cerevisiae* that has been stabilized in formaldehyde–glutaraldehyde medium of 550 milliosmoles. A piece of displaced cell wall, the inner surface of which is covered by remnants of the plasmalemma and the endoplasmic reticulum, is seen. The neighboring cell shows the obverse situation in which part of the cell wall has been broken away, revealing the inner fracture face of the plasmalemma with a few peepholes through which underlying endoplasmic reticulum can be seen. The plasmalemma shows the structures normally found, i.e., invaginations and areas with particles arranged hexagonally (Moor and Mühlethaler, 1963), and it adheres closely to the cell wall. The situation appears similar to that of unfixed frozen cells (Fig. 1b). Nečas et al. (1969) observed the same tendency of the fracture plane to skip from the plasmalemma level down to the endoplasmic reticulum level in protoplasts fixed with aldehyde.

Glutaraldehyde contains significant amounts of α, β-unsaturated aldehydes, oligomers, and polymers which are thought to give stable Michael-type adducts with amines (Richards and Knowles, 1968), so leading to

---

FIG. 3. (a) Ultrathin section of a yeast cell fixed with aldehydes at a hypertonic concentration. $KMnO_4$, uranyl acetate, and lead citrate were used for staining. The plasmalemma is deeply folded. $\times 65,000$. (b) A freeze-etched yeast cell fixed hypertonically with aldehydes and then prepared in distilled water. Because of the fixation the plasmalemma is convoluted. Fracturing occurred within the membrane, revealing inner fracture face areas covered by particles. In other areas fracturing occurs between plasmalemma and cell wall, thus revealing patches of the smooth outer surface (PL OS) of the membrane (cf. Fig. 7). The cross-fractured cell wall as well as its outer surface (CW OS) can be seen. $\times 40,000$.

cross-linkage of proteins. It is suggested that, among the lipids, phosphatidylethanolamine and phosphatidylserine are fixed by the glutaraldehyde. This has been shown by Gigg and Payne (1969) for the former substance.

The dense cytoplasm of yeast cells is stabilized with glutaraldehyde to a rigid gel. Even after drastic extractions (chloroform–methanol–water) or treatments with a detergent (Triton X-100), the overall structure of isolated protoplasts stayed intact. Changes in membrane ultrastructure due to such treatments could easily be investigated, since the fixed plasmagel served as a support for the membranes (Kopp, 1972).

We generally obtained good results with protoplasts using acrolein–glutaraldehyde fixation in cacodylate buffer as recommended by Hess (1966), when we proceeded in the following way. One part of the fixative was added to 9 parts of the last medium containing the cells (isolated organelles, membranes). This was done in the cold. After 15 minutes the suspension was mixed with the fixative 1:1 and then stored overnight in the refrigerator. Before any further staining, the cells were washed three times for 15 minutes with cacodylate buffer.

In the following section a few of the multitude of staining procedures (cf. Geyer, 1973) are briefly discussed.

B. Permanganate Fixation and Staining

Most electron microscope studies with yeasts are carried out on material fixed with permanganate (e.g., see Gay and Martin, 1971; Yoo et al., 1973; Deshusses et al., 1970; Oulevey et al., 1970). The fixation is often carried out in a 2% solution of $KMnO_4$ in water for 1–2 hours at room temperature. In this simple procedure fixation and staining are performed in one step, and samples can be observed without any further staining (Gay and Martin, 1971). Poststaining with a uranyl salt (acetate or nitrate) either in the 70% step of the dehydration or on the section, and/or a lead salt such as lead citrate (Reynolds, 1963) or lead oxide (Karnovsky, 1961), enhances the contrast considerably (Fig. 3a). Cell walls and membranes are stained

FIG. 4. A yeast cell after isotonic fixation with aldehydes and freeze-etching. Through the breaking away of a piece of the cell wall, the inner fracture face of the plasmalemma is revealed. Part of the cell wall is shown from the inside. This view corresponds to the outer fracture face seen in Fig. 1b. Because of the fixation the fracture plane is different; in some areas the fracture skips down to the level of the endoplasmic reticulum. The corresponding plasmic remnants were broken off together with the cell wall. At the top left a small area of the cell wall outer surface (CW OS) has been revealed by the sublimation of ice. ×23,000.

distinctly, but the nucleoplasm and the groundplasm often seem partly cleared. This is possibly due to oxidative degradation and dissolution of plasma components. Ribosomes are not seen.

According to Riemersma (1973), part of the contrast is due to the reaction of permanganate with unsaturated membrane lipids. It is assumed that the addition of permanganate to the double bond results in an anionic product which moves to the plane of the polar groups. The end product is thought to consist of a grainy precipitate of MnO_2.

The peptide bonds of proteins are not destroyed, but side groups such as amino and sulfhydryl groups may be oxidized by $KMnO_4$. However, steric effects may influence the stainability (Hake, 1965). Riemersma (1973) suggests the use of permanganate at lower concentrations and low temperature to reduce the violence of the reagent. Gay and Martin (1971) used 2 and 4% $KMnO_4$ at $0\,°C$ without poststaining with good results.

C.   Osmium Tetroxide Fixation and Staining

The fixation of whole .yeast cells with osmium tetroxide alone or after a preceding fixation with aldehyde does not give good results. Satisfactory penetration of the fixatives and stains seems to be made impossible by the barrier of the cell wall. Several attempts have been made to overcome this problem.

In an impressive article on the fiber apparatus in the nucleus of yeast, Robinow and Marak (1966) reported an elegant method to facilitate the penetration of osmium tetroxide. Between fixation with aldehyde and staining with osmium tetroxide the cell wall was partly digested with snail gut juice. The groundplasm of such preparations appears well preserved. In contrast to permanganate staining, it shows the ribosomes clearly. Ultrastructural details of the nucleus are in good agreement with those found in freeze-etching (Moor, 1966, 1967; Matile et al., 1969).

Schwab et al. (1970) used a mixture of 2% glutaraldehyde and 2% acrolein in 50% aqueous dimethyl sulfoxide to facilitate penetration. The fixation took 7 hours in the cold and was followed, after a short wash, by a 3% osmium tetroxide staining at pH 4. These investigators also obtained good preservation of cytoplasmic features.

Another interesting approach to the fixation problem was recently published by Hereward and Northcote (1972b). They froze their specimens in Freon 22 and freeze-substituted them in a 2% solution of osmium tetroxide in acetone. The substitution took place within 3 days at $-78\,°C$. After the specimens were brought up to room temperature in stages (1 hour at $-40\,°C$; 1 hour at $0\,°C$), they were embedded and poststained with uranyl acetate and lead hydroxide. As also in freeze-etching, this method should allow investigation of the time sequence of rapidly occurring events, since the fixation time is well defined.

It is generally accepted that osmium tetroxide reacts primarily with the olefinic part of lipid molecules. There is some controversy about the stability of the resulting compounds. Riemersma (1968) assumes that lipid polar groups are involved in the reactions of membrane lipids during tissue fixation. He postulates a primary formation of cyclic osmic esters. The resulting anions are attracted by lipid cations and shifted to the polar part of the lipid layer. A further reduction, mediated by proteins or dehydration media, leads to osmium dioxide as the final product. Korn (1966) however, demonstrated, by isolation of bis(methyl-9,10-dihydroxystearate) osmate from amebas, fixed with osmium tetroxide, that osmium is almost certainly covalently bound to the hydrocarbon moiety of lipids.

Lateral polymerization of unsaturated lipids in membranes seems possible. This would account for the stability of the lipid part of the plasmalemma fixed with osmium and extracted with ethanol–acetone (Kopp, 1972). After fracturing in the frozen state, a plastic deformation of the lipid layer treated in this manner was seen, supporting the idea of a strong lateral connection between molecules. In the same publication it was shown that osmium staining, after extraction of aldehyde-fixed protoplasts with chloroform–methanol–water, no longer led to the appearance of the "unit" aspect. A similar result was obtained with osmium staining after a saturation of double bonds by reduction of the mercury adduct with sodium borohydride. In this experiment also no "unit" aspect could be observed, indicating again the importance of the double bonds as the primary reaction site (Kopp, 1972).

In proteins the reaction with osmium tetroxide does not lead to degradation of the peptide bonds (Hake, 1965), but oxidation of side groups, influenced by the configuration of the protein molecules, must be expected. The protein-coagulating effect of osmium tetroxide might be the reason for the good preservation of the groundplasm, in contrast to permanganate staining which leads to partial extraction of proteins.

Good results in staining are obtained by using an 1% solution of osmium tetroxide in cacodylate buffer (Hess, 1966). Fixation should be carried out in an ice bath (high vapor pressure of osmium tetroxide) and should not last longer than 2 hours to avoid "overoxidation" (lit.cit. Geyer, 1973). Usually, poststaining on the section with a uranyl salt and/or a lead salt is necessary (see Section II,B).

For the ultrastructural localization of polysaccharides, the following procedures may be of interest.

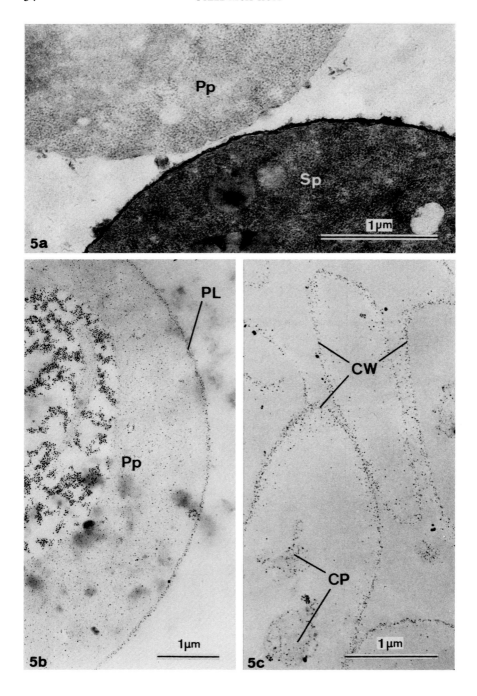

D. Staining with Silver

The procedure of Picket-Heaps (1967) is followed, but instead of block staining, the different steps are performed in suspension. Figure 1a shows a budding yeast cell as revealed after acrolein–glutaraldehyde fixation and staining with silver–hexamethylenetetramine complex. Before staining, oxidation with periodic acid was carried out. Periodic acid is a specific reagent for glycols. The oxidation leads to formation of aldehyde groups which then reduce the silver salt. The method can be employed to distinguish different cell wall components (Picket-Heaps, 1967). Unspecific staining is introduced by the presence of sulfhydryl and aldehyde groups which occur naturally, or aldehyde groups which are introduced by fixation. Several control steps therefore have to be carried out. In the protoplast shown in Fig. 5b before oxidation and staining, aldehyde groups already present were blocked with sodium bisulfite, and sulfhydryl groups with iodoacetic acid (Picket-Heaps, 1967). A great part of the background staining in the ground-plasm has disappeared, but along the plasmalemma a distinct band of silver grains can be observed. The cell wall in Fig. 1a shows a stained outer stratum and a narrow, stained stratum close to the protoplast. We were not sure whether the inner stained stratum belongs to the cell wall or whether it corresponds to the stained line limiting the isolated protoplast. Therefore cell walls were isolated and stained the same way. They show only one stratum (Fig. 5c). Thus the smaller inner stratum shown in Fig. 1a may correspond to a polysaccharide constituent of the plasmalemma.

Matile et al. (1967) isolated a globular mannan protein from the plasmalemma of yeast, and Cortat et al. (1973) demonstrated the incorporation of GDP-^{14}C-mannose into isolated plasmalemma vesicles. A correlation between the stainable substance in the plasmalemma and the mannan–protein is as yet hypothetical. Results of experiments with enzymatic degradation by phosphomannanase support the idea that the outer layer of the cell wall contains mannan (McLellan et al., 1970). This observation is confirmed

FIG. 5. (a) Two cells of S. cerevisiae stained with ruthenium red and osmium tetroxide after treatment with snail gut enzyme. One cell is covered with heavily stained remnants of the cell wall (spheroplast, Sp). The other is devoid of such material and may therefore be called a protoplast (Pp). ×31,0000. (b) A yeast protoplast stained with silver. After blocking of aldehyde and sulfhydryl groups followed by oxidation with periodic acid, the plasmalemma shows a distinct stain when treated with silver–hexamethylenetetramine complex. ×16,000. (c) An isolated cell wall stained with silver as the protoplast in Fig. 5b. Only the outer stratum is stained. Sometimes cytoplasmic material (CP) is left in the interior of the cell wall. From the staining we obtained with whole cells, isolated protoplasts, and isolated cell walls, we concluded that the narrow, stained stratum delimiting the inside of the cell wall belongs to the plasmalemma. ×24,0000.

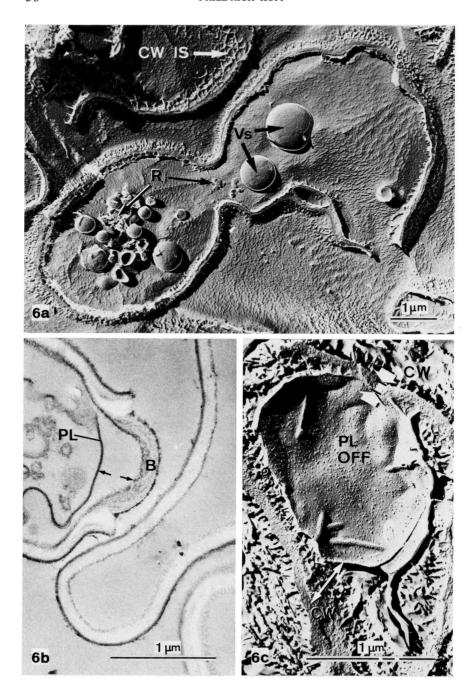

by more recent results of Gerber *et al.* (1973), who labeled the surface of *Candida utilis* with colloidal gold coated with antimannan antibodies, as well as by the use of a mercury-labeled concanavalin A as a marker for the localization of mannan (Horisberger *et al.*, 1971). Mannan is accumulated especially in the bud scars (Bauer *et al.*, 1972).

E.   Staining with Ruthenium Red

Ruthenium red has been used because of its reaction with extracellular acidic mucosubstances. It penetrates the cell only if the plasmalemma is damaged. As a hexavalent cation, ruthenium red reacts with anionic substances. A sufficient contrast in the electron microscope is produced only after a subsequent reaction with osmium tetroxide (Luft, 1971). To improve the staining of the cell wall, we added 1 ppt ruthenium red to both the acrolein–glutaraldehyde fixative and the osmium tetroxide stain in cacodylate buffer as used by Hess (1966). Isolated cell walls (Fig. 6b) revealed an inner and an outer zone of higher density, separated by a more transparent zone. Note the difference in the silver-stained cell wall (Fig. 5c). Enclosed remnants of cytoplasmic membranes display a prominent triple-layered appearance after staining with ruthenium red. The bud scar has a higher affinity for the stain than the original cell wall.

During the formation of protoplasts of *S. cerevisiae* with snail enzyme, two types of spherical cells were easily distinguished by their difference in color in the light microscope when stained with ruthenium red. In the electron microscope the same suspension proved to consist of two types of cells (Fig. 5a). In one type the plasmalemma was coated by a layer of heavily stained material, whereas in the other type this was not the case. Staining with ruthenium red therefore could serve as a method for the recognition of protoplasts in a population of spheroplasts on the light microscope level. The coat of spheroplasts stainable with ruthenium red may be related to the substance in the prospheroplast envelope isolated by Darling *et al.*

FIG. 6.   (a) An isolated cell wall of a budding cell of *S. cerevisiae.* Also, after extensive washing a considerable number of cell walls in the sample contained membrane vesicles (Vs) and other cytoplasmic materials such as clusters of ribosomes (R). A view of the inside of the cell wall is shown (CW IS). ×12,000. (b) Isolated cell walls stained with ruthenium red and osmium tetroxide show intensively stained inner and outer strata. The newly synthesized wall in the bud scar (B) has a greater affinity for the stain. If any plasmalemma was left, we always found it to be detached from the cell wall. ×28,000. (c) The outer half of the split plasmalemma shown from the inside. The plasmalemma has been torn away from the cell wall during rupturing of the cell. In this way it is demonstrated that this outer fracture face does not belong to the cell wall but is in fact an outer part of the split membrane. ×34,000.

(1972), which is stainable with phosphotungstic acid (PTA). These investigators isolated a alkali-soluble glucan as a major component of the part
of the cell wall that is in close apposition to the plasmalemma. It should be
noted that the actual inner surface of isolated cell walls of *S. cerevisiae*
(without the adhering outer half of the split plasmalemma as revealed by
freeze-fracturing; see Section III,A) shows a fibrillar texture (Fig. 6a) [Kopp,
1972 (see p. 293), 1973 (see p. 183)]. These fibrils could well be a glucan,
but the PTA- and ruthenium red–stainable material is more likely to be an
intercalating matrix substance of as yet unknown composition.

F.   Embedding

It can be advantageous to perform all the staining and fixation procedures
in suspension. However, it is difficult to embed a sediment of cells, since
most resin monomers are very viscous liquids. For easier handling of the
material during embedding, it is useful to have the cells (isolated organelles,
membrane vesicles, etc.) concentrated in small gel blocks. Some investigators use agar or gelatine for this purpose. If one wishes to avoid unnecessary heating of the sample, one can use bovine serum albumin (BSA) in a
procedure similar to the one described by Shands (1968). After washing
three times for 15 minutes, the fixed and stained cells (organelles, membranes) are resuspended in a 1.5–3% solution of BSA in tris–HCl buffer
(pH 7.5). The suspension is allowed to stand for 10 minutes. Then the cells
are spun down into a pellet. The supernatant BSA solution is discarded,
and 1 or 2 drops of glutaraldehyde (25%) are carefully added on top of the
pellet. The tubes are closed with a piece of Parafilm and stored overnight in
the refrigerator. The resulting gel is then cut into pieces smaller than 1 mm³,
washed, dehydrated, and embedded in resin in the usual manner.

We have most satisfactory results with the Epon–Araldite embedding
mixture described by Mollenhauer (1964), but for special purposes other
resins may be better (cf. Luft, 1973).

III.   Freeze-Fracturing and Freeze-Etching, Interpretation

The three-dimensional appearance of micrographs of freeze-fractured
and freeze-etched cells is mostly due to cytoplasmic membranes which serve
as predetermined rupture zones where fracturing preferentially occurs. The
following discussion should lead to a better understanding of the spatial
relations between the ultrastructural details revealed.

A. Fracture Faces

Fracturing along the yeast plasmalemma in the frozen state reveals two different face views, a convex inner fracture face (IFF) on the side of the protoplast, and a concave outer fracture face (OFF) on the inside of the cell wall (Figs. 1b and 4). The fact that the two faces show two different but typical ultrastructural aspects speaks in favor of only one possible fracture plane in the yeast plasmalemma, i.e., between the convex inner fracture face, regularly invaginated, and always covered by a large number of particles, and the concave outer fracture face, always with a few particles and the same number of humps as the number of invaginations mentioned. The correspondence of the two complementary fracture faces of the yeast plasmalemma was shown by double-replica techniques by Sleytr (1970) and Hauenstein (cf. Mühlethaler, 1971).

The position of this fracture plane along the membrane was controversial. The question arose whether the two complementary faces correspond to an internal fracture plane (Branton, 1966) or whether the fracture plane follows one of the two membrane surfaces. [The possibility that the fracturing passes along both surfaces can be excluded, since in that case four (twice two complementary) fracture faces would be expected, each face showing another texture, which is not the case.]

Hereward and Northcote (1972a) localized the internal fracture plane in split tonoplast membranes by thin-sectioning of embedded, freeze-substituted drops of frozen and fractured yeast. The cleavage plane along the plasmalemma was judged to be not clearly localized, since after staining the split material a faint line appeared parallel to the inner dark line of the membrane limiting the cytoplasma. At the invagination sites the situation was reversed. In places where the cell wall shows humps corresponding to the invaginations of the plasmalemma, a similar line was found, parallel to the stained line limiting the cell wall on the inside. These investigators suggested as a possible explanation, that "this indistinct third line could have resulted from nonspecific adsorption of stain in which case the membrane would have split."

If we assume, as is done in most freeze-etch work on yeasts, that the fracture plane follows the outer surface of the plasmalemma, the inner fracture faces in Figs. 1b and 4 are views of the plasmalemma surface as seen from outside, whereas the outer fracture faces in the same micrographs correspond to views of the cell wall inner surface. When yeasts are mechanically disintegrated, the plasmalemma is regularly pulled away from the cell wall (Fig. 6). The cell wall inner surface, exposed by such treatment, reveals a fibrillar texture which looks quite different from the outer fracture face of the same area in freeze-fractured yeast (Kopp, 1972, 1973). If the fibrillar

7   1 µm

face belongs to the cell wall, we must conclude that the cell wall inner surface in freeze-fractured cells is covered by an additional layer, i.e., an outer part of the split plasmalemma.

A situation that clearly demonstrates that the outer fracture face is not the inside of the cell wall but an internal face of the split plasmalemma is shown in Fig. 6c. The outer fracture face seen in this micrograph belongs to a plasmalemma that has been partly torn away from the cell wall in the course of cell disintegration by shaking with glass beads. The morphological aspect of this face is similar to that of the outer fracture face exposed on the inside of the cell wall in Figs. 1b and 4.

In this context it should be mentioned that we regularly observed that the plasmalemma collapses away from the cell wall as is shown in Fig. 6, and this is also supported by the fact that the inside of the cell wall fails to stain with silver (Fig. 5c). These results confirm the findings of Dubé et al. (1973), who determined the fate of the plasma membrane during fractionation. The cell wall very often contains a considerable number of trapped vesicles and other cytoplasmic material (Fig. 6a) which are difficult to wash out. Therefore earlier work concerning cell wall composition should be judged cautiously. We have strong doubts that the plasmalemma remained attached to the cell wall during isolation by the method of Suomalainen et al. (1967).

B.   Etching and Membrane Surface

Most of the components added to a culture medium for nutrition, osmotic stabilization, or cryoprotection lead to the formation of a eutectic during the course of the freezing process. During the freezing of a cell suspension in such a medium, the growing ice crystals push the cells into the remaining solution which becomes more and more concentrated. At a certain temperature this solution solidifies to an amorphous state.

If isolated protoplasts are frozen in an osmotically stabilizing medium, they are covered by an eutectic layer of variable thickness. After fracturing the protoplasts show similar convex and concave fracture faces as are revealed in the case of whole cells. Nečas et al. (1969) interpreted the inner convex fracture face as corresponding to the protoplast surface, whereas

---

FIG. 7. Naked protoplasts fixed with aldehydes, treated with 0.02% Triton X-100, and freeze-etched in distilled water. The surface structures after fracturing and etching are seen. Where the fracture plane through the ice touched the protoplast, an outer part of the plasmalemma was broken off, revealing the inner fracture face. After sublimation of some of the ice, the actual outer surface (OS) of the plasmalemma was uncovered. It shows a fairly smooth texture compared to the inner fracture face. Some of the surrounding ice can be seen (I). ×55,000.

the concave face was considered to be only a print complementary to the protoplast surface in the nonetchable envelope. Since we assume that yeast plasmalemma splits internally, we tried to reveal the true outer surface. Exponentially growing yeasts were incubated for 2 hours with snail gut enzyme. After fixation with glutaraldehyde and acrolein, the naked protoplasts were washed with twice-distilled water and stored overnight in water to permit diffusion of all soluble substances. After this extensive washing, which was thought to prevent the formation of a detectable eutectic, the protoplasts were frozen. Then they were fractured and etched for 90 seconds under a vacuum of at least 5.10^{-6} torr, with a cold finger cooled by liquid nitrogen above the specimen. In this way some ice was sublimed, revealing a additional smooth layer (Fig. 7). This layer, which is considered to represent the true outer surface (OS), was proved to be accessible to various lipid solvents (Triton X-100, ethanol–acetone, chloroform–methanol–water) (Kopp, 1972, 1973). In addition, a relation between changes in the appearance of the black lines of the "unit" aspect and changes in the smooth outer surface due to extraction was found, supporting the idea of the lipid nature of the plasmalemma outer surface. Streiblova (1968) found a similar smooth layer on the surface of isolated protoplasts and on protoplast surfaces of cells that were partly plasmolyzed. But she interpreted them as "innermost wall layers." Observations made on isolated plasmalemma vesicles, isolated vacuoles (cf. Frey-Wyssling, 1973, p. 61), and protoplast portions in disintegrated cells (F. Kopp, unpublished results) all show a similar additional layer after sublimation of ice, and lead us to the conclusion that the smooth layer found on the plasmalemma surface belongs to the membrane.

The situation represented in Fig. 3b can now be explained as follows. Because of the hypertonicity of the fixation medium, part of the protoplast has plasmolyzed. This results in a convoluted appearance of the protoplast surface that was uncovered by splintering off part of the cell wall during freeze-fracturing. In areas where the plasmalemma was attached to the cell wall, normal fracturing occurred, revealing the inner fracture face covered by particles. In other areas where the plasmalemma was detached from the cell wall, the fracture plane passed between the cell wall and the plasmalemma. Here the true outer surface of the plasmalemma is visible (compare Fig. 3b with Fig. 7). Some breaks through the plasmalemma to the level of the endoplasmic reticulum can be seen.

From these and other results, we suggested a dynamic membrane concept in which both surfaces of the membrane are limited by a lipid layer. Hydrophobic proteins (particles) are thought to be situated in between, sometimes forming a continuous internal layer. The proportion of the membrane in the form of a lipid bilayer is assumed to be variable. Differences in amount of bilayer areas could explain the different fracture behavior in plasmalemma and tonoplast found by Hereward and Northcote (1972). Protoplasts were treated with various enzymes by Nečas and Svoboda (1974), but no changes in particle structure and distribution were observed. This can partly be explained if one assumes the particles to be protected by an overlaying lipid layer.

ACKNOWLEDGMENT

This work was supported by the Swiss National Foundation for Scientific Research Grant 5252.3.
I express my thanks to Miss Sonia Türler for help with the manuscript.

REFERENCES

Bauer, H., Horisberger, M., Bush, D. A., and Sigarlakie, E. (1972). *Arch. Mikrobiol.* **85**, 202–208.
Branton, D. (1966). *Proc. Nat. Acad. Sci. U.S.* **55**, 1048–1056.
Bullivant, S. (1973). *In* "Advanced Techniques in Biological Electron Microscopy" (J. K. Koehler, ed.), pp. 67–112. Springer-Verlag, Berlin and New York.
Cortat, M., Matile, P., and Kopp, F. (1973). *Biochem. Biophys. Res. Commun.* **53**, 482–489.
Darling, S., Theilade, J., and Birch-Anderson, A. (1972). *J. Bacteriol.* **110**, 336–345.
Deshusses, J., Oulevey, N., and Turian, G. (1970). *Protoplasma* **70**, 119–130.
Dubé, J., Setterfield, G., Kiss, G., and Lusena, C. V. (1973). *Can. J. Mikrobiol.* **19**, 285–290.
Frey-Wyssling, A. (1973). *Protoplasmatologia* IIIG, 1–105.
Gay, J. L., and Martin, M. (1971). *Arch. Mikrobiol.* **78**, 145–157.
Gerber, H., Horisberger, M., and Bauer, H. (1973). *Infec. Immunity* **7**, 487–492.
Geyer, G. (1973). "Ultrahistochemie." Fischer, Stuttgart.
Gigg, R., and Payne, S. (1969). *Chem. Phys. Lipids* **3**, 292–295.
Hake, T. (1965). *Lab. Invest.* **14**, 470–474.
Hereward, F. V., and Northcote, D. H. (1972a). *J. Cell Sci.* **10**, 555–561.
Hereward, F. V., and Northcote, D. H. (1972b). *Exp. Cell Res.* **70**, 73–80.
Hess, W. M. (1966). *Stain Technol.* **41**, 27–35.
Horisberger, M., Bauer, H., and Bush, D. A. (1971). *FEBS* (*Fed. Eur. Biochem. Soc.*) *Lett.* **18**, 311–314.
Karnorsky, M. J. (1961). *J. Biophys. Biochem. Cytol.* **11**, 729–732.
Kopp, F. (1972). *Cytobiologie* **6**, 287–317.
Kopp, F. (1973). *In* "Freeze-etching Techniques and Applications" (E. L. Benedetti and P. Favard, eds.), pp. 181–186. Soc. Fr. Microsc. Electron., Paris.
Korn, E. D. (1966). *Biochim. Biophys. Acta* **116**, 325–335.
Luft, J. H. (1971). *Anat. Rec.* **171**, 347–416.
Luft, J. H. (1973). *In* "Advanced Techniques in Biological Electron Microscopy" (J. K. Koehler, ed.), pp. 1–34. Springer-Verlag, Berlin and New York.
McLellan, W. L., McDaniel, L. E., and Lampen, J. O. (1970). *J. Bacteriol.* **102**, 261–270.
Matile, P., Moor, H., and Mühlethaler, K. (1967). *Arch. Mikrobiol.* **58**, 201–211.
Matile, P., Moor, H., and Robinow, C. F. (1969). *In* "The Yeast" (A. H. Rose and J. S. Harrison, eds.), pp. 219–302. Academic Press, New York.
Mollenhauer, H. H. (1964). *Stain Technol.* **39**, 111–114.
Moor, H. (1964). *Z. Zellforsch. Mikrosk. Anat.* **62**, 546–580.
Moor, H. (1966). *J. Cell Biol.* **29**, 153–155.

Moor, H. (1967). *Protoplasma* **64**, 89–103.

Moor, H. (1969). *Int. Rev. Cytol.* **25**, 391–412.

Moor, H. (1971). *Phil. Trans. Roy. Soc. London, Ser. B* **261**, 121–131.

Moor, H. (1973a). *In* "Freeze-etching Techniques and Applications" (E. L. Benedetti and P. Favard, eds.), pp. 11–19, Soc. Fr. Microsc. Electron., Paris.

Moor, H. (1973b). *In* "Freeze-etching Techniques and Applications" (E. L. Benedetti and P. Favard, eds.), pp. 27–30. Soc. Fr. Microsc. Electron., Paris.

Moor, H., and Mühlethaler, K. (1963). *J. Cell Biol.* **17**, 609–628.

Mühlethaler, K. (1971). *Int. Rev. Cytol.* **31**, 1–19.

Mühlethaler, K. (1973). *In* "Freeze-etching Techniques and Applications" (E. L. Benedetti and P. Favard, eds.), pp. 1–10. Soc. Fr. Microsc. Electron., Paris.

Nečas, O., and Svoboda, A. (1974). *Folia Microbiol. (Prague)* **19**, 81–87.

Nečas, O., Kopecka, M., and Brichta, J. (1969). *Exp. Cell Res.* **58**, 411–419.

Oulevey, N., Deshusses, J., and Turian, G. (1970). *Protoplasma* **70**, 217–224.

Picket-Heaps, J. D. (1967). *J. Histochem. Cytochem.* **15**, 442–455.

Reynolds, E. S. (1963). *J. Cell Biol.* **17**, 208–213.

Richards, F. M., and Knowles, J. R. (1968). *J. Mol. Biol.* **37**, 231–233.

Riemersma, J. C. (1968). *Biochim. Biophys. Acta* **152**, 718–727.

Riemersma, J. C. (1973). *Acta Histochem., Suppl.* **13**, 125–134.

Robinow, C. F., and Marak, J. (1966). *J. Cell Biol.* **29**, 129–151.

Schwab, D. W., Janney, A. H., and Scala, J. (1970). *Stain Technol.* **45**, 143–147.

Shands, J. W. (1968). *Stain Technol.* **43**, 15–17.

Sleytr, U. (1970). *Protoplasma* **70**, 101–117.

Steere, R. L. (1957). *J. Biophys. Biochem. Cytol.* **3**, 45–60.

Streiblova, E. (1968). *J. Bacteriol.* **95**, 700–707.

Suomalainen, H., Nurminen, T., and Oura, E. (1967). *Suom. Kemistilehti B* **40**, 323–326.

Yoo, B. Y., Calleja, G. B., and Johnson, B. F. (1973). *Arch. Mikrobiol.* **91**, 1–10.

Chapter 3

Methods in Sporulation and Germination of Yeasts

JAMES E. HABER AND HARLYN O. HALVORSON

Department of Biology and Rosenstiel Basic Medical Sciences Research Center
Brandeis University, Waltham, Massachusetts

Sporulation of yeasts such as *Saccharomyces cerevisiae* and *Schizosaccharomyces pombe* provides an attractive system for the study of unicellular differentiation, especially the processes of meiosis and genetic recombination. Yeasts are easily cultured, sporulated, and germinated, and are quite amenable to genetic analysis. Nevertheless, the experience of many workers in recent years has been that study of sporulation and germination involves several technical problems which have slowed the progress of understanding fundamental mechanisms of cellular control. In this article we try to identify areas toward which major attention has been directed and to focus on procedures that so far have proved most successful.

I. Sporulation

Sporulation has been examined in several different yeasts, and the majority of work has been focused on *S. cerevisiae*. Unless otherwise noted, all methods cited below have been developed and used primarily with *S. cerevisiae*.

Sporulation occurs when cells preadapted to oxidative growth are incubated in a medium such as 1% potassium acetate in the absence of a nitrogen source (Miller, 1963; Croes, 1967a; Fowell, 1969). Sporulation occurs in diploids or cells of higher ploidy which carry both *a*- and α-mating-type alleles (Roman and Sands, 1953; Gunge and Nakatomi, 1972); diploids of genotype *a/a* or α/α do not sporulate. Sporulation is distinctly different from the vegetative cell cycle, as the conditions that support vegetative growth do not support sporulation of strains of *S. cerevisiae* (Fowell, 1969; Mills, 1972). Sporulation occurs under pseudostarvation conditions in which the cells do not proliferate. During sporulation cells undergo meiosis rather than mitosis and yield four haploid genetic complements enclosed in four refractile spores within an ascus.

It is not clear to what extent sporulation requires the expression of new genetic information. Very few sporulation-specific events have been elucidated. There are, however, substantial changes in the physiology of sporulating cells, many of which make routine examination of biochemical events more difficult than the same studies in vegetative cells. For example, as cells reach stationary phase in vegetative growth medium (prior to sporulation), the cell wall becomes more difficult to remove by enzymatic digestion (Deutch and Parry, 1974). There is a significant increase in the activity both of proteases (Chen and Miller, 1968; A. Singh, personal communication) and RNases (C. A. Saunders and J. E. Haber, unpublished observations).

Sporulation is a relatively slow event. The vegetative cell cycle takes 2–3 hours, while sporulation is not completed until nearly 24 hours. The main stages of this process are illustrated in a generalized form in Fig. 1; however, the reader is directed to several recent reviews for more detail (Fowell, 1969; Haber and Halvorson, 1972a; Tingle *et al.*, 1973).

The major period of protein synthesis during sporulation occurs in the first 6 hours (Esposito *et al.*, 1969), while DNA synthesis and the synthesis of most RNA occurs somewhat later, beginning at 4–6 hours and ending by 10 hours (Croes, 1967a; Mills, 1972). There is no net increase in either RNA or protein during sporulation, because of extensive turnover (Croes, 1967a; Esposito *et al.*, 1969). During sporulation, however, there is a significant increase in carbohydrates and a significant increase in dry weight (Croes, 1967a; Esposito *et al.*, 1969; Roth, 1970).

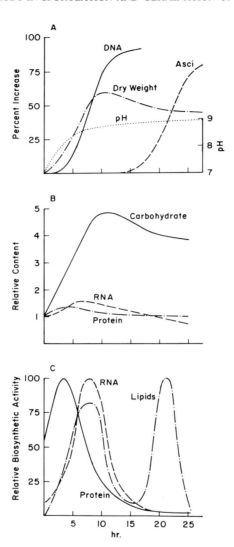

FIG. 1. An idealized representation of major events during sporulation of *S. cerevisiae*. The formation of ascospores takes about 24 hours, DNA synthesis being followed by two meiotic divisions at approximately 12 and 16 hours. Although there is considerable *de novo* synthesis of protein and RNA, extensive turnover reduces the total protein and RNA content to initial levels.

Total cellular carbohydrate content and dry weight both increase significantly during the first 10 hours. These figures are based on data from several laboratories, including Croes (1967a), Esposito *et al.* (1969), Sando and Miyake (1971), Mills (1972), Henry and Halvorson (1973), and Kane and Roth (1974).

A. Improvement of Sporulation Efficiency

1. MEDIA AND CULTURE CONDITIONS

Sporulation is an obligate aerobic process which is carried out in the absence of an external source of nitrogen. Most commonly, diploid cells are grown at 30 °C in a rich glucose medium such as YEPD (1% yeast extract, 2% Bacto peptone, 2% dextrose) to stationary phase—approximately 2×10^8 cells/ml—and then harvested, washed in water, and resuspended in 1% potassium acetate (pH 7) at a density of approximately 2×10^7 cells/ml. Fowell (1969) demonstrated that the efficiency of sporulation is dependent on cell density and that the observed increase in pH of sporulation medium from pH 7 to 9 in the first 3–5 hours is necessary for sporulation to proceed. Cells maintained in a buffered acetate medium at pH 7 are unable to complete sporulation.

One approach to the improvement of sporulation efficiency has been to change the presporulation vegetative growth medium. Croes (1967b) showed that the growth of cells in glucose medium to stationary phase was necessary in order for the cells to become adapted to oxidative growth. Roth and Halvorson (1969) found that cells pregrown on acetate medium containing yeast nitrogen base and yeast extract, buffered at pH 5.5 with potassium biphthalate, sporulated directly from logarithmic growth. This method was limited in that vegetative cells had to be harvested at a relatively low cell density (approximately 1.5×10^7 cells/ml). Furthermore, not all strains grow in this medium. More recently, Küenzi et al. (1974) modified this medium so that cells may be harvested at a density of approximately 6×10^7 cells/ml. The medium used by these workers contains 1% potassium acetate, 0.6% yeast nitrogen base (without amino acids), 0.5% yeast extract, 0.5% peptone, and 1.02% potassium biphthalate (pH 5.5). It is important in using acetate growth medium to provide vigorous aeration which can be achieved by incubation on a rotary shaker in flasks with a volume 10 times greater than the amount of liquid. Fast (1973) has also extended the study of growth media to other carbon sources, such as glycerol and galactose, which do not repress tricarbocylic acid cycle enzymes. Fast uses a growth medium which consists of 1% yeast extract, 2% peptone, and either 1% potassium acetate, 3% glycerol, or 2% galactose. These rich media support growth of all strains tested. Growth in media that do not repress tricarboxylic acid cycle enzymes results in better synchrony of sporulation (Roth and Halvorson, 1969; Fast, 1973). In some cases the percent of cells forming spores is increased. Furthermore, the sporulation process is often considered shortened from approximately 24 to 18 hours or less. However, even under these conditions mature ascospores are formed over relatively long intervals of 6–8 hours.

Even using acetate growth medium and optimal cell density and aeration during sporulation, the process is not synchronous. Haber and Halvorson (1972b) demonstrated that, even for cells growing in acetate presporulation medium, there is a considerable asynchrony which reflects the different ability of cells at different stages of the cell cycle to sporulate. It was concluded that certain cells in the population, notably newly formed daughter cells, sporulate poorly if at all. Similar conclusions for cells sporulated from stationary phase in dextrose medium have been reported by Yanagita *et al.* (1970). The study of sporulation is thus complicated by the fact that within the population there are some cells that do not complete sporulation.

2. Strains

Sporulation efficiency is strain-dependent, ranging from 50 to 95%. Table I lists some of the strains used in recent studies of sporulation. Although several strains used to study sporulation are derived from stock cultures at the Yeast Stock Center, Donner Laboratory, University of California, Berkeley, California, many studies are carried out on strains that are homothallic diploids of unknown origin. Indeed, some of the most efficient sporulating strains are the least well characterized genetically. Several mutants unable to complete sporulation have also been derived and are discussed below.

B.   Identification of Stages during Sporulation

1. Morphological Events

There are very few morphological phases that can be monitored during sporulation. Under the light microscope, without staining, only the appearance of refractile ascospores at the end of sporulation can be seen. Stages of meiosis from mononucleate cells to binucleate and tetranucleate cells can be visualized by the use of Giemsa stain. Several related staining techniques, based on the work of Pontrefact and Miller (1962) or Robinow and Marak (1966), have been published (Miller, 1968; Esposito *et al.*, 1970; Hartwell, 1970). In these methods the cells are fixed in 0.3% Formalin or Carnoy's fixative (Gurr, 1957), fixed to slides or cover slips, and treated to remove RNA. Several alternative treatments have been used, including incubation in 1 N NaOH for 1 hour (Esposito *et al.*, 1970), treatment with 1 N HCl at 60 °C for 10–15 minutes (Miller, 1968), or RNase digestion (Hartwell, 1970). After hydrolysis of RNA, the cells are washed in 0.1 M phosphate buffer, (pH 7) and stained with Giemsa stain for periods up to 3 hours. The cells are destained in 95% ethanol for 30 seconds and may be dehydrated by xylene treatment. Visualization of the nuclei may be improved if the cover slip is pressed very firmly against the slide to flatten the cells slightly.

TABLE I

REPRESENTATIVE a/α DIPLOID STRAINS OF *S. cerevisiae* USED IN THE STUDY OF SPORULATION

Strain	Phenotype	Genotype	Sporulation efficiency (%)	Reference
Y185	Heterothallic, wild type	$\dfrac{a\ ade2\ his5\ his8}{\alpha\ \ +\ \ +\ \ +}$	70	Haber and Halvorson (1972b); Henry and Halvorson (1973)
Y55	Homothallic, wild type	$\dfrac{a}{\alpha}$	90	Haber and Halvorson (1972b); Rousseau and Halvorson (1973a)
S41	Homothallic, arginine-requiring, cycloheximide resistant	$\dfrac{a\ HO\alpha\ HM\ arg4\ cyh1}{\alpha\ HO\alpha\ HM\ arg4\ cyh1}$	80	Esposito and Esposito (1969); Esposito et al. (1969, 1970); Fast (1973)
SK-1	Homothallic, wild type	$\dfrac{a}{\alpha}$	95 (very rapid sporulation)	Kane and Roth (1974)
4579	Heterothallic; adenine-, methionine-uracil-, leucine-requiring	$\dfrac{+\ \ a\ leu2\text{-}27\ \ +}{thr4\ \alpha\ leu2\text{-}1\ his4}$ $\dfrac{ade2\text{-}1\ met2\ ura3}{ade2\text{-}1\ met2\ ura3}$	60	Roth (1973)
D649	Heterothallic, wild type	$\dfrac{MAL2\ \ +\ \ a\ \ +\ \ +}{mal\ \ thr4\ \alpha\ leu2\ his4}$ $\dfrac{+\ \ trp1\ pet6}{adel\ \ +\ \ +}$ $\dfrac{ade2\ lys2}{+\ \ \ +}$	60	Sogin et al. (1972); Haber and Halvorson (1972b)

Electron microscope studies of thin sections (Moens, 1971; Moens and Rapport, 1971a,b), or by freeze-etching (Guth *et al.*, 1972), reveal considerable detail about the meiotic process and the formation of ascospores. These techniques have not yet been widely used to monitor temporal events during sporulation, although Moens *et al.* (1974) used the electron microscope to study alterations in normal sporulation in temperature-sensitive conditional mutants (see Section D,1).

2. GENETIC EVENTS

Meiotic events may also be followed by genetic techniques. As first shown by Sherman and Roman (1963), intragenic recombination of noncomplementing heteroalleles occurs at nearly the same time as premeiotic DNA synthesis. It is therefore possible to use diploids auxotrophic for an amino acid (e.g., leucine-requiring, genotype *leu2-1/leu2-27*) and determine the number of prototrophs arising by recombination, which appear when plated cells are spread on agar plates not containing the amino acid. The proportion of prototrophic cells arising by recombination varies widely with the pairs of alleles selected but may reach 0.5% of the cells (Roth and Fogel, 1971). By taking samples at different times during sporulation, the period of recombination (and presumably DNA synthesis) can be found. Similarly, one can determine the time of haploidization of the genome following meiosis II by the appearance of canavinine-resistant colonies in a diploid which is heterozygous for the recessive allele canavanine resistance (*can^r*). A combination of these two genetic markers (i.e., *leu2-1/leu2-27* and *CAN^S/can^r*) enables the investigator to relate biochemical events during sporulation to the appearance of the haploid state. As seen in Section D,2, the same techniques have been used quite successfully in the isolation of mutants unable to complete sporulation.

C. Measurement of Macromolecular Synthesis

To date, the study of macromolecular synthesis has centered on the changes in the major constituents of the cell during sporulation. Techniques for the measurement of changes in nucleic acid, protein, carbohydrate, and lipid content during sporulation have all been developed. An analysis of the changes in subclasses of lipids, carbohydrates, and a few RNA species has also been achieved. Virtually no work has been reported concerning the appearance of sporulation-specific proteins, mRNA species, or other macromolecules.

One major problem confronting all this work is the general lack of permeability of the sporulating cell to both inhibitors of macromolecular synthesis and radioactively labeled precursors. As shown by Mills (1972), cells undergoing sporulation show a dramatically reduced ability to incorporate amino acids or adenine. Mills found that the reduced permeability reflected a change in the external pH of the sporulation medium, which rises from approximately pH 7 to a maximum of about pH 9 or 8.5. The uptake of both amino acids and adenine is much greater at pH 6 than at pH 9. As a general solution to this problem of permeability, Mills proposed that short-term labeling can be carried out in pH 6 medium for as long as 1 hour without any serious effect on the ability of cells to sporulate if returned to the high-pH medium. Cells maintained at a low pH, however, are unable to sporulate. This method has made it possible to label especially RNA at a much greater specific activity. However, P. J. Wejksnora and J. E. Haber found that transferring cells to fresh medium at pH 6.5 during the period of labeling alters not only the incorporation of adenine-^{3}H or ^{32}P into RNA, but also the kinetics of synthesis. After 1 hour at pH 9, virtually all the label is found in 35 S ribosomal precursor RNA, and almost no labeled 18 or 26 S rRNA is formed. However, at pH 6.5 a very significant fraction of the RNA labeled in this interval is found to be processed to 26 and 18 S RNA. It appears that the method of lowering the pH of sporulation medium to increase the efficiency of labeling must be regarded as a very useful method with some serious potential liabilities. It is not possible from present data to determine whether all the material labeled during a 30-minute or 1-hour interval is of the same type, as well as amount, that would have been found under normal sporulating conditions.

1. DNA SYNTHESIS

a. Calorimetric measurements. The measurement of DNA synthesis during sporulation in yeasts has been carried out almost exclusively by colorimetric rather than radioisotope methods.

i. Diphenylamine test. The use of diphenylamine reactivity with DNA was first described by Burton (1956). For use in yeasts the procedure of pretreatment described by Ogur and Rosen (1950) is generally adopted.

At least 2×10^8 cells are harvested and washed in cold 70% ethanol. Cells are then successively washed with 0.1% perchloric acid (PCA) in 70% ethanol, 0.2 M PCA, and 1 M PCA. Cells are then extracted twice at 70 °C with 1 ml of 1 M PCA for 20 minutes. For DNA determinations involving late stages of sporulation or spores, a third PCA extraction is often used. The colorimetric test is carried out by mixing 1 ml of the extract with 1 ml of diphenylamine reagent (2% diphenylamine in glacial acetic acid and containing 0.08 mg/ml acetaldehyde). The samples are mixed and incubated at room temperature or at 30 °C for 18–24 hours and monitored at 595 nm.

The number of cells generally needed for an adequate assay of DNA is quite large, usually about 40 ml of the sporulation culture (i.e., 8×10^8 cells). The method also suffers from problems involving accuracy, as the extraction of DNA from cells in the later stages of sporulation and from ascospores is more difficult. Recently, Hatzfeld (1973a) reported that as cells enter stationary phase there is a serious fluctuation in the diphenylamine test. There is a 25% decrease in absorbance at 595 nm with no change in cell number as cells remain at stationary phase. These results suggest that diphenylamine-positive material may appear at the end of exponential growth, and may seriously affect the measurement of the percent increase in premeiotic DNA synthesis.

Recently, R. Roth and R. Silva-Lopez (personal communication) devised a micromethod based on the Burton procedure. At intervals during sporulation 5- to 10-ml samples are collected in 15-ml centrifuge tubes and washed once with sterile distilled water; the pellets are frozen until used. All steps take place in the same tube, and the diphenylamine reagent is added directly to the cell pellets at the end. The frozen pellets are thawed and well-suspended in 8 ml of 0.2 M PCA in 50% ethanol (ETOH) for 40 minutes at room temperature. Samples are centrifuged (5000–6000 rpm for 7–10 minutes), and the pellets washed with 4 ml of PCA-ETOH. A small amount of glass beads (120 μm, acid-washed) is mixed with the pellet, using a glass rod, until the pellet is evenly disrupted. The tubes are cooled to 4°C in ice, and the pellet resuspended in 4.0 ml of ETOH–ether (3:1 v/v) at 60°C for 5 minutes. The samples are centrifuged. The supernatant is removed by careful suction, and the lipid extraction with ETOH–ether is then repeated. After centrifugation the pellets are dried at 60°C in an oven (4–8 hours; but up to 24 hours in an oven will not harm the samples). After drying, 1.0 ml of 1.0 M PCA solution is added, and the pellets are resuspended. Tubes are incubated with periodic shaking at 70°C for 30 minutes and allowed to cool and to each tube is added exactly 1.0 ml of diphenylamine reagent. After incubation at 30°C for 17–19 hours for color development, tubes are centrifuged and the supernatant collected. All samples and standard DNA samples are measured spectrophotometrically at 595 and 650 nm. The difference between the two readings is proportional to the DNA concentration in each tube.

ii. Diaminobenzoic acid method. A second method for the measurement of DNA involves the use of diaminobenzoic acid. This method is far more sensitive, and can be carried out on approximately one-tenth the number of cells used in the standard Burton method. This technique, adapted from a general method of Kissane and Robins (1958) by Wehr and Parks (1969), includes the following essential steps.

Cells are precipitated in an equal volume of 10% trichloracetic acid (TCA) and washed four times with 0.1 N potassium acetate in 95% enthanol. The precipitate is then twice extracted with 95% ethanol at 60 °C. The pellet is then evaporated to dryness and resuspended in 0.2 ml of 1.0 N HCl containing 0.3 gm/ml of 3,5-diaminobenzoic acid dihydrochloride (Aldrich Chemical Co., Milwaukee, Wisconsin) and heated at 60 °C for 30 minutes. This sample is then diluted with at least 1 ml of 0.6 N perchloric acid, centrifuged to remove debris, and measured in an Aminco-Bowman spectrofluorimeter by excitation at 408 nm and emission at 520 nm. This assay is linear in the range of 0.5–5 mg of DNA. It is important that fluorometry be carried out within approximately 1 hour of time of incubation, as the color is not stable. In a modification of this procedure developed by Milne (1972), the cells are first incubated in 1 N NaOH for 24 hours before precipitation with TCA and subsequent steps.

   b. Radioisotope Methods. Measurement of DNA synthesis by incorporation of radioisotopes is complicated by the facts that much more RNA than DNA is synthesized and that most yeast strains are unable to take up labeled thymidine or deoxythymidine monophosphate (dTMP) which is incorporated only into DNA (Grivell and Jackson, 1968). The incorporation of radioisotopes into DNA can be measured by determining the amount of alkali-resistant, DNase-sensitive material (Bücking-Throm *et al.*, 1973). Cells are hydrolyzed with 1 N NaOH for 16 hours and then dialyzed against 0.1 M tris-HCl buffer (pH 7.4) containing 0.3 M MgCl$_2$. Two aliquots are then taken, and one is treated with 80 μg/ml DNase. Both samples are precipitated with an equal volume of 20% TCA to which 400 μg of unlabeled high-molecular-weight DNA is added as carrier. Precipitates are collected on nitrocellulose filters and counted in a scintillation counter. The amount of DNA is determined from the difference between the amount of radioactivity before and after treatment with DNase. Hatzfeld (1973b) has reported that adenine-^{3}H can be found in non-DNA-containing material after a somewhat less extensive treatment, and suggests uracil-^{3}H be used.

   Recently, several mutants of yeasts have been isolated that are able to take up dTMP into DNA (Brendel and Haynes, 1972; Wickner, 1974). Wickner has reported four complementation groups of dTMP uptake mutants *(tup)*. It is interesting to note, however, that of 15 alleles of *tup1* studied by Wickner, only diploids *(tup1/tup1)* containing the allele *tup1–108* are able to sporulate. Whether *tup1–108* diploids or diploids homozygous for *tup2*, *tup3*, or *tup4* are able to take up dTMP during sporulation has not yet been reported. These strains nevertheless offer exciting prospects for specific labeling of DNA in the study of meiosis and recombination.

   c. DNA Synthesis Inhibitors. To inhibit DNA synthesis, R. Roth (personal communication) used hydroxylurea, as described by Slater (1973), for yeasts growing vegetatively. Concentrations of 100 mM are needed to block DNA synthesis and sporulation.

It should also be possible to use diploids homozygous for several different temperature-sensitive cell division cycle mutations (*cdc*) affecting initiation of DNA synthesis (Hartwell, 1973). Simchen (1974) has obtained such strains by ultraviolet-induced mitotic crossing-over and found that strains unable to synthesize DNA at 34 °C (e.g., *cdc* 4) do not sporulate.

2. RNA SYNTHESIS

a. Measurement of Synthesis.   Changes in the amount of total RNA during sporulation are generally measured by the method of Ceriotti (1955). Measurement of *de novo* synthesis is accomplished by monitoring increases in the accumulation of radioisotopes into whole RNA or particular electrophoretically identifiable species.

At the stationary phase, and during sporulation, there is a significant increase in the amount of RNase activity found in cells. Such RNase activity poses serious problems for the isolation of intact RNA from sporulating cells. RNA is obtained by several extractions with an equal volume of phenol after breaking the cells either in frozen buffer in a French pressure cell or Eaton press (Bhargava and Halvorson, 1971), or in a Bronwill homogenizer using glass beads (Mills, 1974). The addition of 0.1% or more sodium dodecyl sulfate is important to reduce RNase activity and denature ribosomes. Water or buffer-saturated phenol may be stabilized against oxidation for periods up to several weeks by the addition of 0.1% 8-hydroxyquinoline. The precipitation of RNA from the aqueous phase after phenol extraction by 67% ethanol is facilitated by the presence of at least 0.1 M NaCl. Residual phenol may be removed by extraction with ether. Some polyphosphates (see below) and carbohydrates remain after this extraction procedure.

Analysis of the relative synthesis of RNA species during sporulation has generally employed polyacrylamide gel electrophoresis of labeled RNA, according to the methods described by Loening (1967). There are problems associated with each of the isotopes commonly employed. Labeling with ³H- or ¹⁴C-labeled compounds is restricted by poor permeability at the high pH characteristic of sporulation. Labeling with ³²P yields RNA of very high specific activity; however, ³²P is incorporated into other TCA- (and ethanol-) precipitable molecules—most notably (and troublesome) polyphosphates. Phosphate is first incorporated into large polyphosphates before utilization in the synthesis of phosphorylated nucleosides. Polyphosphates are apparently preferentially used even when large amounts of inorganic phosphate are added to the medium, so that the addition of unlabeled phosphate slows but does not prevent continued incorporation of ^{32}P into RNA (Sogin *et al.*, 1972; Chaffin *et al.*, 1974).

The amount of *de novo* RNA synthesized during sporulation can be measured against an internal reference by first labeling cells with adenine-^{14}C or uracil-^{14}C during vegetative growth. Major RNA species (rRNA and RNA) labeled prior to sporulation are found in RNA extracted during sporulation. There is nevertheless a significant degradation of RNA during sporulation (Croes, 1967a; Esposito *et al.*, 1969), so that the amount of RNA labeled during vegetative growth decreases by as much as 30–50% after 24 hours of sporulation. The breakdown of rRNA is such that the labeled rRNA remaining is still found in an equimolar ratio. There is some confusion about the turnover of RNA during sporulation. If cells are labeled with ^{14}C during vegetative growth, one finds very little ^{14}C incorporation into 20 S RNA (which normally accumulates only during sporulation) (Sogin *et al.*, 1972). Thus RNA is degraded during sporulation, but not effectively recycled into new RNA. Hopper *et al.* (1974) found that most of the labeled adenine liberated by RNA degradation is found in the medium in the form of 2′- and 3′-AMP.

b. Polyribosomes and mRNA.   Recently, attempts were initiated to study mRNA in sporulating cells. Mills and Frank (1973) used the method of Martin (1973) to examine the proportion of ribosomes and polyribosomes during sporulation. This method involves the dissociation of ribosomes not attached to mRNA into subunits, while at the same time leaving ribosomes attached to mRNA as intact 80 S monosome units. The isolation of intact mRNA requires, however, the formation of polysomes not degraded by RNase.

Mills (1974) developed a method in which the majority of polyribosomes appear to be isolated intact. Cells undergoing sporulation are broken either by grinding them in a mortar with sand or granular Dry Ice for approximately 1 minute, or by disruption with glass beads in a Bronwill homogenizer for 15 seconds. Homogenization is carried out in the presence of a small volume of lysing buffer containing 0.05 M triethanolamine (pH 7.4), 0.5 M KCl, and 0.01 M MgCl$_2$. After disruption the samples are diluted with lysing buffer and centrifuged at 12,000 g for 10 minutes to remove debris. Polysomes are separated by layering a 0.5- to 0.9-ml aliquot of the supernatant on an 11-ml, 10–40% sucrose gradient made with lysing buffer and containing a 0.65-ml cushion of 50% sucrose. Gradients are centrifuged in a Spinco SW41 rotor at 150,000 g for 90 minutes at 4°C. In order to stabilize polysomes during extraction, 100 μg/ml cycloheximide is added to samples 5 minutes prior to collection and breaking. Mills reported that the fraction of ribosomes in polyribosomes is greatest (about 60%) if cycloheximide is added to cells transferred to fresh sporulation medium at pH 6. The effect of this pH shift on translation has not been assessed, however.

   c. *Inhibitors of RNA Synthesis.*   Several inhibitors of RNA synthesis, including lomofungin (Cannon *et al.*, 1973), cordycepin (Roth and Anderson, 1972), and thiolutin (Tipper, 1973), have been reported in studies of vegetative cells. To date, no use of these has been reported with sporulating cells.

3.   Protein Synthesis

   Measurement of the synthesis of total protein during sporulation has generally been monitored by the incorporation of radioactively labeled amino acids into a fraction that is insoluble in 5% TCA and resistant to boiling for 30 minutes (Rodenberg *et al.*, 1968). A study of changes in individual protein components of the cell can be monitored using poly-acrylamide gels. One major problem in the analysis of proteins during sporulation is that the specific activity of several proteases increases greatly during this period (A. Singh, personal communication). Protein samples can be protected, at least in part, by the addition of 0.3 mg/ml phenylmethyl-sulfonyl fluoride (PMSF, protease inhibitor); however, at least one of the three major proteases that appear during sporulation is not inhibited by PMSF (Hata *et al.*, 1967; A. Singh, personal communication).

   The importance of adding PMSF to buffers throughout the isolation of enzymes cannot be stressed too greatly. For example, in the absence of PMSF during some stages of the isolation of RNA polymerases of *S. cerevisiae* during sporulation, J. Sebastian and J. E. Haber (unpublished observations) found a significant change in the chromatographic behavior of α-ammanitin-sensitive RNA polymerase II. Subsequent attempts to repeat this observation when the addition of PMSF was more carefully controlled demonstrated that this alteration in RNA polymerase II during sporulation was an artefact caused by a high level of proteases during extraction. PMSF is only sparingly soluble in aqueous solutions. If added directly to aqueous buffers, it tends to remain insoluble, forming a film along the edges of a flask or tube. PMSF may be solubilized in ethanol or dimethy sulfoxide (DMSO). Even under these conditions it is important to add the PMSF to aqueous solution slowly with vigorous stirring.

   Very little work has been reported on the isolation of proteins that appear only during sporulation.

   Protein synthesis is easly inhibited by the addition of 100 mg/ml cyclo-heximide. The labeling of proteins during sporulation can also be enhanced by lowering the pH of the sporulation medium to 6.0 (Mills, 1972).

4. CARBOHYDRATE SYNTHESIS

Methods for the analysis of carbohydrates synthesized or utilized during sporulation have been described by Kane and Roth (1974), using modifications of the methods of Berke and Rothstein (1957) and Trevelyan and Harrison (1952, 1956).

5. LIPID SYNTHESIS

Lipid synthesis during sporulation in yeasts has been described by Henry and Halvorson (1973), using the methods of Letters (1966) as modified by Getz *et al.* (1970).

D. Isolation of Mutants Specific for Sporulation

Mutants that grow normally but are unable to sporulate have been isolated by several techniques.

1. HOMOTHALLIC STRAINS

Homothallic strains of yeasts have been used in order to isolate diploid yeasts homozygous for temperature-sensitive sporulation mutations. Homothallic strains sporulate to produce four haploid spores, however, these spores diploidize on germination and yield a/α diploids homozygous for all other markers. One can then mutagenize haploid spores of such a strain, allow the cell to germinate, and then test these new homothallic diploids for alterations in sporulation. In this way, Bresch *et al.* (1968) using *Schiz. pombe*, and Esposito and Esposito (1969) using *S. cerevisiae*, isolated a very large number of mutants temperature-sensitive for sporulation, and characterized several of them for defects of spore formation and events during meiosis (M. S. Esposito *et al.*, 1970; R. E. Esposito *et al.*, 1972; *Moens et al.*, 1974).

2. DISOMIC STRAINS

An alternative approach for obtaining mutants, especially in the early stages of sporulation (including meiosis and recombination), was developed by Roth and Fogel (1971). These investigators took advantage of the fact that a haploid strain disomic for chromosome III and heterozygous (a/α) for the mating-type allele undergoes most of the events of sporulation up to actual formation of the spore wall. Mutations generated on all chromosomes except chromosome III will be hemizygous and thus expressed. By this technique, Roth and Fogel isolated several mutants defective in DNA synthesis or recombination (Roth, 1973).

3. DOMINANT MUTATIONS

It is also possible to generate dominant mutations simply by treating diploid cells with mutagen. Very little work has been reported to date on such mutants, although various laboratories have indicated that several of such dominant mutations exist (Esposito *et al.*, 1972; Hopper and Hall, 1973; J. E. Haber, unpublished).

E. "Sporulation-Specific" Events

One of the major themes found in the recent literature on sporulation in yeasts is the attempt to demonstrate which events are sporulation-specific. Thus biochemical events found in cells able to undergo sporulation are compared with events in cells genetically unable to complete the process [i.e., diploid cells homozygous for mating-type (*a/a* or *α/α*) or haploid strains]. There are some quite distinct differences between *a/α* cells and these nonsporulating types. For example, Roth and Lusnak (1970) showed that DNA synthesis preceding meiosis occurred only in *a/α* diploids. Similarly, Kadowaki and Halvorson (1971) described the appearance of stable 20 S RNA in sporulating *a/α* cells, but not in haploids or *a/a* and *α/α* diploids under the same conditions. However, recent studies on changes in lipids (Henry and Halvorson, 1973), carbohydrates (Kane and Roth, 1974), and major protein species (Hopper *et al.* 1974) show virtually no differences during the first half of sporulation. That there are no observable differences in these processes does not necessarily mean that there are no sporulation-specific events. It is certainly not necessary that all sporulation events be controlled by the *a/α* genotype which is necessary for DNA synthesis and meiosis. Many biochemical events may in fact be specific responses to the pseudostarvation conditions in which the cells find themselves regardless of mating genotype.

II. Germination

A. Spore Suspension

1. PREPARATION OF LARGE QUANTITIES OF ASCI AND SINGLE SPORES

Fungal spores free of their ascus or of the outer sporangial wall respond better to a germinant after removal of the permeability barrier (Brierley, 1917; Gwynne-Vaughan, 1934; Stuben, 1939; Machlis and Ossia, 1953;

Lowry *et al.*, 1956). Following sporulation the ascus walls of *Saccharomyces fragilis* and *Schiz. pombe* rapidly lyse, liberating the spores. In *S. cerevisiae*, the ascus wall remains intact until the spores germinate. Spores of *Schiz. pombe* are released from the asci after incubation at low temperature and can be enriched by the killing of parental cells with alcohol (Leupold, 1957). Susceptibility to rupture is ascribed to the presence of high level of α-glucanase activities toward $(1 \rightarrow 3)$ and $(1 \rightarrow 6)$-glucans (Phaff, 1971).

Resistant asci can be disrupted by pressure with the tip of a glass needle, which may damage the spores, or by the use of enzymes that dissolve the ascus wall: extracts of *Bacillus polymyxa* (Wright and Lederberg, 1957), digestive juice of the snail *Helix pomatia* (Johnson and Mortimer, 1959), the protease fraction of *Streptomyces griseus* Rousseau and Halvarson, 1969), or others (Kitamura, 1971). However, assays that require a micromanipulator to dissociate the spores makes the method tedious and time-consuming.

Biochemical analysis requires quantities of single spores not obtainable with a micromanipulator and which are free of contaminations from the ascus. Oil-phase separation (Emeis and Gutz, 1958) is a technique in which the lipophilic spores are taken up on the oil phase and the debris and the vegetative cells remain in the aqueous phase. Ascus-free spores tend to clump, thereby lowering the yield. This technique does not allow enough single spores to be recovered in the oil phase, since too many asci contaminate the spore preparations (Enebo *et al.*, 1968; Fowell, 1969). Resnick *et al.* (1967) introduced stable-flow, free-boundary electrophoresis to harvest an almost pure suspension of spores, based on the difference in electrophoretic properties of the vegetative cells and the spores, however, this technique is limited in the yield of single spores that can be obtained. For the same reason the oil-phase or liquid paraffin (Windisch, 1961) procedure is unsatisfactory for biochemical studies of germination. Siddiqui (1971) improved the paraffin technique, using ultrasound to disperse the spores. Partsch (1969) has described a procedure involving superposed filters to isolate single spores.

In general, the methods for preparation of single spores of *S. cerevisiae* give poor yields, are unable to prevent clumping, and lack simplicity. The use of spores, free of their asci, is required to avoid the presence of any activator or inhibitor of germination from the mother cell, to improve the synchrony of a germinating spore suspension, and to prevent cell agglutination and cell fusion.

Rousseau and Halvorson (1969) have described a method for the preparation of yeast spores, which is convenient, reproducible, avoids clumping, and gives a final preparation which is suitable for studies on germination. A diploid *S. cerevisiae* was grown first on slants of YMA medium

[3 gm of yeast extract, 3 gm of malt extract, 5 gm of peptone, 10 gm of glucose, 20 gm of agar, and 1 liter of water (pH 5.5)] for 2 days at 30°C. The cells were washed off the slant with sterile water and inoculated at a density of 5×10^4 cells/ml into YEPD medium. After 24 hours of aeration at 30°C, the culture was harvested by centrifugation for 15 minutes at 550 g at 25°C, washed once with distilled water, and inoculated at a final level of 5×10^7 cells/ml of sporulating medium into 2-liter Erlenmeyer flasks containing 800 ml of medium (0.1% glucose, 0.25% yeast extract, and 2% potassium acetate) containing 20 μg/ml tetracycline to prevent contamination. The flasks were aerated on a shaker at 300 rpm at 30°C for 6 days. After 132 hours sporulation was over 97%. The asci were then harvested by centrifugation and washed twice with cold distilled water. The yield from 10 liters of sporulation medium was 85 gm of wet asci.

To prepare a suspension of single spores, a suspension of asci (about 5–10 OD at 600 nm) was incubated overnight at 30°C on a shaker with 5 mg/ml pronase in 0.02 M citrate–phosphate buffer (pH 7). Spores do not germinate in this medium. To disperse the spores after the incubation, Tween 80 was added to a final concentration of 1%, and the suspension was passed through a French press (10°–15° C) at 7000 psi by using a flow rate of 20-25 ml per minute. The suspension was then centrifuged at 25 g for 20–25 minutes, and the pellet was suspended in medium containing 1% Tween 80. Occasionally, aggregates of single spores occur which are easily removed by allowing the suspension to settle for a few minutes and discarding the sediment.

By the above procedure suspensions of single spores can be obtained which remain dormant for several days when stored as a thick suspension at 4°C. The treatments employed (pronase, Tween 80, or pressure to disrupt the ascus) affect neither their ability to germinate nor their viability. Higher centrifugation speeds of single-spore preparations should be avoided, since they give rise to clumps of spores which are difficult to disperse.

2. PREPARATION OF SPORES LACKING MITOCHONDRIAL DNA

Sporulation is an aerobic process, and established petite mutants, resulting from alterations in mitochondrial DNA, are respiratory-deficient and do not sporulate. We designed an approach (Küenzi et al., 1974) to determine the contribution of the mitochondrial system to germination and outgrowth that takes advantage of the unusual effect of the intercalating agent ethidium bromide (EthBr).

Short exposure of a vegetative culture of S. cerevisiae to EthBr results in the complete conversion of wild-type rho+ cells to petite rho− cells, caused by the rapid destruction of mtDNA (Goldring et al., 1970; Perlman and Mahler, 1971). After removal of the EthBr, the culture is in a transient state, being genotypically petite but retaining respiratory activity for a long time (Mahler and Perlman, 1972). Such a culture is capable of sporulating (Küenzi et al., 1974). Yeast strains are grown on medium containing 1% potassium acetate, 0.6% yeast nitrogen base (without amino acids), 0.5% yeast extract, 0.5% peptone (Difco, Detroit, Michigan), and 1.02% potassium biphthalate. During the exponential-growth phase, EthBr is added to a concentration of 25 μg/ml, and growth continued for one generation of growth. Almost the entire population is converted to the rho$^-$ genotype without loss in viability, and mtDNA is not detected. The cells are centrifuged at room temperature, washed twice with sterile, distilled water, resuspended in sporulation medium (1% potassium acetate buffered to pH 7) to 4.4–5.0 OD and incubated at 30°C. The percentage of asci formed is only slightly below that in the control culture. The control cultures produce asci with three to four spores, whereas two-spored asci are predominant in EthBr-treated culture. The spores are harvested by centrifugation, washed, and stored as an aqueous suspension at 4°C.

The asci and unsporulated cells are suspended in 0.067 M potassium phosphate buffer (pH 6.8) containing 0.1% of a saturated solution of mercaptoethylamine, 1% Tween 80, and 1 mg of "protoplasting enzyme" per milliliter (Kirin Brewery Co., Takasaki, Japan). After incubation for 1 hour at 30°C, all the unsporulated cells are lysed and asci walls are digested. The spores are washed five times with 0.067 M potassium phosphate buffer (pH 6.8) containing 1% Tween 80. The spores are suspended in ice-cold yeast extract (1%), peptone (2%), and Tween 80 (1%) medium and passed through a French pressure cell at 7000 psi. The resulting single-spore suspension is adjusted to a concentration of 3×10^7 spores/ml before use in germination experiments.

3. Storage of Spore Preparation

Large-scale preparative methods require storage of the spores at some stage of their isolation (Rousseau and Halvorson, 1969). Lyophilization of asci led to a 50% loss in viability. Also, rapid freezing of asci with Dry Ice and acetone in citrate–phosphate–mercaptoethylamine hydrochloride buffer, 5% glycerol, or 5% DMSO led to a significant decrease in viability. When liquid asci suspensions were placed in the freezer at -15°C (slow freezing), excellent recoveries were observed even after 70 days of storage.

B. Germination Requirements

It has been known since for some time that the germination of ascospores of S. cerevisiae could occur either in a nitrogen or an oxygen atmosphere. In 1961, Palleroni investigated qualitatively the nutritive requirements of spores isolated with a micromanipulator and found that only glucose, fructose, and mannose could trigger the germination process. Rousseau and Halvorson (1973b) extended this study and found that only D-glucose and sucrose (1%) could trigger a rapid decrease in absorbance from the spore population. D-Mannose, D-galactose, and D-maltose could support the germination, but the process was delayed and highly asynchronous as measured microscopically or by the decrease in absorbance. The same kinetics of germination was observed for the first 5 hours of incubation, but thereafter a faster rate of growth was observed with mannose than with galactose, and a lesser rate for maltose. The yeast spores germinated poorly in the presence of D-trehalose.

The negative results (Rousseau and Halvorson, 1973b) obtained with other substrates (amino acids and nucleotides) strongly support the critical role of a carbohydrate such as glucose, as a germination stimulant. The phosphate concentration in the germination medium is important. The optimal conditions for a synchronous germinating spore population require a temperature of 30°C, glucose in the medium, and a pH within the range 5.4–8.2; carbon dioxide concentration does not seem to be an important factor.

C. Methods for Measuring Germination and Outgrowth

1. CYTOLOGICAL

Skinner et al. (1951) and Hashimoto et al. (1958) contend that the inner spore wall in yeasts serves as the cell wall of the new vegetative cell. Electron microscope study of germinating yeast spores (Hashimoto et al., 1958) showed that the inner spore coat became very electron-dense during germination and outgrowth but did not clearly suggest that the surface of the cell was changing with time of incubation. The exact chemical nature of the outer spore coat and inner spore coat has not yet been investigated; only a few (contradictory) clues to its nature exist at the present time (Brigley et al., 1970; Miller and Hoffman-Ostenhof, 1964; Snider and Miller, 1966). According to Miller (1969), the surface of the spore is hydrophobic in nature and contains lipids. However, Brigley et al. (1970) showed that the electrophoretic mobilities of the spores changed when they were treated with proteolytic enzymes, suggesting the presence of a protein layer on the surface.

During the first 90 minutes of germination, the percentage of germinated cells, indicated by semirefractility and phase darkness, increases from 55% at 30 minutes to 95% at 90 minutes. By 120 minutes nearly 100% of the population has germinated. All germinated spores are swollen by this time, and slightly elongated.

A typical ascus contains four ungerminated ascospores arranged in the shape of a tetrahedron. In phase-contrast microscopy most asci appeared to be planar (Rousseau *et al.*, 1972). The outer surface of the intact ascospores appears smooth in the scanning microscope, but an individual spore free of both the ascus wall and outer spore wall exhibits some surface irregularity. The most intense modification of the spore surface occurs between 30 and 60 minutes, which concurs with the greatest loss in refractility (bilayer). The spore at 60 minutes is rough, and has continuous ridges over the entire surface. At 90 minutes maximum phase darkness is reached; also, the spore swells and becomes pear-shaped. A scanning electron micrograph reveals heightened surface irregularity at the beginning of outgrowth. Further outgrowth (120 minutes) is seen as an extension of one end of the cell, which continues to 150 minutes by which time initial bud formation has occurred.

2. Spectrophotometric

During germination (Fig. 2) a rapid decrease in light absorbance of the spore suspension occurs (Rousseau *et al.*, 1972). A strong correlation appears between the decrease in light-refractile spores or the increase in nonrefractile spores. The swelling of the individual cells begins at 90 minutes. At 120 minutes the spores start to increase in absorbance and continue to 240 minutes. On initial bud formation at 150 minutes, the cells clump; thereafter, it becomes difficult to determine accurately the percentage of the various morphological cell forms. Cell weight, expressed as percent dry weight, decreases about 10% during the first 60 minutes of incubation, after which it increases (mainly as a result of increases in KOH-soluble carbohydrates) (Northcote and Horne, 1952; Pazonyi and Markus, 1955; Trevelyan and Harrison, 1956; Rousseau *et al.*, 1972).

3. Biochemical and Physiological Properties

Several biochemical and physiological properties can be used to follow the kinetics of germination and outgrowth.

a. Respiration.   Studies of the rate of carbon dioxide evolution and oxygen uptake in SSM medium and in citrate–phosphate buffer disclose very intersting relationships (Fig. 3) (Rousseau and Halvorson, 1973a). There is a burst of carbon dioxide production in the first 15–30 minutes of germination. After this the rate of carbon dioxide changes, depending on the germination medium; in the synthetic medium there are two different and linear rates of increase (from 1 to 3 hours, and from 3 hours on) (Fig. 3A), whereas in the buffer only one low rate occurs following the burst (Fig. 3B). Oxygen uptake experiments also demonstrate a break in the rates of increase at about 210 minutes after the onset of germination in SSM medium

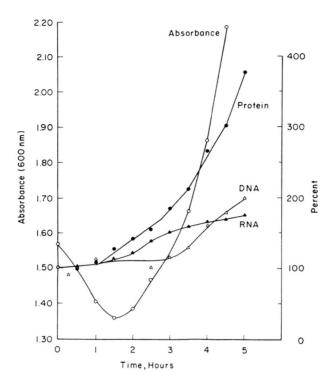

FIG. 2. Kinetics of net macromolecular synthesis during *S. cerevisiae* spore germination and outgrowth in succinate synthetic medium. Absorbance (open circles) of the suspension was also determined. Values for macromolecules are given as percent of T_0 values: open triangles, DNA; solid triangles, RNA; circles, protein. (From Rousseau and Halvorson, 1973c.)

(Fig. 3A). In the buffer there is almost linear oxygen consumption but, after 120–150 minutes of germination, the rate follows the same as found in the case of carbon dioxide production.

The burst of carbon dioxide in SSM medium or in buffer at the beginning of germination is more evident when the CO_2/O_2 ratio (respiratory quotient, RQ) is plotted against time. In complete SSM medium spores exhibit a high RQ of 18.5 after 15 minutes of germination, decreasing to 4 after 140 minutes and increasing slowly thereafter. However, suspended in the buffer supplemented with glucose, spores exhibit also a high RQ of 5 within 30 minutes after the onset of germination. This increase is less than that observed when spores are suspended in SSM medium. In buffer an RQ of 1 is obtained about 4 hours after the initiation of germination.

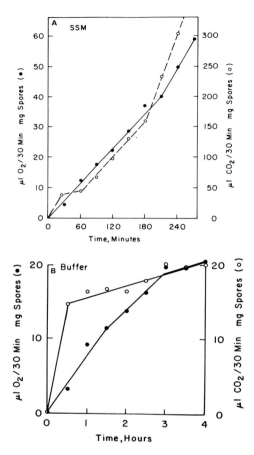

FIG. 3.   Kinetics of oxygen uptake and carbon dioxide evolution during germination and outgrowth. By using conventional Warburg procedures, the respiratory activity of yeast spores was studied in SSM medium and in citrate–phosphate buffer. (From Rousseau and Halvorson, 1973a.) Reproduced by permission of the National Research Council of Canada from the *Can. J. Microbiol.* **19**, 547–555 (1973).

b. Decrease in Lipids and Amino Acid Release.   During the first 4 hours of germination and outgrowth, the total lipids decrease continuously 75%. During this same period amino acids accumulate in the medium.

Before 60 minutes of incubation, amino acids cannot be detected in the medium. By 90 minutes proline, threonine, glycine, cystine, serine, and valine are found in the medium. Other amino acids are undetectable.

c. Macromolecular Synthesis.   Figure 2 shows the kinetics of net RNA, DNA, and protein synthesis during *S. cerevisiae* spore germination and outgrowth (Rousseau and Halvorson, 1973c). The increases observed are normalized to the initial levels in the spore. As can be seen, proteins increase first, followed soon after by RNA, and 2 hours later by DNA. Proteins increase continuously, whereas RNA synthesis shows a step increase about 2 hours after the onset of germination.

The level of RNA increases very slowly during the first 2 hours, then shows a significant increase at 2.5 hours, and decreases thereafter. During the first 2 hours of germination, the accumulation of protein parallels the accumulation of RNA.

Between 1 and 3 hours increases in protein occur in the absence of net DNA synthesis. After 3 hours both DNA and protein levels rise. From the data in Fig. 2, it appears that RNA transcription precedes replication.

ACKNOWLEDGMENTS

Part of the work described above was supported by NIH Grants GM 20056 (James E. Haber) and AI 10610–04 (Harlyn O. Halvorson).

REFERENCES

Berke, H. L. and Rothstein, A. (1957). *Arch. Biochem. Biophys.* **72**, 380.
Bhargava, M., and Halvorson, H. O. (1971). *J. Cell Biol.* **49**, 423.
Brendel, M., and Haynes, R. H. (1972). *Mol. Gen. Genet.* **117**, 39.
Bresch, C., Miller, G., and Egel, R. (1968). *Mol. Gen. Genet.* **102**, 301.
Brierley, W. B. (1971). *Ann. Bot. (London)* **31**, 127–132.
Brigley, M. S., Illingworth, R. F., Rose, A. H., and Fisher, D. J. (1970). *J. Bacteriol.* **104**, 588–589.
Bücking-Throm, E., Duntze, W., Hartwell, L. H., and Manney, T. R. (1973). *Exp. Cell Res.* **76**, 99.
Burton, K. (1956). *Biochem. J.* **62**, 315.
Cannon, M., Davies, J. E., and Jimenez, A. (1973). *FEBS (Fed. Eur. Biochem. Soc.) Lett.* **32**, 277.
Ceriotti, G. (1955). *J. Biol. Chem.* **214**, 59.
Chaffin, W. L., Sogin, S. J., and Halvorson, H. O. (1974). *J. Bacteriol.* **120**, 972.
Chen, A. W., and Miller, J. J. (1968). *Can. J. Microbiol.* **14**, 957.
Croes, A. F. (1967a). *Planta* **76**, 209.
Croes, A. F. (1967b). *Planta* **76**, 227.
Deutch, C. E., and Parry, J. M. (1974). *J. Gen. Microbiol.* **80**, 259.
Emeis, C. C., and Gutz, H. (1958). *Z. Naturforsch. B* **13**, 647.
Enebo, L., Johnsson, E., Nordstrom, K., and Möller, A. (1960). Quoted in "The Yeasts" (A. H. Rose and J. S. Harrison, eds.), Vol. 1, pp. 303–383. Academic Press, New York, 1969.
Esposito, M. S., and Esposito, R. E. (1969). *Genetics* **61**, 79.
Esposito, M. S., Esposito, R. E., Arnaud, M., and Halvorson, H. O. (1969). *J. Bacteriol.* **100**, 180.
Esposito, M. S., Esposito, R. E., Arnaud, M., and Halvorson, H. O. (1970). *J. Bacteriol.* **104**, 202.
Esposito, R. E., Frink, N., Bernstein, P., and Esposito, M. S. (1972). *Mol. Gen. Genet.* **114**, 241.

Fast, D. (1973). *J. Bacteriol.* **116**, 925.
Fowell, R. R. (1969). In "The Yeasts" (A. H. Rose and J. S. Harrison, eds.), Vol. 1, pp. 303–383. Academic Press, New York.
Getz, G. S., Jakovcic, S., Heywood, J., Frank, J., and Rabinowitz, M. (1970). *Biochim. Biophys. Acta* **218**, 441.
Goldring, E. S., Grossman, L. I., Dikrupnick, S., Cryer, D. R., and Marmur, J. (1970). *J. Mol. Biol.* **52**, 323–335.
Grivell, A. R., and Jackson, J. F. (1968). *J. Gen. Microbiol.* **54**, 307.
Gunge, N., and Nakatomi, Y. (1972). *Genetics* **70**, 41.
Gurr, G. T. (1957), "Biological Staining Methods," 6th ed. G. T. Gurr Ltd., London.
Guth, E., Hashimoto, T., and Conti, S. (1972). *J. Bacteriol.* **109**, 869.
Gwynne-Vaughan, H., and Williamson, H. S. (1934). *Ann. Bot. (London)* [N.S.] **48**, 261–272.
Johnson, J. R., and Mortimer, R. K. (1959). *J. Bacteriol.* **78**, 292.
Haber, J. E., and Halvorson, H. O. (1972a). *Curr. Top. Develop. Biol.* **7**, 61–83.
Haber, J. E., and Halvorson, H. O. (1972b). *J. Bacteriol.* **109**, 869.
Hartwell, L. H. (1970). *J. Bacteriol.* **104**, 1280.
Hartwell, L. H. (1973). *J. Bacteriol.* **115**, 966.
Hashimoto, J., Conti, S. F., and Naylor, H. B. (1958). *J. Bacteriol.* **76**, 406–416.
Hata, T., Hayashi, R., and Doi, E. (1967). *Agr. Biol. Chem.* **31**, 150.
Hatzfeld, J. (1973a). *J. Cell. Physiol.* **83**, 159.
Hatzfeld, J. (1973b). *Biochim. Biophys. Acta.* **299**, 34.
Henry, S. A., and Halvorson, H. O. (1973). *J. Bacteriol.* **114**, 1158.
Hopper, A. K., and Hall, B. D. (1973). *Genetics* **74**, Suppl., S119.
Hopper, A. K., Magee, P. T., Welch, S. K., Friedman, M., and Hall, B. D. (1974). *J. Bacteriol.* **119**, 619.
Kadowaki, K., and Halvorson, H. O. (1971). *J. Bacteriol.* **105**, 826.
Kane, S. M., and Roth, R. M. (1974). *J. Bacteriol.* **118**, 8.
Kissane, J. M., and Robins, E. (1958). *J. Biol. Chem.* **233**, 184.
Kitamura, K. (1971). *Arch. Biochem. Biophys.* **145**, 402–404.
Küenzi, M. T., Tingle, M. A., and Halvorson, H. O. (1974). *J. Bacteriol.* **117**, 80.
Letters, R. (1966). *Biochim. Biophys. Acta* **116**, 489.
Leupold, U. (1957). *Allg. Pathol. Bacteriol.* **20**, 535.
Leoning, V. E. (1967). *Biochem. J.* **102**, 251.
Lowry, R. J., Sussman, A. S., and Boventer-Heidenhain, B. V. (1956). *Mycologia* **48**, 241.
Machlis, L., and Ossia, E. (1953). *Amer. J. Bot.* **40**, 358.
Mahler, H. R., and Perlman, P. S. (1972). *Arch. Biochem. Biophys.* **148**, 115–129.
Martin, T. E. (1973). *Exp. Cell Res.* **80**, 496.
Miller, J. J. (1963). *Can. J. Microbiol.* **9**, 259.
Miller, J. J. (1968). *In* "Effects of Radiation on Meiotic Systems," pp. 177–184. IAEA, Vienna.
Miller, J. J. (1969). "Spectrum, Recent Trends in Yeast Research," pp. 73–79. Georgia State University, Athens, Georgia.
Miller, J. J., and Hoffman-Ostenhof, O. (1964). *Z. Allg. Mikrobiol.* **4**, 273–294.
Mills, D. (1972). *J. Bacteriol.* **112**, 519.
Mills, D. (1974). *Appl. Microbiol.* **27**, 944.
Mills, D., and Frank, K. (1973). *Genetics* **74**, Suppl., S182.
Milne, C. (1972). M.A. Thesis, University of Washington, Seattle.
Moens, P. B. (1971). *Can. J. Microbiol.* **17**, 507.
Moens, P. B., and Rapport, E. (1971a). *J. Cell. Biol.* **50**, 344.
Moens, P. B., and Rapport, E. (1971b). *J. Cell. Sci.* **9**, 665.
Moens, P. B., Esposito, R. E., and Esposito, M. S. (1974). *Exp. Cell Res.* **83**, 166.
Northcote, D. H., and Horne, R. W. (1952). *Biochem. J.* **51**, 232–236.

Ogur, M., and Rosen, G. (1950). *Arch. Biochem.* **25**, 262.

Palleroni, N. J. (1961). *Phyton (Buenos Aires)* **16**, 117–128.

Partsch, G. (1969). *Appl. Microbiol.* **17**, 925.

Pazonyi, B., and Markus, L. (1955). *Agrokem. Talajtan* **4**, 225–235.

Perlman, P. S., and Mahler, H. S. (1971). *Nature (London), New Biol.* **231**, 12–16.

Phaff, H. J. (1971). *In* "The Yeasts" (A. H. Rose and J. S. Harrison, eds.), Vol. 2, pp. 135–210. Academic Press, New York.

Pontrefact, R. D., and Miller, J. J. (1962), *Can. J. Microbiol.* **8**, 573.

Resnick, M. A., Tippets, R. D., and Mortimer, R. K. (1967). *Science* **158**, 803–804.

Robinow, G. F., and Marak, J. (1966). *J. Cell Biol.* **29**, 129.

Rodenberg, S., Steinberg, W., Piper, J., Nickerson, K., Vary, J., Epstein, R., and Halvorson, H. O. (1968). *J. Bacteriol.* **96**, 492.

Roman, H., and Sands, S. M. (1953). *Proc. Nat. Acad. Sci. U.S.* **39**, 171.

Roth, R. (1970). *J. Bacteriol.* **101**, 53.

Roth, R. (1973). *Proc. Nat. Acad. Sci. U.S.* **70**, 3087.

Roth, R., and Anderson, J. M. (1972). *Abs., Annu. Meet. Amer. Soc. Microbiol.* p. 176.

Roth, R., and Fogel, S. (1971). *Mol. Gen. Genet.* **112**, 295.

Roth, R., and Halvorson, H. O. (1969). *J. Bacteriol.* **98**, 831.

Roth, R., and Lusnak, K. (1970). *Science* **168**, 493.

Rousseau, P., and Halvorson, H. O. (1969). *J. Bacteriol.* **100**, 1426–1427.

Rousseau, P., and Halvorson, H. O. (1973a). *Can. J. Microbiol.* **19**, 547–555.

Rousseau, P., and Halvorson, H. O. (1973b). *Can. J. Microbiol.* **19**, 1311–1318.

Rousseau, P., and Halvorson, H. O. (1973c). *J. Bacteriol.* **113**, 1289–1295.

Rousseau, P., Halvorson, H. O., Bulla, L. A., Jr., and St. Julian, G. (1972). *J. Bacteriol.* **109**, 1232–1238.

Sando, N., and Miyake, S. (1971). *Growth Develop. Diff.* **12**, 273.

Sherman, F., and Roman, H. (1963). *Genetics* **48**, 255.

Siddiqui, B. (1971). *Hereditas* **69**, 67–76.

Simchen, G. (1974). *Genetics* **76**, 745.

Skinner, C. E., Emmons, C. H., and Tsuchiya, H. M. (1951). *In* "Molds, Yeasts, and Actinomycetes" (A. T. Henrici, ed.), 2nd ed., pp. 264–304. Wiley, New York.

Slater, M. L. (1973). *J. Bacteriol.* **113**, 263.

Snider, I. J. and Miller, J. J. (1966). *Can. J. Microbiol.* **12**, 485–488.

Sogin, S. J., Haber, J. E., and Halvorson, H. O. (1972). *J. Bacteriol.* **112**, 806.

Stuben, J. (1939). *Planta* **30**, 353.

Tingle, M., Singh Klar, A. J., Henry, S. A., and Halvorson, H. O. (1973). *Soc. Gen. Microbiol.* **23**, 209–243.

Tipper, D. J. (1973). *J. Bacteriol.* **116**, 245.

Trevelyan, W. E., and Harrison, J. S. (1952). *Biochem. J.* **50**, 298.

Trevelyan, W. E., and Harrison, J. S. (1956). *Biochem. J.* **63**, 23.

Wehr, C. T., and Parks, L. W. (1969). *J. Bacteriol.* **98**, 458.

Wickner, R. B. (1974). *J. Bacteriol.* **117**, 252.

Windisch, S. (1961). *Wallerstein Lab. Commun.* **24**, 316–325.

Wright, R. E., and Lederberg, J. (1957). *Proc. Nat. Acad. Sci. U.S.* **43**, 919.

Yanagita, T., Yagisawa, M., Oishi, S., Sando, N., and Suto, T. (1970). *J. Gen. Appl. Microbiol.* **16**, 347.

Chapter 4

Synchronous Mating in Yeasts

ELISSA P. SENA, DAVID N. RADIN, JULIET WELCH,
AND SEYMOUR FOGEL

Genetics Department,
University of California,
Berkeley, California

I. Introduction

The development of methods for hybridizing yeasts has concerned investigators in fungal biology and fermentation technology for several decades. For the most part early experimental procedures were directed toward producing and isolating hybrids in conjunction with genetic analysis. The requirements for such analyses are conveniently satisfied by obtaining mating frequencies up to several percent. Low yields, although adequate for genetic purposes, severely limit biochemical studies of mating. The important advances in hybridization procedures have been reviewed by Fowell (1969).

Recently, broadened interests in yeast sexuality stimulated efforts to develop large-scale methods for synchronizing yeast mating. The pivotal position of the zygote in the yeast life cycle increases the value of these methods for several different research programs. Large-scale mating techniques can be used either to study (1) the control and expression of developmental changes associated with mating itself, or (2) interactions between parental components specifically initiated in the zygote. Synchronously mating yeast cell populations allow both approaches to be used simultaneously at the morphological, physiological, biochemical, and genetic levels.

Matings between haploid yeast cells generate zygotes which typically produce diploid buds by mitosis. Direct micromanipulative pairings of unbudded a and α cells starved for 48 hours on buffered solid media mate with a frequency $> 95\%$ (Mortimer, 1955). These observations imply that cells competent to mate must be critically regulated as regards nuclear DNA replication, since the regular alternation of haploidy and diploidy is maintained by meiosis and mating, respectively. For these reasons currently successful procedures for synchronizing the yeast mating reaction sequence utilize two approaches. The first approach employs cells taken from stationary cultures and subsequently manipulated or phased to enhance the mating reaction. A second method depends on selecting unbudded cells from logarithmically growing cultures.

Our basic synchronous mating procedure (Sena *et al.*, 1973a) utilizes log-phase unbudded a and α cells grown in supplemented minimal medium and isolated via zonal gradient centrifugation. On resuspension in supplemented minimal medium, the unbudded cells of opposite mating type rapidly initiate a round of synchronous mating. Mating mixtures taken at various times during incubation, and centrifuged on sorbitol gradients, can be fractionated to yield suspensions of zygotes or unmated cells. Suspensions containing 90–95% zygotes can be obtained in quantities suitable for physiological, biochemical, and genetic analysis.

Modifications of the above procedure are presented in Sections II,A,1, b and c, and II,A,3. The principal changes involve (1) isolating unbudded a and α fractions separately before mixing, rather than after, and (2) resuspension of purified zygote preparations in growth medium to allow first zygotic bud production. The former modification allows more precise analysis and control of the early events in mating reactions, especially cell pairing and agglutination, while the latter procedure permits the study of zygote maturation as well as the controls regulating the initiation, maturation, and separation of the first diploid zygotic bud.

While this article deals only with heterothallic mating systems at the haploid-diploid level, the methods presented can be applied to homothallic strains or other ploidy levels. Presented here are our methods for producing synchronously mating populations of *Saccharomyces cerevisiae*. The experimental approaches of other researchers in this organism and in *Hansenula wingei* are briefly reviewed. A detailed comparison of all procedures as regards methodological approaches and zygotic yields is included. Modifications are suggested which might circumvent variability arising from differences in sex factor production, agglutinability, or response to factors. Included in the discussion section are suggestions concerning new approaches to achieving synchronized yeast mating and the application of such systems to the analysis of mating itself, or as a tool for analyzing other developmental sequences.

II. Mating Procedures

A. Density Gradient Methods and New Approaches

1. CELL ISOLATION

a. Premix Method. As previously published (Sena *et al.*, 1973a), the premix method utilizes early- to mid-log-phase haploid *a* and α cells grown overnight in minimal medium (pH 4.5) containing 0.145% YNB, 1.0% glucose, 0.3% $(NH_4)_2SO_4$, and auxotrophic nutrients as required. Such cultures represent eight or nine cell generations and are harvested at a concentration of 5–7.5 $\times$ 10^6 cells/ml. Liquid cultures containing equal cell numbers are mixed and centrifuged, and the pelleted cells resuspended in 20–25 ml of water. The mixture, containing about 10^{10} cells, is sonicated and layered on a linear, 8 to 35% sorbitol gradient in an MSE zonal rotor (Measuring Scientific Equipment, Ltd., London, England). A detailed description of zonal rotor methodology for yeast cell separations has been published (Halvorson *et al.*, 1971). After centrifuging the gradient at 1200 rpm for 10 minutes, 15-ml fractions are collected from the top of the gradient using a displacement pump and an automatic fraction collector. The fractions are examined microscopically, and those containing unbudded cells are pooled. After the pooled cells are isolated by centrifugation and resuspended in fresh minimal media for the mating incubation, a sample is diluted and plated on selective media to determine the a/α cell ratio obtained from the rotor.

This method is quite satisfactory when the particular *a* and α strains exhibit equivalent size distributions among the unbudded fractions. However, if the unbudded cells differ in size, the a/α input ratio is not recovered. Instead, an excess of *a*s is usually obtained. This observation implies that unbudded *a* cells are generally smaller than comparably grown α cells.

Attempts to achieve equivalence of input/output by starting with excess α cells yielded inconsistent results. Although the above procedure produced synchronous zygotes, the method was modified because the highest zygote yield is consistently produced when a and α cells are present in a $1:1$ ratio.

  b. *Separate Fractionation Method—Large Scale.* If two zonal rotors are available, it is possible to circumvent the problem of controlling the a/α cell ratio in the isolation procedure. The separate fractionation method is similar to the premix method, with the important exception that the a and α overnight cultures are harvested separately. The a and α pellets are each resuspended in 20–30 ml of water, sonicated, and layered onto their respective 8 to 35% sorbitol gradients in two MSE zonal rotors. Fractions are taken from each rotor, after centrifugation at 1200 rpm for 10 minutes, and those containing unbudded cells are pooled. The pooled cells are pelleted and resuspended in fresh minimal medium at a concentration of 5–7×10^{6} cells/ml. The separate a and α cell suspensions are then mixed in a $1:1$ ratio for mating.

  Any isolation procedure yielding high levels of unbudded cells may be used. The Sorvall SZ-14 reorienting density gradient zonal rotor can be used for fractionating logarithmically growing yeast cell cultures (Wells, 1974). We found that a linear, 8 to 35% sorbitol gradient in the SZ14 rotor separated single from budded cells less effectively than the MSE rotor. Consequently, the single-cell yield was decreased, a disadvantage that might be eliminated with other gradients. We also fractionated a and α cells sequentially in the same MSE rotor. Sequential processing imposes significant differentials on the two mating types. Clearly, after passage through the first gradient on the rotor, the unbudded cells harvested cannot be mixed with cells of the opposite mating type until the latter are processed in a subsequent second gradient. To discourage growth we retain the cells isolated first in sorbitol while the other mating type is partitioned.

  In addition to modifying the isolation procedure in the premix method, we also instituted media changes based on the objectives of different experiments. Cultures are grown in 0.05% glucose for mitochondrial studies, and 0.005% $(^{15}NH_4)_2SO_4$ for DNA studies (Sena 1972, 1975). The mating medium contains 0.05% glucose, a low nonrepressive sugar level which, nonetheless, maintains normal zygote production.

  c. *Small Gradient Fractionation.* Small 25- or 50-ml, linear, 8 to 35% sorbitol gradients are adequate for single-cell isolations if relatively few cells are required; e.g., 10^6 to 10^8 (adapted from Mitchison and Vincent, 1965). The 25-ml gradients are made in 30-ml (2×10 cm) Corex tubes and centrifuged in a Sorvall HB4 rotor at 1000 rpm for 2 minutes. To accommodate 50-ml gradients, 100-ml Kimax tubes (2.5×16 cm) are cut to fit into Mistral 50-ml tube holders. Gradients are centrifuged in the Mistral

Universal swing-out head at 800–1000 rpm for 4–6 minutes and then fraction-ated from the top into 1-ml samples using a Buchler Densi-Flow gradient collector and a Hughes Hilo flow pump. The unbudded cell fractions are identified, pooled, and utilized as above (Section A,1,b).

2. MATING PROCESS

The mixed-cell suspensions from the isolation procedures are shaken at 250 rpm at 30°C. Surface-to-volume factors are considered in Section III,B,1,c. If cell concentrations are increased above 10^7 cells/ml, distinctive pear-shaped cells ("schmoos") are formed and zygote frequency is greatly reduced. The morphological change (Duntze et al., 1970) is attributed to the action of an extracellular peptide produced by α cells on cells of the opposite mating type (Duntze et al., 1973). Similar effects are generated when liquid cultures are allowed to stand without shaking, or if stationary-phase cultures are used in the premix method (Section A,1,a).

The mating kinetics are relatively insensitive to differences in the isolation procedure when the initial time of a and α mixing is taken into account. The depicted time lag for initial zygote production is shorter in the premix method (Figs. 1a and b), probably because of the occurrence of a and α cell interactions before resuspension of cells in mating medium. With premixed

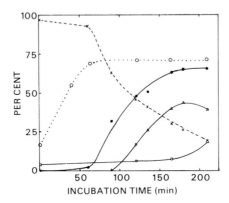

FIG. 1a. Kinetics of the mating reaction—premix method. A mixture of a and α unbudded cells isolated from exponentially growing cultures by zonal gradient centrifugation was incub-ated at 30°C in liquid YNB medium. Sexual interactions between mating types probably began at the initial time of mixing, typically 1 hour before zero incubation time. Budded and un-budded cells that failed to mate are counted as one unit, while zygotes, budded or unbudded, are considered two units. Solid circles, total zygotes; triangles, budded zygotes; squares, budded cells; circles, agglutination; ×, unbudded cells. Percent agglutination is estimated by using a Coulter counter to determine the number of nonagglutinated cells present in a diluted sample of the mating culture before and after brief sonication.

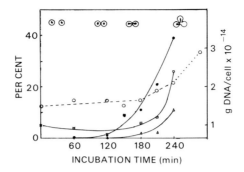

FIG. 1b.   Kinetics of the mating reaction—separate fractionation method. Conditions and counting are the same as Fig. 1a, but note that *a* and *α* cells cannot initiate any interactions until they are mixed at time zero. Solid circles, total zygotes; triangles, budded zygotes; squares, budded cells. The values for DNA per cell (open circles, scale on right) were determined by the DABA method. (Sena *et al.*, 1975, adapted from Kissane and Robbins, 1958.) See Section II,A,2 for a description of the last point. Drawings of Giemsa-stained cells at various stages of the mating reaction appear at the top of the figure.

cells agglutination is evident at the beginning of incubation (Figs. 1a and 2a). After 60 minutes marked cell agglutination is clearly apparent macroscopically as a particulate suspension and microscopically as clumps of cells (Figs. 1a and 2b). Although quantitative measurements were not taken, agglutination comparable to that in the premix method was observed using the separate fractionation method. However, the reaction was delayed 30–60 minutes. Both cell pairs and agglutinated clumps can be disrupted sonically. Such treatment does not disrupt cell pairs that have initiated zygotic fusion. Sonicated clumps reagglutinate within a few minutes. Extensive mating occurs within agglutinated clumps between 60 and 160 minutes using the premix method (Fig. 1a), or between 140 and 240 minutes with the separate fractionation method (Fig. 1b). Budded zygotes appear about 30 minutes after the onset of cell fusion. Giemsa-stained cells taken at various points during the mating reaction are depicted at the top of Fig. 1b.

   Figure 1b also depicts the increase in total DNA per cell during mating as determined by chemical measurement [diaminobenzoic acid (DABA) procedure]. DNA per cell shows an initial increase between 3 and 4 hours corresponding with first zygotic bud initiation, and increases rapidly as zygotes mature. Measurements of percent nuclear and mitochonrial DNA, as well as ^{14}N incorporation into ^{15}N-labeled nuclear and mitochondrial DNA in mating cells, show that both *a* and *α* cells synthesize mitochondrial DNA continuously during early stages of the mating process, and nuclear DNA synthesis is delayed until zygotic bud initiation (Sena *et al.*, 1973b, 1975). Presumably, nuclear DNA synthesis occurs after zygotic spindle

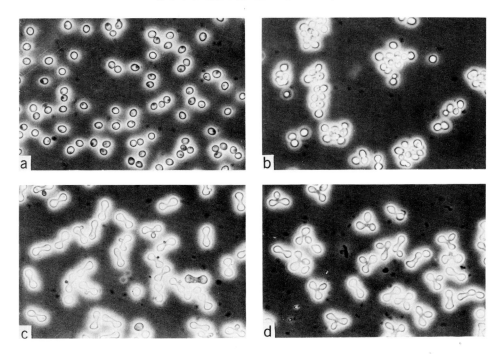

FIG. 2. Sequence of events during synchronous mating derived from the premix method. (a) Cells after 20 minutes. (b) Cells after 60 minutes. (c) Purified zygotes after 120 minutes. (d) Purified zygotes after 150 minutes. Micrographs were taken with phase optics at a magnification of $1100 \times$. Reproduced from Sena *et al.* (1973a).

plaque fusion (Byers and Goetsch, 1973). Cryer and associates (1973) reported that Storm and Lam observed the persistence of mitochondrial DNA synthesis in *a* cells exposed to concentrations of α factor inhibiting nuclear DNA synthesis. Thus the previously reported observations that *a* and α cells produce extracellular substances which arrest cells of the opposite mating type at G_1 and inhibit DNA synthesis (Throm and Duntze, 1970; Bücking-Throm *et al.*, 1972; Hartwell, 1973; L. E. Wilkinson and N. R. Pringle, personal communication) refer only to nuclear DNA synthesis. The final point on the DNA curve is a measurement of DNA per zygote in the mating culture and is therefore connected with a dotted line. The value was determined from zygotes isolated to 98% purity (see Section II,A.3) at 240 minutes of incubation, followed by a further 45-minute incubation in fresh medium.

The mating reaction in the premix method is essentially complete 150 minutes after the cells are resuspended in conjugation medium. Zygotes are harvested by the separate fractionation method (see Section II,A,3) before mating reaction completion. For both procedures cell number per milliliter remains constant during mating, and unmated cells subsequently display a round of budding. The timing of zygote formation may be delayed by several factors including excessive handling, strain differences in agglutinability, and mating factor levels.

3. ZYGOTE ISOLATION

Biochemical experiments are facilitated by the use of pure reagents. By the same token purified zygote preparations facilitate biochemical experiments concerned with the dynamics of the mating process. To raise the zygote levels in mating mixtures to 90–95%, we devised the following zygote isolation procedure. After the zygotes attain the desired developmental stage as determined by microscopic observation, the mating mixtures are harvested by centrifugation and washed once with distilled water. This step is essential for effective separation on the sorbitol gradient. After resuspension in 0.5–4 ml of water, the cells are sonicated for 5 seconds and immediately layered on 25-ml, linear, 8 to 35% sorbitol gradients. The gradients are centrifuged at 1000 rpm in a Sorvall HB4 rotor for 4 minutes. Overloading the gradient leads to streaming and pellet formation. The zygotes may be obtained by removing successive fractions from a gradient with either a syringe (Fig. 3) or a fraction collector. Alternatively, if bands are clearly defined, the lower zygote-containing band may be collected with a syringe.

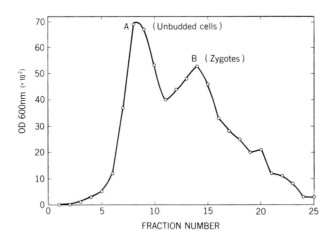

FIG. 3. Purification of zygotes on a sorbitol gradient. Cells from a synchronously mating culture containing 25% zygotes were partitioned on a 25-ml, 8 to 35% sorbitol gradient as described in Section II,A,3. The 600-nm OD of 1-ml fractions was determined on a Gilford spectrophotometer. The top of the gradient is at the left. Reproduced from Sena *et al.* (1973a).

Zygote purification is approximately 2-fold after one sorbitol gradient. If further zygote purification is required, the procedure may be repeated after centrifugation, washing, and resuspension in water. Dilution of zygote-containing sorbitol gradient fractions with water facilitates centrifugation. When heterogeneous zygote mixtures are fractionated, partial separation between budded and unbudded zygotes can be realized. A purified zygote preparation (95%) isolated from a 120-minute synchronous mating culture containing 65% unbudded zygotes and 30% zygotes with small buds is illustrated in Fig. 2c. The zygote preparation (90%) in Fig. 2d was recovered from a 150-minute incubation mixture in which 70% of the zygotes had large buds. Viability and sporulative capacity of purified zygotes is excellent. Genetic segregations derived from tetrad analysis of zygotic asci are normal.

The method's success depends on the fact that the mating mixture contains only zygotes and single cells or cells with very small buds at the time of zygote isolation. The separation of zygotes is incomplete if the mixture contains cells with large buds, because these cosediment with the zygotes. Extremely large samples of mating mixtures containing more than 5×10^9 cells are readily partitioned on the zonal rotor.

The density gradient centrifugation method can also be used to isolate the first zygotic buds. If purified zygotes are resuspended in fresh medium and further incubated until mature zygotic buds emerge, these buds will either drop off or can be detached from the zygote sonically. The mixture of zygotes and released diploid buds can be resolved on a 8 to 35% sorbitol gradient. In this manner, we obtained an 95% pure first zygotic bud fraction; the buds were 100% viable.

The various aspects of the methods presented in Section II,A are summarized diagrammatically in Fig. 4. Isolation of unbudded cells via zonal rotor or tubes, mating, and gradient separation of zygotes from unmated cells are depicted.

B. Other Published Methods

Jakob (1962) devised a technique for synchronous zygote production in *S. cerevisiae*. She grew cells for 48 hours to stationary phase at 30°C on agar medium containing 0.5% yeast extract and 3% dextrose (YED 3% medium). Cells of both mating types were mixed 1:1 at a concentration of 1.6×10^6 cells/ml in YED 3% medium. After 2 hours at 25°C the cells were centrifuged and allowed to remain pelleted for 30 minutes at 25°C. The supernatant was then decanted, and the cells resuspended in fresh medium. After 3.5–4.5 hours, zygotes were generated. They represented about 20% of the total cell population.

Using *H. wingei* grown in YED–KH$_2$PO$_4$ to early stationary phase, Brock

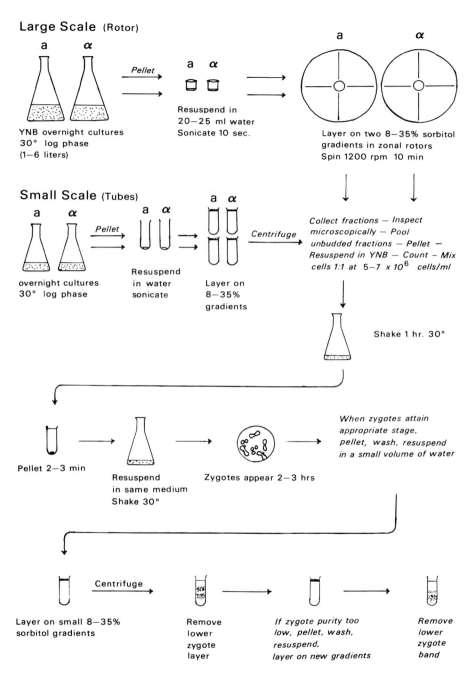

Large Scale (Rotor)

a   α

Pellet →

a   α

→

YNB overnight cultures
30° log phase
(1—6 liters)

Resuspend in
20—25 ml water
Sonicate 10 sec.

a   α

Layer on two 8—35% sorbitol
gradients in zonal rotors
Spin 1200 rpm 10 min

Small Scale (Tubes)

a   α

Pellet →

a   α

→

a   α

Centrifuge →

overnight cultures
30° log phase

Resuspend
in water
sonicate

Layer on
8—35%
gradients

*Collect fractions — Inspect
microscopically — Pool
unbudded fractions — Pellet —
Resuspend in YNB — Count - Mix
cells 1:1 at 5–7 x 10^6 cells/ml*

Shake 1 hr. 30°

Pellet 2—3 min

→

Resuspend
in same medium
Shake 30°

→

Zygotes appear 2—3 hrs

→

*When zygotes attain
appropriate stage,
pellet, wash, resuspend
in a small volume of water*

Layer on small 8—35%
sorbitol gradients

Centrifuge →

Remove
lower
zygote
layer

→

*If zygote purity too
low, pellet, wash,
resuspend,
layer on new gradients*

→

Remove
lower
zygote
band

FIG. 4. Procedures for mating synchrony production and zygote isolation using alternative methods of harvesting unbudded *a* and *α* cells.

(1965) and M. A. Crandall (personal communication) mixed equal numbers of each mating type and incubated 3×10^8 cells/ml in mating medium [0.01 M KH$_2$PO$_4$, 0.1% MgSO$_4$, 0.5% glucose (pH 5.5)]. After a 1-hour incubation at 30°C with shaking, the cells may be deagglutinated by two washings with water and a 10-second sonication. However, this procedure was not utilized routinely. Cells were resuspended in fresh mating medium. Conjugation began almost at once and a 70–80% zygote level was attained after 3 hours.

Biliński et al. (1973) grow S. cerevisiae cells in rich medium to early stationary phase. 5×10^7 cells/ml of each mating type are mixed and incubated 1 hour at 30°C in YED 3% (pH 4.5). After centrifugation the cells are resuspended in YED 3% (pH 8.5) and incubated 1 hour at 30°C. The cell mixture is pelleted and resuspended in YED 3% (pH 4.5). After sonication and dilution to 10^7 cells/ml, the mixture is centrifuged, and incubated as a pellet for 30 minutes at 30°C. The cell pellet is then disintegrated by gentle shaking. After another 30 minutes it is gently shaken again and reincubated without shaking. About 30–40% zygotes are produced in 4.5–5.5 hours.

All the procedures described in Section II yield synchronously mating yeast cell populations. However, the various methods differ considerably in equipment requirement, cell culture manipulation, and zygote yield. Each method possesses unique limitations regarding its usefulness for specific purposes. For example, studies centered on the mating reaction as a sequential process within the total yeast life cycle pose different requirements compared to those aimed at determining the fate of parental mitochondrial DNA molecules within the zygote.

III.   Discussion

A.   Comparison of Published Methods

No single mating synchrony method currently available is uniquely suited to all research situations. Thus, as an aid to investigators of yeast mating attempting to decide which method might be most suitable for a specific mating experiment, we constructed Table I. It briefly summarizes and compares the major published methods for mating synchrony based on the following criteria.

1. Yeast genera used for mating (Hansenula, Saccharomyces).
2. Premating culture growth medium (complex versus minimal). Use of a defined minimal medium may be required for experiments utilizing radioactive or density label incorporation.

TABLE I: COMPARATIVE ASPECTS OF

Method[a]	Yeast species	Growth medium	Growth cycle stage	Mating medium	Cell concentration for mating and ratio of mating types	Maintenance of cell contact
A	S. cerevisiae	Complex	48-hour plate, stationary	YED 3%	1.6×10^6 ml, 1:1	Pellet
B	H. wingei	Complex	Early stationary	KH_2PO_4, $MgSO_4$, glucose (pH 5.5)	3×10^8 ml, 1:1	Agglutination
C	S. cerevisiae	Complex	Early stationary	YED 3% different pH values	1×10^8 ml, 1:1	Pellet
D	S. cerevisiae	Defined minimal YNB	Log phase (unbudded cells isolated)	Defined minimal YNB (pH 4.5)	$5-7 \times 10^6$ ml, 1:1	Agglutination

[a] A, Jakob (1962); B, Brock (1965); C, Bilinski *et al.* (1973); and D, Sena *et al.* (1973a).

3. Growth cycle stage of initial mating cells (late-log-phase, unbudded cells versus early- or late-stationary phase). Actively growing cells may incorporate labeled compounds more efficiently than less actively growing cells during early incubation in mating medium.

4. Mating medium (complex versus minimal medium, and medium supporting or not supporting cell growth). A defined medium and/or cell growth may be necessary if specific label uptake is required during mating.

5. Cell concentration per milliliter of mating medium and ratio of opposite-mating-type cells (1.6×10^6–3×10^8/ml; 1:1). Mating efficiency may be enhanced at other input ratios if either cell type were a weak or an overproducer of specific mating factors (see Section III,B,2).

6. Maintenance of cell contact (either physically forcing opposite mating types together in a pellet and/or relying on the agglutinative properties of the mating culture). Since strains vary in agglutinative ability, this property may dictate the use of one method over another. Alternatively, certain other procedural variations can be utilized (see Section III,B,1).

7. Mating mixture manipulations (centrifugation, sonication, medium changes).

Manipulations	Time of zygote forma- tion	Percent of zygotes	Final zygote yield	Zygote purifica- tion (%)	First zygotic bud purifica- tion
Centrifuge at 2 hours; stand 30 minutes	3.5–4.5 hours	20	4×10^4	—	—
Transfer to fresh medium after 1 hour	2–3 hours	70–80	2.4×10^{11}	—	—
Multiple transfers	4.5–5.5 hours	30–40	3.6×10^9	—	—
Zonal or other gradients; centri- fuge at 30 minutes to 1 hour	2.5–3.5 hours	40–50	5×10^9	90–95	95%

See Section II,B for descriptions of methods A, B, and C; see Section II,A for method D. For further remarks on comparisons of these methods, see Section III,A.

8. Timing of zygote formation (3–5 hours after mixing).

9. Zygote percentage produced in the mating mixture (20–80%).

10. Final zygote yield (4×10^4–2.4×10^{11}).

11. Zygote purification from unmated cells. This step requires that un-mated cells in a mating mixture be either unbudded or carry only very small buds at the time of zygote isolation (see Section II,A,3).

12. Isolation of first zygotic buds. For isolation of zygotic buds, purified zygote preparations are necessary (see Section II,A,3).

B. Problems in Mating Procedures

1. AGGLUTINATION

Vigorous and rapid agglutination is necessary for mating in an agitated liquid culture (Sakai and Yanagishima, 1971). Although initial cell aggrega-tion occurs in non-growth-supporting medium (Campbell, 1973), agglutina-tion requires medium containing essential growth nutrients (Radin et al., 1973) and is optimal at pH 4.5 (Biliński et al., 1973). Although all laboratory strains tested thus far exhibit some agglutination, different strains vary in their capacity to agglutinate. Strongly agglutinating strains can be handled as described in our basic mating procedure. Problems with agglutination may be circumvented as follows.

1. To enhance agglutination an initial slow shaking period (as low as 80–100 rpm) can be maintained until agglutination is established. If desirable, the shaker speed may then be increased.

2. A 3- to 5-minute, low-speed (500 rpm) pelleting of the mating culture between $\frac{1}{2}$ and 1 hour after cell mixing forces the cells into close contact and increases agglutination between weakly agglutinating strains. The pellet should then be resuspended gently in the supernatant medium.

3. The ratio of flask volume to mating mixture volume influences the agglutination reaction. A 20:1 ratio is nearly optimal when cultures agglutinate poorly.

4. Although many experiments require minimal defined medium, others do not. An enriched medium (e.g., supplemented with a full complement of amino acids and bases) enhances the agglutination response of poorly agglutinating strains.

5. Preincubation of cells in the late-log-phase filtrate (pH adjusted to 4.5) taken from cultures of the opposite mating type, before resuspension in conjugation medium, promotes and hastens agglutination. This effect may be caused by the presence of agglutination inducers in the growth medium (Yanagishima *et al.*, 1972; Radin and Fogel, 1975). This technique works well when it is combined with immediate pelleting of the mating mixture.

6. Normally, diploids do not agglutinate. Agglutination and mating are reduced as the percentage of a/α diploids in the mating reaction rises. Since cultures often undergo spontaneous diploidization after several transfers or growth cycles, the genetic integrity of the strains should be monitored regularly.

2. DELAY OF MITOSIS IN MATING CULTURES

The delay in initiation of mitosis in unmated cells in a mating culture is primary to the usefulness of many experiments. Results of mating experiments can be confused by the superimposition of effects due to mitotic activity of unmated cells. Mitotic delay is also a central consideration in purifying zygotes via the gradient fractionation technique. This delay is probably caused by the presence of *a* and α factors in the medium (Hartwell, 1973; Sena *et al.*, 1973a). Mitotic delay can be affected by the following techniques.

1. Mixing log-phase cultures prior to gradient fractionation decreases the lag time for zygote formation, while the kinetics of bud initiation for nonmated cells remains constant (see Fig. 1a and b). Accordingly, premixing increases the delay between mating reaction completion and the onset of

mitotic bud initiation. Preincubating cells in filtrate from opposite-mating-type cultures adjusted to pH 4.5 prior to mixing also increases this delay.

2. If the strains employed are deficient in factor production or activity, addition of exogenous a and/or α factors could postpone the onset of mitosis. The success of such additions is variable and depends on establishing the optimal concentration for added sex factors. The appropriate concentration range seems to be narrow. For example, addition of excess α factor inhibits mating.

3. A very rich medium may hasten the initiation of mitosis without comparably hastening the mating reaction. Therefore using rich medium to enhance agglutination and mating must be weighed against this undesirable result.

4. Low a or α factor production or activity may reduce mitotic delay and mating efficiency. This effect may be overcome by varying the cell ratio in the mating mixture to favor the inefficient strain. Biliński and his colleagues (1973) reported maximal mating with a/α cell ratios other than 1:1, an observation that probably reflects this phenomenon.

C. Future Experimental Approaches for Producing Synchronously Mating Populations of Yeasts

Any technique employing cells either arrested, or selected, at a stage capable of immediate mating initiation can be used to produce populations of synchronously mating cells. The following represents heuristically useful speculative approaches.

1. Appropriate concentrations of purified a or α factor (Duntze et al., 1973; L. E. Wilkinson and J. R. Pringle, personal communication) or other antimitotic agents may be used to arrest yeast cells at the mononucleate, unbudded stage. Such treatments accumulating unbudded a and α cell populations, after washing to remove factor and resuspension in mating medium, can synchronize the production of zygotes. We employed phased a cells treated with purified α factor (Duntze et al., 1970), at a concentration just inhibitory to mitotic initiation, in matings with gradient-isolated, unbudded α cells. While zygote yields were approximately 30–40%, zygote synchrony was poor. The poor synchrony can probably be attributed to the fact that α-factor-arrested cells show differential recovery from mitotic inhibition (Bücking-Throm et al., 1973). If differential recovery could be minimized by a procedure selective for cells displaying uniform recovery times, sex factor phasing of a and α cells could represent a rapid, efficient method for producing massive populations of synchronously mating cells distinguished by high zygote yields.

2. Phasing might also be accomplished by incorporating cell division

cycle mutations into mating strains. Reid found (L. H. Hartwell, personal communication) that *cdc 28* cells are arrested at the restrictive temperature prior to the initiation of mitotic DNA synthesis, and retain the ability to mate at the restrictive temperature. It is conceivable that *a* and α cultures arrested in this manner could produce synchronously mating cell populations.

3. Possibly, homothallic systems could yield abundant synchronously mating yeast populations. Certain homothallic yeast strains display highly synchronous efficient mating among the early sister cells of an ascosporal clone. Thus twin zygotes are commonly observed (Oshima and Takano, 1972). Although parental genotypes are rendered homozygous under these circumstances and the utility of segregating markers is lost, such populations may be useful for many biochemical studies of mating.

4. A stable-flow, free-boundary migration and fractionation (STAFLO) apparatus can be utilized to fractionate cell mixtures. Mel's STAFLO (1964) system has been used to prepare pure spore suspensions (Resnick *et al.*, 1967) and log- or stationary-phase unbudded single cells from haploid or diploid cultures (S. Fogel, unpublished). These selected spore or cell fractions could yield synchronously mating cell populations.

D.   Applications

Consequent to developing methods for producing large, synchronously mating yeast cell populations and pure zygote preparations, a previously eclipsed phase of the yeast life cycle has become accessible to investigation. Studies of mutants blocked at specific stages in the mating reaction (Crandall and Brock, 1968; MacKay and Manney, 1974a,b) may now be initiated on a large scale. The entire mating process and its individual component reactions can be analyzed simultaneously at the ultrastructural, genetic, and molecular levels. Since a zygote is the direct product of cell fusion, zygotic parental inputs can be manipulated directly by specific experimental treatment of either parent. Initial interactions between parental cytoplasmic organelles, membranes, and especially nuclear and mitochondrial DNAs are prime targets for study. The dual and pivotal alternatives of zygotic maturation via first mitotic bud production or immediate zygotic meiosis and sporulation may well allow novel analyses of the interactions between parental components.

Because mating involves cell wall dissolution concomitant with cell fusion, molecules and organelles not ordinarily incorporated by yeast might be assimilated into the zygote at the time of its formation (Lhoas, 1972). Also, an understanding of the molecular events associated with cell recognition could illuminate the basis for mating incompatibilities between previously nonhybridizable yeast species. The availability of such hybrids, possibly by protoplast fusion or other procedures, might be extremely useful technologically and at the same time improve our understanding of speciation and evolution in yeasts.

ACKNOWLEDGMENTS

We gratefully acknowledge the excellent technical assistance of Patricia Leslie and Karin Lusnak. This research was supported by National Institutes of Health Research Grant GM-17317 to S. Fogel and National Institutes of Health Training Grant 5TO1-GM-00367-15 to the Department of Genetics, University of California, Berkeley.

REFERENCES

Biliński, T., Litwińska, T., Żuk, J., and Gajewski, W. (1973). *J. Gen. Microbiol.* **79**, 285–292.
Brock, T. D. (1965). *J. Bacteriol.* **90**, 1019–1025.
Bücking-Throm, E., Duntze, W., Hartwell, L. H., and Manney, T. R. (1973). *Exp. Cell Res.* **76**, 99–110.
Byers, B., and Goetsch, L. (1973). *Cold Spring Harbor Symp. Quant. Biol.* **38**, 123–131.
Campbell, D. (1973). *J. Bacteriol.* **116**, 323–330.
Crandall, M. A., and Brock, T. D. (1968). *Bacteriol. Rev.* **32**, 139–163.
Cryer, D. R., Goldthwaite, C. D., Zinker, S., Lam, K. B., Storm, E., Hirschberg, R., Blamire, J., Finkelstein, D. B., and Marmur, J. (1973). *Cold Spring Harbor Symp. Quant. Biol.* **38**, 17–29.
Duntze, W., MacKay, V., and Manney, T. R. (1970). *Science* **168**, 1472–1473.
Duntze, W., Stotzler. D., Bücking-Throm, E., and Kalbitzer, S. (1973). *Eur. J. Biochem.* **35**, 357–365.
Fowell, R. R. (1969). *In* "The Yeasts" (A. H. Rose and J. S. Harrison, eds.), Vol. 1, pp. 303–383. Academic Press, New York.
Halvorson, H. O., Carter, B. L. A., and Tauro, P. (1971). *In* "Methods in Enzymology" (K. Moldave and G. Grossman, eds.), Vol. 21, Part D. pp. 462–470. Academic Press, New York.
Hartwell, L. H. (1973). *Exp. Cell Res.* **76**, 111–117.
Jakob, H. (1962). *C. R. Acad. Sci.* **254**, 3909–3911.
Kissane, J. M., and Robins, E. (1958). *J. Biol. Chem.* **233**, 184–188.
Lhoas, P. (1972). *Nature (London), New Biol.* **236**, 86–87.
MacKay, V., and Manney, T. R. (1974a). *Genetics* **76**, 255–271.
MacKay, V., and Manney, T. R. (1974b). *Genetics* **76**, 273–288.
Mel, H. C. (1964). *J. Theor. Biol.* **6**, 159–180.
Mitchison, J. M., and Vincent, W. S. (1965). *Nature (London)* **205**, 987–989.
Mortimer, R. K. (1955). *Radiat. Res.* **2**, 361–368.
Oshima, Y., and Takano, I. (1972). *In* "Fermentation Technology Today" (G. Terui, ed.), pp. 847–852. Soc. Ferment. Technol., Kyoto, Japan.
Radin, D. N., and Fogel, S. (1975). In preparation.
Radin, D. N., Sena, E., and Fogel, S. (1973). *Genetics* **74**, s222.
Resnick, M. A., Tippetts, R. D., and Mortimer, R. K. (1967). *Science,* **158**, 803–804.
Sakai, K., and Yanagishima, N. (1971). *Arch. Mikrobiol.* **75**, 250–265.
Sena, E. P. (1972). Ph.D. Dissertation, University of Wisconsin, Madison.
Sena, E. P., Radin, D. N., and Fogel, S. (1973a). *Proc. Nat. Acad. Sci. U.S.* **70**, 1373–1377.

Sena, E., Welch, J., Radin, D., and Fogel, S. (1973b). *Genetics* **74**, s248–s249.
Sena, E., Welch, J., and Fogel, S. (1975). In preparation.
Throm, E., and Duntze, W. (1970). *J. Bacteriol.* **104**, 1388–1390.
Wells, J. R. (1974). *Exp. Cell Res.* **85**, 278–286.
Yanagishima, N., Sakai, K., and Shimoda, C. (1972). *In* "Fermentation Technology Today" (G. Terui, ed.), pp. 853–856. Soc. Ferment. Technol., Kyoto, Japan.

Chapter 5

Synchronous Zygote Formation in Yeasts

T. BILIŃSKI, J. LITWIŃSKA, J. ŻUK, AND W. GAJEWSKI

Institute of Biochemistry and Biophysics,
Polish Academy of Sciences,
Warsaw, Poland

I. General Characteristic of the Conjugation Process

Although genetic analysis of the yeast *Saccharomyces cerevisiae* based on sexual fusion and tetrad analysis is one of the most advanced as regards eukaryotic organisms, knowledge of sexual processes in this organism is still very limited. It was only recently that the processes occurring during the sexual cycle were experimentally studied. For any kind of physiological or biochemical study of the sexual cycle in yeasts, synchronous mass production of zygotes is a first prerequisite.

The sexual processes in haploid strains of opposite mating type (a and α) leading to zygote formation consist of many characteristic consecutive stages. Three main stages of the sexual cycle are courtship, cell fusion, and karyogamy. In each stage characteristic genetic and biochemical processes occur which prepare the cells for the next sexual reaction. As already established by Hartwell (1973), a and α cells are competent for sexual fusion only at an appropriate stage of the cell cycle; competent cells are unbudded and in a cycle position just before the initiation of DNA replication.

Duntze *et al.* (1970) and Yanagishima (1971) demonstrated that interaction between haploid cells of opposite mating type is mediated by hormonal substances released into the culture medium. According to Throm and Duntze (1970) and Bücking-Throm *et al.* (1973), α cells produce an α factor (peptidic in nature) which is released into the medium. This diffusible α factor inhibits the initiation of DNA replication in *a* cells. In this way synchronization of *a* cells is achieved; they stop budding and become competent for sexual fusion.

Analogous processes of synchronization probably also occur in α cells. We showed recently (Biliński *et al.*, 1974) that, in α cells separated from the conjugation mixture on Dowex resin, DNA synthesis and budding are inhibited. The lag in DNA synthesis preceding sexual fusion of *a* and α cells is not only due to swelling but, according to our unpublished data, the dry synthesis and budding are resumed.

The mixing of haploid cells of opposite mating type results not only in mutual synchronization, but also in other changes characteristic of the conjugation process. The cell volume increases, and agglutination of cells takes place (Sakai and Yanagishima, 1971). The increase in the cell volume is not only due to swelling but, according to our unpublished data, the dry mass content also increases owing to intensive RNA and protein syntheses. Agglutination is a decisive factor ensuring intimate contact between cells of opposite mating type. Agglutination, however, is nonspecific, it occurs among cells of different as well as the same mating type and results in the formation of large clumps. In *S. cerevisiae* agglutination is much weaker than, for instance, in *Hansenula wingei* (Crandall and Brock, 1968).

In the next step cells of opposite mating type form pairs and undergo fusion and plasmogamy. As a result, young zygotes are formed in which nuclear fusion (karyogamy) takes place. Diploid zygotes proliferate mitotically or undergo meiosis and ascospore formation, depending on nutritional conditions.

The capacity for sexual reaction between *a* and α cells is controlled by one allelic pair in the mating-type locus in yeasts. This locus is probably multicistronic and regulates the synthesis of many products involved in mating reactions. It is known that copulation and efficiency of mating reactions are highly variable, depending on the yeast strain used. Most probably the efficiency of the mating reaction depends on the interaction of many different genetic factors present in different strains.

For a detailed study of the biochemical processes occurring in consecutive stages of sexual cell fusion in yeasts, the use of genetic mutants with specific blocks for different stages is necessary. To obtain such mutants, however, a general knowledge of culture conditions for each stage and of synchronization of mating processes must first be obtained. Only in this way can the genetic mechanisms governing the sexual cycle in yeasts be determined.

II. Basic Principles for Synchronous Mass Production
of Zygotes

The classic method of Lindegren gives a yield of zygotes usually of the order of a few percent of the haploid cells in the conjugation mixture. Moreover, the period of zygote formation is extended to several hours. Jakob (1962) greatly increased the yield of zygotes by enforcing cellular contacts between competent haploid cells by centrifugation. This procedure raised the yield of zygotes to about 15%. However, the synchronization in Jakob's procedure is still rather poor, and the zygotes are formed in the course of more than 2 hours.

Since the initial a- and α-cell cultures are unsynchronized and the synchronization through mutual interaction in the conjugation mixture is only partial and transitory, the first and most important step is the preparation of a homogeneous mixture of competent cells of opposite mating type. This can be accomplished either by separation of homogeneous a and α cells in suitable stages for sexual fusion, as was done by Sena et al. (1973), or by differentiating the mating procedure in such a way that conditions are optimal for synchronization but unfavorable for cell fusion (Biliński et al., 1973). Homogeneous populations of competent cells ensure a maximal yield of zygotes for any given pair of haploid strains, whatever the method used for purification of competent cells.

Synchronization of zygote formation can be achieved in several ways. We usually obtain the highest efficiency of conjugation in cultures from late logarithmic phase when the initial cell population is highly asynchronous. The time required for cells from different phases of the cell cycle to achieve competence varies greatly, and therefore a basic condition for synchronization of cell fusion is the prevention of premature zygote formation before the whole cell population becomes competent. This can be achieved by creating conditions favorable to quick competence development and by disrupting early contacts between competent cells by sonication. In this way it is possible to obtain a practically pure competent cell suspension (95%) without premature zygote formation.

The next condition for synchronous zygote formation is the enforcement of contact between competent cells by centrifugation as applied by Jakob (1962). A contact enforced in a pellet during $\frac{1}{2}$ hour results in restriction of the time of fusion between cells to about 1 hour, this being a much higher degree of synchronization as compared with that obtained by other methods. At the end of efficient conjugation, sonication should be applied again to disintegrate clumps and to stop formation of new zygotes.

III.   Method for Poorly Synchronized Zygote Formation for Genetic Analysis

The cultures of a and α strains in late log phase in YPG medium (yeast extract, 1%; Bacto peptone, 1%; and glucose, 2%) are centrifuged and resuspensed in conjugation medium containing 1% yeast extract and 3% glucose with pH 4.5 (adjusted with acetic acid). The concentration of both strains should be nearly identical within the range of 10^7 to 10^8 cells/ml.

The cells of both strains are mixed in a 1:1 ratio, and the mixture is incubated for 2 hours at 28 °C without shaking. After incubation the mixture is diluted 10 times with fresh medium, immediately centrifuged for 2 minutes at 3000 g, and left for 30 minutes at 28 °C. Then the pellet is gently agitated by manual shaking, and the preparation left for another 30 minutes at 28 °C without shaking. After this time the pellet is disintegrated by more vigorous manual shaking without formation of a homogeneous suspension. The shaking is applied only to provide better access of nutrients to the pellet; more thorough disintegration decreases markedly the efficiency of conjugation. The pellet is then further incubated at 28 °C. For most of the strains tested, this procedure gives maximum zygote formation 5 hours after the mixing of a and α cells.

IV.   Method of Synchronous Zygote Mass Production

The basic medium for zygote formation is 3% YE consisting of 3% of glucose and 1% of yeast extract. For cultivation of haploid strains, YPG medium consisting of 1% yeast extract, 1% Bacto peptone, and 2% glucose is used. The haploid strains are cultivated in 250-ml Erlenmeyer flasks containing 100 ml of YPG medium for 18 hours at 30 °C on an orbital shaker (120 rpm). Cells are harvested at the end of the log phase by centrifugation for 5 minutes at 3000 g (Sorvall SS1) in 50-ml steel tubes sterilized with 70% ethanol. Suspensions of 10^8 cells/ml of each strain in the conjugation medium (3% YE) are prepared. The pH of the medium is adjusted to 4.5 with 1 M acetic acid.

The suspensions of cultures of both mating types are mixed together in a 1:1 ratio. The samples containing 35 ml of mixture are incubated for 1 hour at 30 °C in 50-ml steel Sorvall centrifuge tubes. After incubation the cells are centrifuged under the same conditions for 5 minutes at 3000 g in a Sorvall SS1 centrifuge, resuspended in 35 ml of 3% YE, adjusted to pH 8.5 with 1 M Na$_2$HPO$_4$, and exposed for a further incubation of 1 hour. The change in pH of the medium from 4.5 during the first hour to 8.5 during the next hour of the mating reaction is very important. It provides optimal conditions for synchronization of cells, but suboptimal conditions for cell fusion and zygote formation.

After 1 hour of incubation at pH 8.5, the cells are centrifuged, resuspended in 35 ml of fresh YE medium (pH 4.5), and sonicated twice for 30 seconds (amplitude from peak to peak 2 μm with an MSE 100-w ultrasonic disintegrator). Sonication ensures maximum efficiency of synchronization without cell fusion.

Samples of 20 ml of sonicated suspension are added to 250-ml steel centrifuge tubes containing 180 ml of fresh 3% YE medium (pH 4.5) and centrifuged immediately on a MSE 18 high-speed centrifuge with an angle rotor (6 × 250 ml) at room temperature for 2 minutes at 3000 rpm. The samples in the centrifuge tubes are then incubated at 30° for 30 minutes. Afterward the pellet is fragmented by gentle shaking and incubated without shaking for another 30 minutes. At the end of incubation, the samples are shaken again to disintegrate the pellet, and further incubated without shaking. With this procedure zygote formation starts after 4.5 hours of incubation following the mixing of a and α cells, and their number rises abruptly within 1 hour. During this time no buds are formed by the zygotes, which are all in the same relatively young stage. The yield of zygotes ranges from 30 to 40%.

The changes observed in the cell population during the conjugation procedure under the conditions described above are shown in Fig. 1. Samples of initial cultures of a and α cells and the conjugation mixture are analyzed microscopically at 15-minute intervals for estimation of budded cells percentage. Also, the number of conjugated cells (zygotes) can be estimated in the conjugation mixture. The first zygotes appear 4.5 hours after cell mixing, and the curve of zygote formation rises very abruptly during the first hour. A further increase in the frequency of diploid cells is due to the separation of diploid buds from the zygotes.

The incubation time for maximal efficiency of zygote production must be determined for each pair of haploid strains used. When the maximum is reached, the conjugation mixture should be sonicated to prevent the formation of new zygotes. If strict synchronization is required and the maximum yield of zygotes is less important, sonication may be applied much earlier.

V.   Limitations of the Method

The described method of synchronous mass production of zygotes is applicable to any pair of haploid strains, provided they are mutually well synchronized in the conjugation mixture so that the majority of haploid cells are competent to fuse and form zygotes. If the a and α cells do not

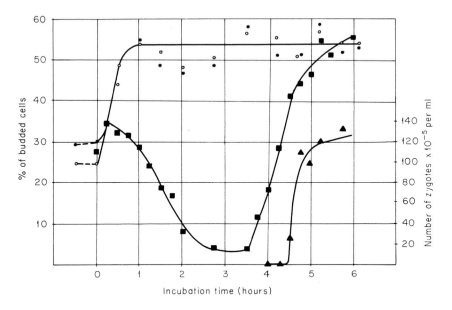

FIG. 1. Numerical relations among different kinds of cells during conjugation. Solid circles, percent of *a* cells with buds in initial culture; open circles, percent of α cells with buds in initial culture; squares, percent of *a* and α cells with buds in conjugation mixture; triangles, number of zygotes in 1 ml of conjugation mixture.

synchronize well and the percentage of competent cells in the conjugation mixture is low, the application of the procedure proposed by Sena *et al.* (1973) is recommended. This method consists of isolation of competent cells from the conjugation mixture by gradient centrifugation. For all technical details of this method, the reader is referred to Sena's publication (Sena *et al.*, 1973). However, if the ability of the cells to fuse is low, it is impossible to raise the yield of the zygotes. If the zygotes are being produced for biochemical or physiological studies, it is important to start with *a* and α strains checked beforehand for high capacity for zygote formation (see our simplified method, Section III).

It should be stressed that the proportion of *a* and α cells in the conjugation mixture is a very important factor for high efficiency of zygote formation. As Sena *et al.* (1973) showed, the optimal proportion is 1:1. Even small changes in this proportion can result in a drastic decrease in zygote production. We have already described (Biliński *et al.*, 1973) the inhibitory effect of increasing proportions of α-mating-type cells in the conjugation mixture on zygote formation. Figure 2 shows the effect of changes in the

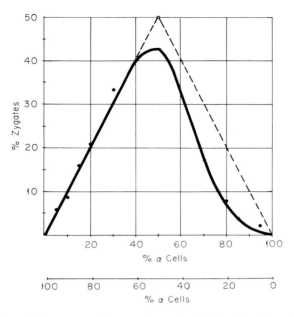

FIG. 2.   Mating efficiency at various a and α ratios. Broken line, theoretical corrected; solid line, observed.

proportions of a and α cells on the efficiency of conjugation. As seen, the efficiency curve is clearly asymmetric on the side of the α surplus and differs markedly from the symmetric theoretical curve. This phenomenon is observed very often, and the degree of asymmetry differs greatly according to the pair of strains used for conjugation.

The absolute concentration of cells in the conjugation mixture is also of paramount importance for cell fusion processes. For mass production of zygotes, concentrations of 10^8 cells/ml in the first phase and 10^7 cells/ml during the conjugation process are recommended (as described in our procedure). High concentrations of cells are very convenient for all manipulations and economize in the use of media. Lower concentrations (10^7 cells/ml) during the first 2 hours and 10^6 cells/ml, but not less, during zygote formation may also be used if a smaller number of zygotes is required. For a high degree of synchronization, higher concentrations are recommended, but for cell fusion itself rather low concentrations are optimal. The composition of the medium also has a very strong influence on the course of conjugation. As a rule, it should be a rich medium, and the most important constituent is glucose. Any other source of carbon or energy that still permits growth practically stops zygote formation completely.

REFERENCES

Biliński, T., Litwińska, J., Żuk, J., and Gajewski, W. (1973). *J. Gen. Microbiol.* **79**, 285.
Biliński, T., Jachymczyk, W., Litwińska, J., and Żuk, J. (1974). *J. Gen. Microbiol.* **82**, 97.
Bücking-Throm, E., Duntze, W., Hartwell, H., and Manney, T. R. (1973). *Exp. Cell Res.* **76**, 99.
Crandall, M. A., and Brock, T. D. (1968). *Bacteriol. Rev.* **32**, 139.
Duntze, W., MacKay, V., and Manney, T. R. (1970). *Science* **168**, 1472.
Hartwell, L. H. (1973). *Exp. Cell Res.* **76**, 111.
Jakob, H. (1962). *C. R. Acad. Sci.* **254**, 3909.
Sakai, K., and Yanagishima, N. (1971). *Arch. Microbiol.* **75**, 260.
Sena, E. P., Radin, N., and Fogel, S. (1973). *Proc. Nat. Acad. Sci. U.S.* **70**, 1373.
Throm, E., and Duntze, W. (1970). *J. Bacteriol.* **104**, 1388.
Yanagishima, N. (1971). *Physiol. Plant.* **24**, 260.

Chapter 6

Continuous Cultivation of Yeasts

A. FIECHTER

Institute of Microbiology,
Swiss Federal Institute of Technology,
Zürich, Switzerland

I. Introduction

A. General Remarks

Continuous-cultivation techniques have been developed in the last few decades. Attempts before World War II failed, because of incorrect understanding of growth kinetics and the lack of efficient equipment. Contamination in long-term runs was usual, and steady states were frequently destroyed by inconsistent operational features of the equipment.

The theoretical background in growth kinetics was developed simultaneously by Monod (1950) and Novick and Szilard (1950). Specific growth rates turned out to be dependent on the substrate, and therefore they are variables. The theory also predicted a very high productivity of continuous-flow systems, and widespread application of such methods in production plants was thought to be only a matter of investment. However, continuously operated plants turned out to economically unfeasible if the underlying biochemical and regulatory features of the processes were not fully understood. Nevertheless, most microbiologists considered chemostat methods primarily a powerful tool for industrial work.

It is the aim of this article to emphasize the usefulness of the methods for microbiological, genetic, and biochemical studies.

Principally, the procedure is based on the facts that:

1. The specific growth rate μ can be set (within the limits of the organism) according to the needs of the experiment. Thus one can select a given state of metabolic activity and control.

2. Cells in this defined state can be obtained independent of time.

3. Comparatively large amounts of defined cell material can be produced with laboratory-scale equipment.

This is in contrast to all kinds of batch methods, which are in principle not controllable reactions in a system of changing parameters. Since substrate concentration S is a function of time, intrinsic regulation represents a transient state with very small time constants.

Continuous-flow methods are particularly suitable for:

1. Kinetic studies on growth and metabolic turnover (degradation and synthetic rates).

2. Determination of metabolic pool concentrations.

3. Production of cell material of defined properties for further examination.

4. Elucidation of effects of external parameters (nutrients, inhibitors, enhancers, etc.).

Bench-scale chemostats have been improved during the last few years, and, although such equipment is more expensive than classic cultivation devices, prices fit reasonably into regular laboratory budgets.

In the following (Sections I,C and II,B), a short technical description of a single chemostat is given to facilitate evaluation of the main parts and accessories recommended by manufacturers.

However, definition of the minimum requirements is not simple. As everywhere, the selection of proper equipment is a question of the aim of the work to be undertaken. Therefore some more sophisticated instrumentation and its use primarily in yeast studies is also mentioned.

B. Classification of Continuous-Culture Systems

A variety of culture systems for continuous cultivation has been described for dealing with different organisms and investigations with different goals. The following classification is an extract from Herbert's scheme and embraces only types of more practical relevance. The arrangement of the list is based on only three criteria. For more detailed information see Herbert (1961).

Criterion I: Feedback of Cells

Open systems	Closed systems
Outflow contains cells: No feedback or partial feedback	No cells in outflow: 100% of the cells feedback to the reactor; "quasi continuous"
Examples	
Single stirred reactor: Completely stirred reactor; cells are homogeneously dispersed (Fig. 1)	Stirred reactor with 100% feedback: X is not proportional to $(S_0 - S)$ (in presence of toxics or if K_S is large)

Criterion II: Mixing

Homogeneous	Heterogeneous
Stirred reactor (homogeneously dispersed cells in a homogeneous mixture of gaseous and liquid phases)	Pipe flow reactor Partitioned tank (organisms are kept back by sedimentation) Thin-layer pan with pellicle at the surface

Criterion III: Number of Stages

Multistage	Single stage
Open systems can be combined to form chains of two or more reactors; feeding of substrate is totally or partly to the first reactor	No serial combination of closed systems is possible Closed systems must be combined with an open reactor (acting as seed tank)

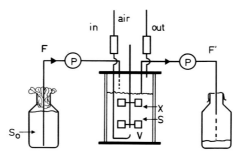

FIG. 1. Homocontinuous single-stage setup. Fresh medium with a substrate concentration S_0 is fed by a peristaltic pump (P) to the reactor at a constant rate (F). Actual volume V is completely mixed and contains cells of concentration X, residual substrate S, and gas bubbles (air). Reaction mixture is removed at a rate (F') equal to medium feed: $F' = F$. Relative feed $F/V = D$ (dilution rate per hour). Specific growth rate μ is variable, depending on the dilution rate: $D \equiv \mu$. $1/D$ = residence time.

For most yeast studies, an open, completely stirred (homogeneous), single-stage reactor is adequate (Fig. 1).

C. Technical and Operational Prerequisites

The prerequisites of continuous-cultivation methods are the attainment and maintenance of a flow equilibrium. During steady-state operations no changes in substrate concentration, cell number, and metabolite pools should occur. As a result of constant environmental conditions (aeration, nutrition, pH, temperature), the regulatory properties of the cell reach stability over long periods of time.

Failures or instabilities of equilibria may be caused by either technical inadequacies or biological effects. Perfect and reliable equipment is therefore a prerequisite for efficient application of the method.

The minimum requirements for a suitable bench-scale unit have been defined as (see Fiechter, 1965): a sterilizable vessel; devices for aseptic coupling of feed lines; filters and sampling devices; mixing and aeration facilities; control of temperature, pH, foam, and "true" working volume (equal to the volume at "rest," i.e., neither agitation nor aeration); transfer elements for medium (in and out), acid and/or alkali.

Because of the long time usually necessary for chemostatic studies, all the numerous building elements must be of good workmanship, featuring high operational reliability. The assembling of ordinary laboratory equipment leads to unsatisfactory improvisations. Many researchers starting this way have abandoned the chemostat method after a frustrating period of unsuccessful trial-and-error work.

Proper equipment is now available at a reasonable price, and although investments may seem high at first, much money and time will be saved afterward.

II. Homocontinuous Single-Stage Systems

In view of the theoretical and experimental complexities of multiple systems, in all their possible variations, the following description of continuous-cultivation technique is confined to the simplest case, namely, the single-stage homocontinuous chemostat (Fig. 1).

A. Theory

Continuous-flow cultivation is strongly related to the kinetics of microbial growth. It makes use of the relationship between metabolic activity and the availability of substrate. Control of the rate of substrate supply to the cell gives rise to definable metabolic turnover, biosynthesis, and regulatory behavior of the cell. The situation can be described by the basic equation of exponential growth and the Monod equation describing the substrate–growth rate relationship. Exponential growth follows the equation

$$N = N_0 e^{\mu t}$$

where N = cell number at time t, N_0 = cell number at time $t = 0$, μ = specific growth rate, and t = time. It occurs when substrate is present in excess, every cell is actually proliferating, and no inhibitors are acting on the cells.

Exponential growth is normally achieved in batch experiments only after a lag during which organization of the cell is directed to unrestricted growth.

μ is the specific growth rate according to

$$\frac{dX}{dt} = \mu X$$

$$\frac{1}{X}\frac{dX}{dt} = \frac{d(\ln X)}{dt} = \mu = \frac{\ln 2}{t_d}$$

where X = biomass and t_d = doubling time. Monod (1942), for the first time, demonstrated that μ is not an inherent constant of the cell. It is a variable depending on the substrate concentration according to the relation

$$\mu = \mu_{max} \frac{S}{K_S + S} \tag{1}$$

where μ_{max} = maximum specific growth rate, S = substrate concentration, and K_S = saturation constant. This relationship is similar to the classic Michaelis-Menten equation for a single enzyme. Cell growth is therefore expressed as the kinetics of the sum of all enzymes involved in growth processes.

All theories on continuous cultivation are based primarily on this model, which has been extended to describe more complicated systems. Extensions of the model should be considered only after critical examination of all experimental data obtained, and of the underlying conditions of the experiment. In the following only enough theory is presented to allow correct use of the flow-culture method, and to allow recognition of its limits in a given system. For detailed information the reader is referred to the work of Monod (1950), Moser (1958), and Fencl (1966); more condensed treatments of the matter are given by Herbert *et al.* (1956), Pfenning and Jannasch (1962), and Kubitschek (1970). An annotated bibliography is compiled by the Prague group and annually published in *Folia Microbiologia* (Ričica, 1973).

In Fig. 1 the basic arrangement of a single-stage chemostat is given. Medium containing the substrate (e.g., glucose) at concentration S_0 is fed at a constant rate F to a completely homogeneous stirred vessel containing a constant-volume V of culture (cells, medium, and gas). A portion of this mixture is taken out at an equal rate $F' = F$. The relative feeding rate F/V is the dilution rate D, which represents the only variable of the system. All external growth parameters are kept constant (stirrer speed, aeration rate, temperature, pH, etc.) In such a system the dilution rate D is identical with the growth rate μ following the equation

$$\frac{dX}{dt} = \mu X - DX \tag{2}$$

Change in biomass = growth − output concentration

During a true steady state, $dX/dt = 0$. Therefore

$$DX = \mu X \tag{3}$$

and

$$D = \mu$$

A sequential increase in D eventually leads to a washout, which occurs at the critical dilution rate D_c which corresponds to μ_{max}.

The resulting concentrations of biomass X and residual substrate S are a function of D, and can be predicted from batch experiments.

For substrate S,

$$\frac{dS}{dt} = DS_0 - DS - \frac{1}{Y}\mu X \tag{4}$$

Change = input − output − uptake

Y is the yield constant (dimensionless) describing the relationship between biomass formed from the substrate:

$$\frac{dX}{dt} = -Y\frac{dS}{dt}$$

or

$$\frac{\text{Weight of cells formed}}{\text{Weight of substrate used}} = Y$$

Equations (2) and (4) both contain μ, which is itself a function of S [Eq. (1)]. Substituting Eq. (1) into these equations gives, from Eq. (2),

$$\frac{dX}{dt} = X\left[\mu_{max}\left(\frac{S}{K_S + S}\right) - D\right] \tag{5}$$

and from Eq. (4),

$$\frac{dS}{dt} = D(S_0 - S) - \frac{1}{Y}X\mu_{max}\left(\frac{S}{K_S + S}\right) \tag{6}$$

Solving Eqs. (5) and (6) for $dX/dt = dS/dt = 0$, the steady-state values of X and S are given as:

$$S = K_S\left(\frac{D}{\mu_{max} - D}\right)$$

and

$$X = Y(S_0 - S) = Y\left[S_0 - K_S\left(\frac{D}{\mu_{max} - D}\right)\right] \tag{8}$$

Steady-state concentrations of cell mass and substrate in the vessel can be predicted for any given value of the dilution rate, providing the organism constants μ_{max}, K_S, and Y are known. Determinations of these constants can be made in batch growth experiments, where μ_{max} is found during maximum slope of the growth curve and K_S is approximately equal to the substrate concentration in the stationary-growth phase (for correct estimation of K_S, see Fencl, 1966). Herbert et al. (1956) calculated the values for

X and S as functions of D for hypothetical organism constants of $\mu_{max} =$ 1.0 per hour, $Y = 0.5$, $K_S = 0.2$ gm/liter, and concentration of the inflowing substrate $S_0 = 10$ gm/liter. Curves for productivity (DX = gm/liter per hour) and doubling time $t_d = \ln 2/\mu$ are also given (Fig. 2).

It is obvious that concentration changes for biomass and substrate are rather small over a wide range of D. Productivity increases linearly to a maximum value at

$$D_{max} = \mu_{max} \left[1 - \sqrt{\frac{K_S}{K_S + S_0}} \right] \qquad (a)$$

Beyond this point washout is observed, and with still rising D biomass finally drops to zero. $D_{max} \equiv \mu_{max}$, and the substrate reaches its maximum concentration ($S \equiv S_0$). Under experimental conditions cell number, although small, never reaches zero at supercritical rates. Only a few data are available from such cell types. The cell wall of *Saccharomyces cerevisiae* is very fragile, the protein content is increased, specific gravity is reduced, and storage carbohydrates and fat contents are close to zero. Cells of this type can more easily be obtained from a two-stage system (see Section V).

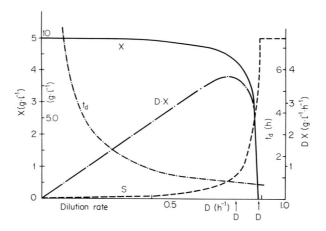

FIG. 2.   Steady state as a function of D. Theoretical values of substrate concentration S, biomass concentration X, productivity XD (output), and doubling time t_d at different dilution rates. Numerical values of Eqs. (7) and (8) for an organism with the following growth constants: $\mu_{max} = 1.0$ per hour, $Y = 0.5$, $K_S = 0.2$ gm/liter, and substrate concentration in the inflowing medium of $S_0 = 10$ gm/liter. Doubling time $t_d \equiv \bar{g}$ (mean generation time): $t_d = \ln (2/\mu)$. D_m denotes dilution rate for maximum productivity, and D_c is dilution rate at washout of the system. D_c corresponds to the maximum specific growth rate (Herbert et al., 1956).

B. Cultivation Techniques

1. VESSELS AND MASS TRANSFER

Microbiologists mostly assume that any kind of mixing device meets the requirement of homogeneous dispersion of air and cells in aequous solution. However, careful examination of data obtained from chemostat experiments revealed inconsistencies due to incomplete mixing, wall growth, foaming, variable volume V, or other side effects from inadequate equipment.

Yeasts are rather high oxygen consumers ($Q_{O_2}^{max} = 14$ mmoles/gm per hour for *Candida tropicalis*). A sufficient supply of this nutrient is therefore necessary for yeast studies.

Figure 3 demonstrates the effect of inadequate mixing in a classic fermenter with two flat-blade turbines (see Fig. 4b). *Saccharomyces cerevisiae* is a glucose-sensitive yeast whose respiration is affected in the presence of even small amounts of glucose (50 mg/liter). Dropwise addition of medium may result in local repression if mass flow in the fermenter is limiting. Even with a high stirrer speed (900 rpm), decrease in biosynthetic activity (biomass formation) and specific respiration (Q_{O_2}) is observed beyond a critical dilution rate (D_R) for purely respirative metabolism.

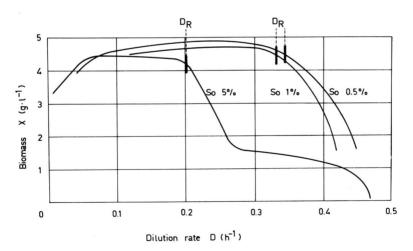

FIG. 3. Continuous cultivation of *S. cerevisiae* (glucose-sensitive). Biomass concentration X as a function of D. Critical dilution rate D_R for purely respiratory turnover is marked by a vertical bar. D_R depends on substrate concentration S_0 in the inflowing medium (0.5, 1, and 5%), indicating limitations of mixing effectiveness. D_R corresponds to maximum productivity (XD). Values for X are calculated for a common value of $S_0 = 1\%$. Yields at dilution rates above D_R are reduced because of repression of respiration.

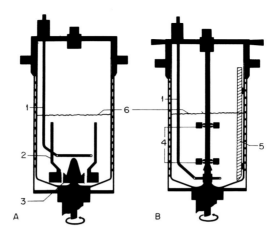

FIG. 4. Mixing systems of bench-scale reactors. (A) Circulation overflow system with draft tube (Einsele *et al.*, 1973). (B) Classic flat-blade turbine with two turbines (geometry according to Finn, 1954). 1, Air inlet with sparger; 2, draft tube with baffles; 3, radial stirrer; 4, flat-blade turbines; 5, baffles; 6, liquid surface.

Figure 3 shows also that D_R is a function of S_0, and therefore overall mixing is the limiting parameter. It was demonstrated (Oldshue, 1966; Einsele, 1972) that blade turbines in baffled vessels have poor mass-flow characteristics. Stagnant zones behind the baffles and slow distribution of the entering glucose give rise to remarkable concentration profiles along the three axes of the reactor.

More uniform mixing is obtained by circulation overflow systems (Fig. 4A). The liquid flows upward in the annular space between the wall and an inserted draft tube and returns to the stirrer inside the tube. High back-mixing features may contribute to good overall transfer rates (Blanch and Einsele, 1973) (Fig. 5).

High values for mass transfer rates are obtained with moderate airflow rates (Table I).

Foam abatement is necessary for proper bulk mixing and simple control of the true actual volume. Cells tend to accumulate in a stagnant layer on top of the liquid, causing incomplete dispersion of the cells. Foam also gives rise to changing (apparent) volumes and therefore causes oscillations of the steady state.

Foam abatement is effected by diluted silicone emulsions, a substrate which has proved to be inert if added in extremely small amounts throughout the experiment. Special control loops consisting of a heated probe which actuates a feed pump for an antifoam agent when touched by the rising foam layer are available. But quite a few of these devices are not free of

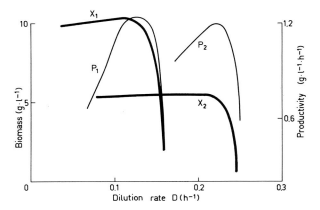

FIG. 5. Continuous cultivation of *C. tropicalis* on hexadecane media (Einsele *et al.*, 1972). Maximum productivity is equal in both runs of $S_0 = 1\%$ and $S_0 = 0.5\%$, but washout of biomass X occurs in the more concentrated system at lower dilution rates (apparent $\mu_{max} = 0.16$). Formation of flocs from oil droplets, cells, and air bubbles is due to limiting mixing effectiveness. For growth characteristics see table on p. 115. $(X_1P_1; X_2P_2 \cong 1.2$ gm/liter per hour.)

drawbacks, and antifoam consumption is in most cases higher than with dropwise addition actuated by a preset timer.

2. ATTAINMENT AND MAINTENANCE OF THE STEADY STATE

Disturbance of the steady state may occur as a result of internal (biological) or external (equipment) reasons (see also Section IV). Figure 6, for example, shows the oscillating response on feed rate change.

TABLE I

MASS TRANSFER IN BENCH-SCALE BIOREACTORS[a]

Air flow rate (vvm)	Revolutions per minute	K_La	Agitation system
0.7	1000	155	Circulation overflow system 20 liters
0.7	2000	328	total and 10 liters actual volume (Meyer, 1974)
0.5	1000	123	Flat-blade turbine 5 liters total and 10 liters actual volume (Robinson,
0.5	2000	238	1971)

[a]K_La is the transfer rate (per hour) of oxygen in an aqueous yeast suspension, indicating the transfer of mass per mass per liter. For assay methods, see Aiba *et al.* (1973). Data show same orders of magnitude of K_La values in all types of agitation systems. For details, see Meyer (1974).

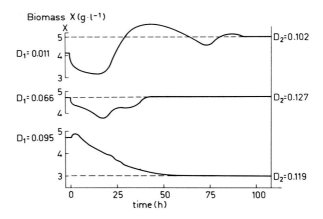

FIG. 6. Transient states in continuous cultivation of *S. cerevisiae* on ethanol (Mor and Fiechter, 1968). New steady states are reached after a period of damped oscillation of biomass concentration when dilution rates are shifted at low growth rates. Oscillation time depends on growth rate and differences in selected dilution rates.

The basic experiment utilized ethanol as the carbon source displaying $\mu_{max} = 0.25$ for *S. cerevisiae*. Damped oscillations of biomass and substrate concentration are observed before stable conditions are attained. Equations describing the transient functions may be of first or second order. The total time to reach a steady state depends on the absolute value of D, and on the ratio between old and new dilution rates. Up to about 100 hours was necessary to reach stability again. Usually, a 5-fold medium volume of the actual volume V has to be fed to attain the new steady state. However, up to 10 times the volume may be necessary, as shown in the example given by Mor and Fiechter (1968).

Experiments of this type reveal the very dynamic nature of continuous-flow cultivation. All kinds of parameters (temperature, pH, feed rate, actual volume, contamination, mutation, etc.) can destroy a given steady state.

It must be mentioned that transient states have a high information potential if observed properly (see Section III,B). However, such a technique is nearly unknown to microbiologists and needs further development.

3. Minimum Requirements for Equipment

Principally, a homocontinuous system is represented by a classic batch fermenter, plus asscessorial devices for transfer of liquids into and out of the vessel.

However, *aseptic operation* is of particular importance in chemostat experiments. Proper design is necessary to prevent contamination at the stirrer shaft by the air or the medium supplied, by leaky gaskets, or as a result of operational manipulations such as inoculation, sampling, or coupling of fresh medium reservoirs. Good results are obtained with mechanical seals for the stirrer shaft, and a membrane closure system for connecting separately sterilized tubings (addition of inoculum, medium, acid or alkali, antifoam agent) or to insert air filters (see Fiechter, 1965). Tubings are fitted with a needle and pushed through the membrane after alcohol is burned at its surface. This system is very versatile and provides safe operation in cases in which more complicated setups are required (e.g., anaerobic cultivations).

The *sampling technique* consists of rapid removal of a small volume of reaction mixture.

The vessel is pressurized by blocking the air outlet. Samples are taken with a needle connected with a rubber tubing, and sampling times are below 0.5 seconds (Weibel *et al.*, 1974). Even with sophisticated sampling techniques, loss of working volume V may be considerable if samples must be taken rather frequently. Stability of the steady state is endangered. Vessel capacity of not less than 5–8 liters total volume should therefore be chosen in order to allow an actual volume of 2–5 liters. Less than 1 liter often gives rise to insurmountable difficulties.

pH control must be used regularly; even this control loop is rather expensive. However, constant hydrogen ion concentration cannot be maintained only by buffering the media, if deflections to extreme values are to be avoided. In addition, either onset or speed of growth in batch experiments or irregularities like synchronization effects are observed (Fig. 7) if the setting of acid or alkali flow is chosen accordingly.

Gas metabolism is an important characteristic of cell metabolism. Assay of the gas exchanged is therefore a very valuable tool in biological studies. Monitoring of gas concentration, in either the liquid or the gas phase, improves the quality of experimental work (see also Section II,C).

C. Analytical Methods

It is the aim of this article to define the minimum requirements for proper chemostat work. Therefore only the most relevant assay procedures allowing proper control of a reaction are given. Apart from control purposes, analytical data can be used for calculation of the metabolic turnover rates and thus for biological characterization of physiological activities. In this way cell material of defined properties can be reproduced at any time and in any quantity.

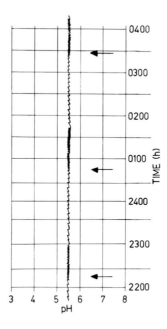

FIG. 7. pH record of a continuous cultivation of *S. cerevisiae* on synthetic glucose medium (von Meyenburg, 1969). Single additions of alkali are recognizable according to growth rate. Irregularities are due to synchronization effects in the cell population. Onset of bud formation with mean generation time $\bar{g}$ = 4.1 hours; oscillation time T = 2.7 hours full period.

The most relevant values in yeast and other studies are X (biomass or number of cells), S (substrate), gas exchange (oxygen, carbon dioxide), and intermediate metabolites (e.g., ethanol). From this metabolic balances (e.g., carbon balance) at specific turnover rates can be calculated. They are an important basis of metabolic and regulation studies.

In chemostat studies a great number of samples are processed. If possible, labor-saving methods should be chosen, preferably ones that can be automated.

1. BIOMASS

The determination of cell mass is based on drying at 105 °C until a constant weight is reached. The method cannot easily be automated at reasonable expense. The cell samples must therefore be centrifuged, washed, and dried separately. Photometry is more suited for monitoring purposes, and instruments are commercially available (Mor *et al.*, 1973). However, to obtain stochiometric material balances, photometric values must be converted to absolute values with the aid of a calibration curve (Pringle and Mor, this volume).

2. GLUCOSE

Older methods (potassium ferricyanide, copper) are unsuitable, because of high manpower costs and poor specificity. Enzymatic tests avoid these disadvantages. Experience shows that glucose oxidase electrodes are not yet sufficiently developed for routine application. Automatic assay of samples collected sequentially is possible (Mor *et al.*, 1973).

3. GAS EXCHANGE

Oxygen and carbon dioxide can be determined with automatic gas monitors. The difference between oxygen and carbon dioxide content in the gas stream between inlet and outlet of the reactor corresponds to the actual gas exchange. The essential advantage of this procedure lies in the fact that no disturbance of the steady state occurs. Further details have been described elsewhere (Fiechter and von Meyenburg, 1968). In lieu of paramagnetic methods for oxygen determination, the less expensive oxygen electrodes can be used. Corresponding pCO_2 sensors are unfortunately not yet available for carbon dioxide measurements. Infrared photometry is still the only reliable method for this purpose.

4. ETHANOL

Enzymatic determination with alcohol dehydrogenase is very specific and easily adapted for automation (Mor *et al.*, 1973). The classic method (i.e, oxidation with potassium dichromate) is inexpensive but requires much labor and lacks specifity.

5. OTHER COMPOUNDS

Assay of numerous compounds is possible by automation (references are available, e.g., Technicon Corp., New York City). However, full automation of a large number of parameters is costly, and strategy for data collection must be carefully evaluated.

6. DATA PROCESSING

Data collected according to paragraphs 1 to 4 permit the calculation of a carbon balance, which gives proof (1) of the quality of the assays, and (2) of the completeness of the stochiometry of the underlying reaction equation. Moreover, from complete assay of substrates and products, the specific values for specific substrate uptake (Q_s), oxygen uptake (Q_{O_2}), carbon dioxide release (Q_{CO_2}), and the specific formation of excreted metabolites

(Q_P) can be calculated:

$$Q_S = \frac{D(S_0 - S)}{X} - \frac{1}{X}\frac{dS}{dt}$$

$$Q_P = \frac{DP}{X} + \frac{1}{X}\frac{dP}{dt}$$

$$Q_{O_2} = \frac{([O_2]_{in} - [O_2]_{out})F_{air}}{VX}$$

$$Q_{CO_2} = \frac{[CO_2]_{out} - [CO_2]_{in}F_{air}}{VX}$$

A simple computer program for rationalization of the calculations has been developed by Mor *et al.* (1973) for batch and chemostat cultivation techniques.

D.   Anaerobic Cultivation Technique

A special cultivation technique is necessary for cultivation of obligate and facultative anaerobes. Additional measures are required for the removal of oxygen and to avoid reentrance of oxygen into the cultivation system (Fig. 8). The complex setup may explain why relatively few data relating to the anaerobic metabolism of yeasts are available.

The main criterion is the removal of oxygen. Oxygen can be eliminated in all parts of the setup by flooding with a highly purified carrier gas. Nitrogen gas is purified in an alkaline pyrogallol solution. Methylene blue indicator is sensitive for 20 ppm of oxygen or more. Extensive work done by Schatzmann (1974) showed that oxygen concentration of less than 0.05 mmoles of oxygen excludes any effects of aerobiosis (absence of cytochromes a to a_3, b, and c to c_1).

The transport of liquid media anaerobically gives rise to particular difficulties. For chemostat systems only peristaltic pumps can practically be adopted. However, no rubber or plastic tubings exist that are gas-impermeable. Glass or stainless-steel conducts are best suited for medium transfer. Peristaltic pumps can be protected from oxygen by placing them in a glove box flooded with purified nitrogen. The evolved carbon dioxide is removed from the fermenter by a flow of oxygen-free nitrogen gas and brought to the carbon dioxide analyzer at a constant rate. Samples are withdrawn either from the reactor or collected from the pump at the outlet medium flow. For *Saccharomyces* species ergosterol, unsaturated fatty acids, and nicotinic acid (Schatzmann and Fiechter, 1974) must be added for anaerobic growth. The capacity of synthesis is much reduced in wild-type strains when com-

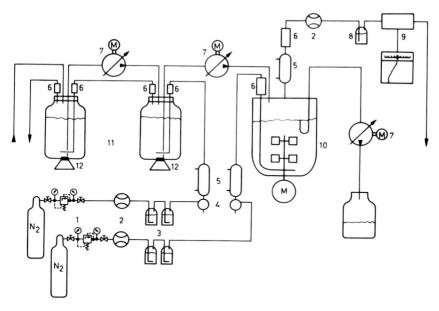

FIG. 8.   Flow sheet of the anaerobic chemostat (Schatzmann and Fiechter, 1974). N_2, Pure nitrogen, <1% oxygen; 1, reducing valve; 2, flowmeter; 3, washing bottles with 6 parts of 60% KOH plus 1 part of 25% pyrogallol; 4, water trap with oxygen indicator; 5, reflux cooler; 6, gas filter (glass wool or ceramic filter); 7, peristaltic pumps; 8, water trap; 9, carbon dioxide infrared monitoring and assay (URAS, Hartmann and Braun, Frankfurt, Germany) plus recorder; 10, reactor 10 l (Chemap, Männedorf, Germany); 11, medium reservoirs 20 l (for deoxygenation); 12, magnetic stirrer.

pared with aerobiosis (yield = 0.1). Ethanol and carbon dioxide are formed in high equimolar quantities (>25 mmoles/gm per hour).

Only a minor reduction in growth rates is observed in anaerobiosis (μ_{max} ~ 0.34 per hour) (Schatzmann, 1974). This metabolic type is identical with a respiration-free petite mutant cultivated at oxygen saturation (*S. cerevisiae* E5, 90; see Fig. 9 and Table III). Despite the nonavailability of oxygen, petite mutants seem to show some mitochondrial residual activity, which might be responsible for the slightly higher yield of 1.3 as against the wild type. Schatzmann (1974) succeeded with the described cultivation method in giving quantitative proof of a glucose effect (also occurring in aerobiosis), which regulates the residual activities of some respiration enzymes (compare Lowdon *et al.*, 1972; Rogers and Stewart, 1973). With the chemostat technique the effects of glucose and oxygen on respiratory functions and glycolysis (Fiechter and Schatzmann, 1972) can therefore unequivocally be separated. Four different regulation states are obtained, which have not

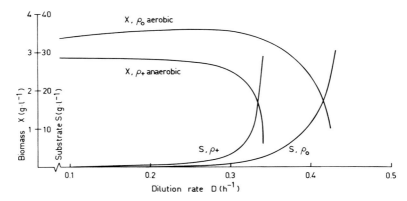

FIG. 9. Continuous cultivation of *S. cerevisiae* on synthetic glucose media (Schatzmann and Fiechter, 1974; Maag, 1973). Biomass formation of the wild type (p^+) in the absence of oxygen, and of the petite mutant E5 (p_0) in the presence of air. The lower values for biomass of the anaerobic wild-type growth indicate adequate removal of oxygen.

yet been described in detail (aerobic and anaerobic repression and derepression).

Respiratory regulation in anaerobiosis results in structural changes in the mitochondria, apart from biochemically recognizable mutations, which assume so-called promitochondrial states.

The described methodology of anaerobic chemostat techniques is not well known as yet. Potentially, it is a very important tool for the study of regulatory and molecular problems.

III. Cultivation of Glucose-Insensitive Yeast Types

The preceding paragraphs have given a short description of the chemostat methods and the attendant theories. Both the theory and the remarks on equipment are principally valid for all kinds of microbial cells and not confined to the biology of yeasts. In contrast, the following sections are devoted to the biology of yeasts and illustrate application of chemostat methods to their study.

A.   Effects of Oxygen and Mass Transfer Limitations on Growth of *C. tropicalis*

Candida tropicalis, like the well-known *Candida utilis*, is a representative of the so-called respiration yeasts *Atmungshefen.*

TABLE II

ORGANISM CONSTANTS AND METABOLIC CHARACTERISTICS OF A
GLUCOSE-INSENSITIVE YEAST (*C. tropicalis*) ON SYNTHETIC MEDIUM
(S_0 = 10 GM/LITER GLUCOSE)[a,b]

Characteristic	Glucose	Hexadecane
Yield, Y	0.55	1.10
μ_{max} (per hour)	0.74	0.28
K_S (mg/liter)	12.0	110.0
$Q_{O_2}^{max}$ (mmoles/gm per hour)	14.0	14.0
Productivity (gm/liter per hour)	3.4	3.9
Q_S	8.0	1.0

[a]Knöpfel, 1972.
[b]Specific growth rates are reduced on hexadecane, but Q_{O_2} is the same as on glucose. Limitation of biosynthetic capacity is given by respiratory activity (Einsele *et al.*, 1972).

With an extensive oxygen supply they show no glucose repression of respiration, high growth kinetics, and high yields. The former type, like many others, is able to adapt to hydrocarbons as growth substrate (see Section III,B). In chemostat studies growth strongly follows Monod kinetics according to the theory depicted in Fig. 2. Equimolar amounts of gas are taken up (Q_{O_2}) and released (Q_{CO_2}), rising linearly with D. Typical growth constants and turnover rates for *C. tropicalis* in glucose media are given in Table II. They show the high potential of this cell type for biosynthetic

TABLE III

ANAEROBIC GROWTH OF *S. cerevisiae* ρ^+ (WILD TYPE)[a]

Growth characteristic	Wild type, ρ^+	Petite mutants E5, ρ_0
Yield, Y	0.10	0.15
μ_{max}	0.34	0.42
K_S (mg/liter)	400	—
Q_S^{max} (mmoles/gm per hour)	18.5	14.4
Q_{CO_2} (mmoles/gm per hour)	32.0	27.2

[a]Synthetic medium supplemented with ergosterol, Tween 80, nicotinic acid (Schatzmann and Fiechter, 1974). Yield or biomass is slightly higher in aerobic chemostat cultivation of *S. cerevisiae* E5, ρ_0 (petite mutant lacking mitochondrial DNA (Nagley and Linnane, 1970). YEPD-Medium.

activity. Interestingly enough, a drastic reduction in maximum growth rate on hydrocarbons is observed.

Identical, maximum respiration rates (Q_{O_2}) on two substrates of different oxidation level seem to support the idea that the biological potentials represent a controlling factor (Table III). Provided that the value of $Q_{O_2} =$ 14 mmoles/gm per hour obtained on hexadecane is corrected for the mitochondria-independent fraction of oxygen for the first oxidation step (approximately 10%), the value becomes close to $Q_{O_2} = 11$ mmoles/gm per hour on glucose (Einsele *et al.*, 1972). Furthermore, the high oxygen demand for the oxidation of hexadecane involves a lower rate of synthesis.

The ratio of demands, calculated on the basis of the material, is: glucose (C_6)/hexadecane (C_{16}) = 1:2.5; and the ratio of maximum growth rates is: $\mu_{max}(C_{16})/\mu_{max}(C_6) = 1:2.5$.

The reported Q_{O_2} values are typical for this *Candida* strain, and no significant enhancement was obtained by higher aeration rates. Nevertheless, it must be kept in mind that these experiments were carried out in reactor systems whose mass transfers seem to be the limiting factor. Even higher respiration rates may occur if this restriction can be overcome. The low value for specific hexadecane uptake (Q_S) suggests that the limitation occurs in either mass transfer in the bulk of the liquid or in the transport and/or oxidation steps (Einsele *et al.*, 1972; Hug, 1973). However, more information is necessary on the mechanism of uptake of insoluble substrates and their transfer into the cell, as well as on the nature of the mitochondria. These organelles are rather insensitive to glucose repression, as can be concluded from Fig. 10. Although repressional effects are observed under conditions of oxygen limitation (Fig. 10), the observed sensitivity to glucose is, on the molecular level, very poorly understood.

B. Transient Experiments

In the chemostat technique, the main criterion is the attainment of a stable steady state. It represents an equilibrium in flow of mass, balanced growth, and constant rates of biosynthetic and degrading activities, and displays the dynamics of regulatory behavior. Instabilities are due to inadequacies of the system, e.g., incomplete media leading to starvation, inconsistent equipment, or differentiation phenomena in the cell population (Sections I,C and IV,D).

It is possible to expand the chemostat technique by introducing a systematic change in one of the given parameters. As a result, the flow equilibrium is disturbed and, according to the given new set of flow rate, nutrient composition, or any other external parameter, a new steady state will be reached. The time course of such a transient state is seldom linear. Occurrence of

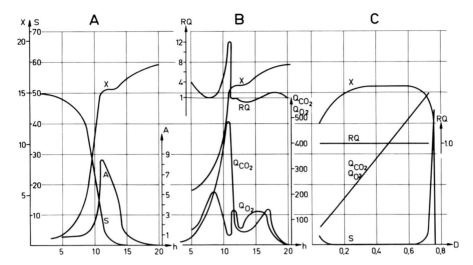

FIG. 10. Continuous cultivation of *C. tropicalis* on synthetic glucose medium (1% glucose). No repression at high dilution rates is observed. No formation of alcohol (C). Diauxic growth (repression) is possible only in the presence of high glucose concentrations ($S_0 \geqq 5\%$) and limiting oxygen supply (air rate = 0.14 vvm). Repression phenomena are the same as with a glucose-sensitive yeast (see Fig. 12). Alcohol is accumulated during the first growth phase. It serves as a substrate for growth in the following phase (A and B) (Knöpfel, 1972).

damped oscillations indicates the presence of regulatory feedback systems. In most cases the mechanisms of the underlying control loop cannot be recognized in full detail. However, the changes in the cell components can be measured and ascribed to the changed parameter.

Figure 11 gives an example of a transient experiment in which glucose was replaced by hexadecane as a carbon source.

It is known that cells of *C. tropicalis*, when grown on hydrocarbons, contain twice as much lipid as when grown on glucose (Hug *et al.*, 1974). In a shift experiment from glucose to hexadecane, an adaptation phase occurs during which the lipid concentration increases greatly. The abrupt change in substrate from glucose to hexadecane involves another transport mechanism and new enzymes in metabolism.

Hexadecane cannot be taken up by glucose-grown cells, and there is no concentration increase during the initial period. Thus either the transport of substrate to the site of enzymatic activity is not possible, or the oxygenase system for alkane degradation is not active. It is more likely that the adaptation phase is due to the requirement for synthesis of the enzymes necessary for oxidation. The duration of 4 hours of the phase corresponds to the generation time (which is 4.6 hours at $D = 0.15$ per hour), suggesting sub-

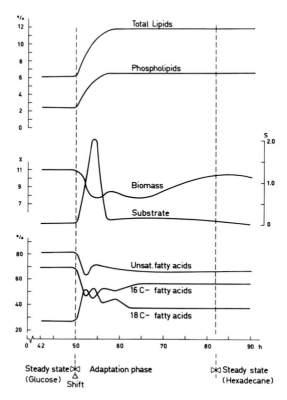

FIG. 11.   Response of cellular lipids and fatty acid composition to a substrate change from glucose to hexadecane in continuous culture (dilution rate $D = 0.15$; generation time, 4.5 hours). Fatty acid concentrations are expressed as percent of total fatty acids detected. Accumulation of the new substrate (hexadecane) occurs during the first phase of adaptation, as a result of restrictive function of the cell envelope (Einsele *et al.*, 1974). Onset of substrate uptake and increase in biomass concentration (oscillation) is initiated only after enrichment of phospholipids (suggesting membrane alteration) and C_{16} fatty acids (Hug *et al.*, 1974).

strate induction of enzyme synthesis. It is assumed that the oxidizing enzyme for hydrocarbon oxidation is bound to some membrane system. If phospholipids are preferentially bound to such membranes, the increase in this lipid fraction indicates the induced buildup of competent membrane structures during the adaptation phase. At least the high cellular lipid concentration seems necessary for hydrocarbon assimilation and does not just reflect the lipophilic nature of the substrate. Proof or disproof of the assumptions made in this study is not possible simply by using the transient technique. However, this example demonstrates perfectly the usefulness of the method for exploration of structural or functional parameters of the cell.

IV.   Cultivation of Glucose-Sensitive Yeast Types

A.   Glucose Effect on Respiration

In contrast to the *Candida* yeast types described in Section III, many other strains are very sensitive to free glucose. Their respiratory metabolism and the synthesis of cell mass are strongly reduced if the concentration of this sugar exceeds 50 $\mu g/ml$.

Uptake and glycolysis are not controlled by glucose itself. However, in the presence of oxygen this initial pathway undergoes inhibition, and uptake of the free sugar is reduced (Pasteur Effect; Sols *et al.*, 1971). This regulation phenomenon is also observed with microorganisms other than yeasts. Among the latter only glucose-sensitive types show the Pasteur effect, indicating that oxygen is not the primary effector (Krebs, 1972).

According to this model, high respiration rates affect glycolysis through ATP and citrate generated by active mitochondria, which act on phosphofructokinase in the form of an allosteric inhibition. As a result, the uptake of glucose is ultimately reduced because of the inhibition of hexokinase by glucose 6-phosphate (Sols *et al.*, 1971). It was found that, in contrast to the hexokinases of higher organisms, the yeast enzyme undergoes allosteric inhibition by glucose 6-phosphate. However, overall transport of glucose across the cell membrane is similar to that in higher organisms. Fully derepressed cells have equal kinetics in both metabolic sequences of glycolysis and the tricarboxylic acid cycle, whereas under repressional conditions metabolic products of the glycolysis are not oxidized quantitatively and ethanol is excreted into the medium.

Obviously, the overall response of a complex control mechanism does not simply follow Monod kinetics (Fig. 12). With higher feed rates, when free glucose is detectable in the bulk of the liquid, less cell mass is produced at the expense of ethanol formation. It is important to note that maximum growth rates are obtained with repressed cells, and that the reduced yield is only due to respiratory repression (Fiechter and von Meyenburg, 1968). Repression is highest at supercritical feed rates ($D > D_c$), even under oxygen saturation. The yield tends to become low, as in anaerobiosis (Fiechter and Schatzmann, 1972), indicating that glucose is a potent effector like oxygen.

In batch cultivation this regulation phenomenon gives rise to diauxic growth behavior. Ethanol is accumulated during the first (repressed) growth phase and serves as the carbon source in a following sequence. A reduction of this growth pattern is represented by a yeast type such as *Schizosaccharomyces pombe*, which is unable to grow on C_2 compounds (Flury, 1973). The anaplerotic pathways are out of function in the presence

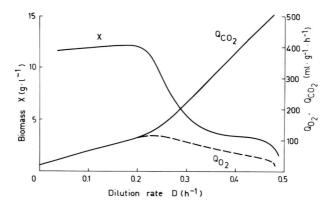

FIG. 12. Continuous cultivation of *S. cerevisiae* in synthetic glucose medium ($S_0 = 3\%$). Respiratory repression at higher feed rates (decrease in specific oxygen and uptake rates; increase in specific carbon dioxide production). Biomass production is reduced according to the degree of repression. This repression phenomenon is exclusively due to the presence of free glucose (excessive oxygen supply). All enzymes involved in the citric acid cycle and the glycoxylic acid bypass are affected (Beck and von Meyenburg, 1968; Knöpfel, 1972).

of ethanol. This substrate is fully oxidized to carbon dioxide and water. Again, specific growth rates under repression are not strongly affected by the absence of the pathway, but respiration, and consequently biosynthesis, are reduced (Table IV).

TABLE IV

ORGANISM CONSTANTS OF GLUCOSE-SENSITIVE
YEASTS[a]

Characteristic	*S. cerevisiae*	*Schiz. pombe*
Yield, Y	0.48	0.2
μ_{max} (per hour)	0.47	0.39
K_S (mg/liter)	35	30
D_R (per hour)	0.30	—
$Q_{O_2}^{max}$	8.0	1.35
Q_S	4.0	1.5
Productivity, DX (gm/gm per hour)	1.25	—

[a]Growth of *S. cerevisiae* is observed during repression and derepression (Fiechter, 1965). Derepressed cells of *Schiz. pombe* are able to oxidize the accumulated ethanol but unable to grow (Flury, 1973).

B.   Regulation of Respiration

Two effectors are responsible for respiration regulation in yeasts. In respiratory yeasts repression appears only if the oxygen supply is limited (Section III,A), whereas in other types, such as *S. cerevisiae*, glucose is a very potent effector of respiratory repression.

Detailed examinations revealed that probably all enzymes of the tricarboxylic acid cycle are regulated (Beck and von Meyenburg, 1968, Knöpfel, 1972; Flury, 1973).

The mechanism of this regulation is unknown. Generally, it is denoted as catabolite repression. But demonstration of the true effector role of glucose or a catabolite of it is still lacking. Catabolite repression in prokaryotes is regulation at the level of RNA transcription. Breakdown products control the amount of cyclic AMP. A specific protein factor as well as cyclic AMP must be present to allow a RNA polymerase to recognize the catabolite-sensitive promotor. The fact that the whole array of enzymes involved in respiration is regulated simultaneously indicates the presence of a more complex control mechanism.

Flury (1973) demonstrated this complexity with one of the participating enzymes of the citric acid cycle. Malate dehydrogenase (MDH) holds a central position in metabolism. It takes part in at least four different pathways of metabolism: the tricarboxylic cycle, the Wood-Werkman reaction, the glyoxylate shunt, and glyconeogenesis. MDH is found in cytoplasm and in mitochondria. Furthermore, it is postulated that MDH is involved in the transport of ions through mitochondrial membranes (Borst, 1963).

The existence of multiple forms of malate dehydrogenase in different organisms is well established. The enzyme occurs mainly in two forms which differ markedly in their amino acid composition and their kinetic properties. In general, one of these forms is localized in the mitochondria and the other in the cytoplasm. In *Schiz. pombe* two isoenzymes have been found. One of them is present only in glucose-repressed cells. It disappears during respiratory derepression. The second form is absent in the presence of glucose, and fully derepressed cells contain this second form exclusively. The activity ratio between the repressed and the derepressed forms is 1:20. Addition of glucose to the derepressed form leads to rapid inactivation, probably through an enzymatically catalyzed chemical modification. Inhibition experiments with cycloheximide reveal that this isoenzyme is synthesized by cytoplasmic and not mitochondrial ribosomes. During the course of a derepression experiment, two new forms are observed before the final derepressed form is fully developed.

These results with *Schiz. pombe* show that regulation of enzyme activity during glucose repression is not simply due to catabolite repression, and views on respiratory regulation in eukaryotes must be expanded.

C. Cell Differentiation Phenomena

In the chemostat the growth rate is determined by the rate of supply of the growth-limiting nutrient. This is essentially true of high feed rates, where the individual cells of the population display indentical features and differential phenomena cannot take place (see Section IV,D). However, Beran (1967), determining the relative distribution of cell age in a random population, demonstrated that the number of bud scars per cell (a characteristic of cell age) differs substantially from the theoretical. Depending on the dilution rate, the number of newly born cells (carrying no bud scar) is always above the theoretical. The individual growth rate is therefore below that of older (mother) cells (Table V). The cell population is consequently heterogeneous with respect to individual growth kinetics. Morphological and physiological differentiation phenomena are also observed at low feed rates.

However, at low dilution rates metabolism is diversified and other effects become visible. Under glucose limitation the synthesis of storage carbohydrate increases strongly with increasing limitation of substrate supply,

TABLE V

RELATIVE AGE DISTRIBUTION IN YEAST POPULATIONS AT DIFFERENT DILUTION RATES[a]

Number of of scars	Cell fraction (%)		
	Theoretical distribution	Dilution rate, D	
		0.15	0.35
0	50,000	59.89	56.98
1	25,000	17.17	21.37
2	12,500	11.17	10.48
3	6,250	5.88	5.01
4	3,125	2.98	2.96
>4	3,125	3.38	2.26

[a]Age is expressed in terms of bud scar numbers (Beran, 1967). *Saccharomyces cerevisiae* bud counts in the fluorescence microscope after primulin staining.

i.e., decreasing growth rate. Repressed (high glucose feed rate) cells accumulate only glycogen, whereas derepressed cells (low substrate feed rate) contain trehalose and high amounts of glycogen.

With *S. cerevisiae*, Küenzi (1970) observed a broad spectrum of cell sizes at dilution rates below $D = 0.2$ (0–180 μm^3), whereas with $D = 0.455$ cells were much more uniform and fell in only two classes (60–90 and 90–120 μm^3). Observation with the phase light microscope showed the occurrence of a class of dark cells in addition to large, bright cells. The dark cells were elliptical or long in contour, and glycogen staining was negative. Bright cells were nearly globular and glycogen-positive. In long-lasting steady states, another class of dark cells occurred. These elongated cells were repressed and excreted ethanol (Küenzi, 1970).

With dilution rates below $D = 0.1$, regular oscillations of gas metabolism were observed. The duration of one oscillation cycle was half the mean generation time, indicating a synchronization effect due to cell differentiation.

Cell size measurements in a population of this kind revealed that the class of largest cell ($> 5 \mu$m diameter) governed the synchronization. Every second oscillation was due to the newly born daughter generation. Very small cells ($< 3.7 \mu$m in diameter) in such populations were not able to bud. Before initiation of the new bud, a burst of carbon dioxide was observed (Figs. 13 and 14A). During the cell cycle of these regularly budding cells, typical changes in carbohydrate composition were observed. Before onset of bud formation, glycogen and trehalose contents decreased considerably and

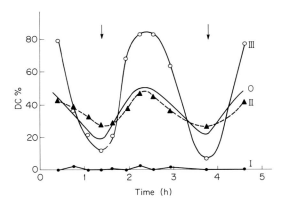

FIG. 13. Stable oscillation in continuous cultivation of *S. cerevisiae*. Mean generation time $\bar{g} = 4.1$ hours; $D = 0.169$ per hour; oscillation time $= 2.4$ hours. Numbers of double cell fractions of size classes I, mean diameter ≤ 3.75 nm; II, mean diameter 3.75–5 nm; and III, mean diameter >5 nm. 0, Fraction of double cells in total population; arrows, maximum release of carbon dioxide. Class III cells maintain the synchronization effect (Küenzi, 1970).

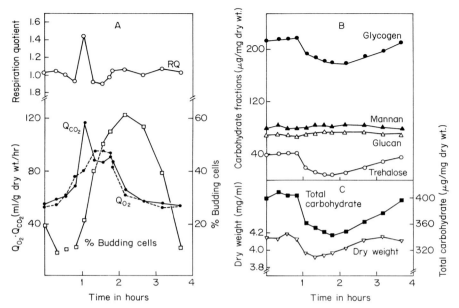

FIG. 14. Partially synchronized population of *S. cerevisiae* cells. Dilution rate $D = 0.10$ per hour. Percent budding cells (A), dry weight (C), gas exchange (A), and changes in carbohydrate fractions (B) as a function of time during a steady state. Onset of budding is possible only after mobilization of storage carbohydrates accumulated during bud formation. Note the peak in carbon dioxide release at the initiation point of a new budding round (Küenzi and Fiechter, 1969).

recovered only after separation of the daughter cells from the mother cells (Fig. 14C). The single-cell phase is the period of carbohydrate formation and storage by the cell. The continuous supply of fresh medium refills the storage carbohydrates. Once a certain maximum value has been reached, bud formation is triggered, and the cell passes into the double-cell phase. The energy generated by glycogen and trehalose degradation is required for budding. Small amounts of ethanol are thereby excreted, serving those cells whose substrate supply (internal or external) is subcritical to initiate synchronous budding. The determinant for joining the first or the second category of budding cells is effected by the ethanol (see also Küenzi and Fiechter, 1972).

D.   Synchronization Effects in Continuous Culture

Synchronization effects are often encountered in chemostat experiments. Several effects may cause synchronization of part of the cells. At high dilution rates, the steady state is normally perfectly stable because differences in individual generation times are rather small. The substrate supply is too high to allow differentiation before initiation of the next round of budding. Cells unable to grow are washed out from the system in a short time.

Under conditions of low substrate supply, individual cells tend to differentiate according to individual substrate availability. As shown in Section IV,C, the substrate (e.g., ethanol) may function as a coordinating agent. To achieve synchronization, according to Küenzi, a chemostat steady state at high dilution rates (e.g., $D = 0.32$) has to be maintained for some time. After six generation times were completed, the supply was interrupted until all the ethanol was used up by the continuing growth. Next, fresh medium was fed at any low dilution rate ($D < 0.1$–0.15), and a cyclic, partially synchronized culture was obtained. In contrast to any other method, this procedure yields synchronization at different growth rates. A drawback of the method is the relative low degree of synchronization. However, sampling of the synchronized cell fraction is possible by sucrose gradient centrifugation (Matile and Bahr, 1968).

Synchronization effects at low dilution rates are also obtained by sudden changes in external parameters such as temperature or pH (von Meyenburg, 1969). Application of this "shock" at regular sequences can maintain perpetual synchronization. However, it is not clear how much the effects of such drastic measures do interfere with the organizational state of the cell.

V. Multistage Systems

Extensive treatment of the theory is given elsewhere (see Fencl, 1966). Only a few remarks are made to highlight the usefulness of multistage chemostat systems for biological studies. Theoretically, the layout of a multistage system can be varied in terms of number and type of reactor (Section I,B). For simplicity only a combination of two homogeneous stirred reactors (a two-stage system) is considered here (subscripts denote corresponding stages).

This system is a combination of two identical reactors depicted in Fig. 1, if the outflow of stage 1 is connected to vessel 2. Additional substrate may be fed to stage 2 (F_{02}). Inflow concentration of X from the first vessel is X_{02}. In such a system dilution rates no longer represent the specific growth rate, e.g., $D_2 > \mu_2$:

$$D_2 = \mu_2 \frac{X_2}{X_2 - X_{02}}$$

and

$$\frac{X_2}{X_2 - X_{02}} > 1$$

The maximum value of μ_2 could reach μ_{max} only if the total feed of the substrate is supplied to stage 2 (F_{02} = total sugar addition), reducing the multiple system to a single-stage chemostat. Furthermore, washout of the higher stages is not possible, since

$$X_2 = X_{02} + \frac{\mu_2}{D_2 - \mu_2} X_{02}$$

and

$$X_{02}, \mu_2, D_2 > 0$$

and

$$D_2 \geq \mu_2, X_2 \geq X_{02}$$

Thus cell concentration in the first stage can never exceed that of the latter stages. A two-stage system therefore offers the opportunity to grow cells in the presence of high amounts of substrate, while maintaining high dilution rates. Cell density is low in the second stage but never zero.

According to the theory, maximum specific growth and turnover rates cannot rise above those reached in batch or one-stage systems. However,

TABLE VI

CULTIVATION OF *C. utilis* IN A TWO-STAGE
CHEMOSTAT SYSTEM[a,b]

	Batch	First stage	Second stage
μ_{max} (per hour)	0.49	0.47	1.0
Concentration (mg/mg)			
DNA	—	0.005	0.005
RNA	—	0.16	0.16
Protein	—	0.5	0.55
Synthetic rate (per hour)			
Q_{DNA}	—	0.002	0.008
Q_{RNA}	0.0024	0.05	0.28
Q_{prot}	—	0.2	0.8

[a](Mian *et al.*, 1971).
[b]Rates of synthesis (DNA, RNA, protein) increase according to specific growth rates. Unusually high synthetic activity is attained under conditions of unrestricted growth.

more recent data from growth studies of *C. utilis* show that kinetics can be raised to unexpected levels in the second stage. Assuming that Monod kinetics are still followed, these findings can be interpreted only as an expression of limiting mass transfer onto the cell surface. It seems that substrate profiles at the interface between liquid (medium) and solid (cell) control the substrate uptake and therefore also the metabolic turnover rates of the cell.

It is seen from Table VI that the maximum growth rate of *C. utilis* is enhanced from 0.49 per hour in the batch system up to 1.0 per hour in the second stage. The doubling time of nearly 1 hour is an unexpected low value for an organism of this size. Specific rates for DNA, RNA, and protein synthesis rise linearly according to the increase in growth rates. The author assumes that under conditions of abundant substrate, as attained in the second stage, specific growth rates and synthetic activities approach a maximum.

VI. Concluding Remarks

The aim of this article was to demonstrate the usefulness of continuous-cultivation methods in microbiological studies. These methods have been considered by scientists primarily as a potential tool in the development of efficient industrial production plants. Indeed, some processes such as single-cell production or effluent treatment plants are operated continuously. But one should also mention that in most cases the old batch cultivation techniques are applied for products of biosynthetic origin.

Advantages of chemostat techniques for scientific work, as indicated in Sections III to V, are obvious. They are particularly suitable for work on growth kinetics and metabolic fluxes, as well as for investigations on dynamics of biological control mechanisms. However, we cannot overlook the fact that many of the data accumulated during the last two decades represent single facts, merely describing interesting biological phenomena. Many of these phenomena are not explainable by the classic theory of Monod. New models of growth and regulation should be developed to provide a more consistent picture of the living cell.

It is also seen from this article that the cost of equipment for chemostat studies is distinctly higher than that for classic techniques. Many scientists also reject such methods because of the apparently enormous labor involved in such studies. They seldom take into account that, with a new and more sophisticated method, available information also increases.

We must therefore conclude that more widespread application of the chemostat technique can contribute effectively to our efforts toward more complete recognition of the nature of the cell.

List of Symbols

C_6		Glucose
C_{16}		Hexadecane
D	per hour	Dilution rate F/V
D_2	per hour	Dilution rate in the second stage
D_c	per hour	Critical dilution rate for washout
D_R	per hour	Maximum dilution rate for derepression
D_{max}	per hour	Dilution rate for maximum productivity
DX	gm/liter per hour	Productivity
F	ml/hour	Flow rate of medium
F_{02}	ml/hour	Flow rate of substrate to the second stage
g	hours	Generation time
$\overline{g}$	hours	Mean generation time
K_S	mg/liter	Saturation constant for growth
MDH		Malate dehydrogenase
N		Number of cells
N_0		Initial number of cells
P	gm/liter	Product
Q_{CO_2}	mmoles/gm per hour	Specific carbon dioxide release
Q_{O_2}	mmoles/gm per hour	Specific oxygen uptake
$Q_{O_2}^{max}$	mmoles/gm per hour	Maximum specific oxygen uptake
Q_P	mmoles/gm per hour	Specific formation of product
Q_S	mmoles/gm per hour	Specific uptake of substrate
Q_S^{max}	mmoles/gm per hour	Maximum specific uptake of substrate
S	gm/liter	Substrate concentration of the reactor mixture
S_0	gm/liter	Initial concentration of the medium
T	hours	Time of oscillation, period
t	hours, minutes	Time
t_d	hours	Doubling time
vvm	liters/liter per minute	Airflow rate
V	liters	Volume of the reactor mixture
X	gm/liter	Concentration of biomass
X_2	gm/liter	Concentration of biomass in the second stage
X_{02}	gm/liter	Concentration of biomass flowing to the second stage
Y		Yield; dimensionless; grams of biomass formed per gram of substrate taken up
ρ		Petite mutant
ρ_0		Petite mutant lacking mitochondrial DNA
μ	per hour	Specific growth rate
μ_{max}	per hour	Maximum specific growth rate
μ_2	per hour	Specific growth rate in the second fermenter

ACKNOWLEDGMENTS

This work was supported by the Schweizerischer Nationalfonds zur Förderung der wissenschaftlichen Forschung (Project No. 3.628.71) and the Kommission zur Förderung der wissenschaftlichen Forschung (Projects Nos. 91 and 91.2).
The author is indebted to Dr. John R. Pringle, Dr. Arthur Einsele, and Dr. Juan-Ramon Mor for their fruitful discussions and suggestions during the preparation of the manuscript.

REFERENCES

Aiba, S., Humphrey, A. E., and Millis, N. F. (1973). "Biochemical Engineering," 2nd ed., Academic Press, New York.
Beck, C., and von Meyenburg, K. (1968). *J. Bacteriol.* **96**, 479.
Beran, K. (1967). *Mitt. Versuchssta. Gaerungsgewerbe Wien* **21**, 101.
Blanch, H. W., and Einsele, A. (1973). *Biotechnol. Bioeng.* **15**, 861.
Borst, P. (1963). *Proc. Int. Congr. Biochem., 5th, 1961* Vol. 2, p. 233.
Einsele, A. (1972). "Die Funktion des Stofftransportes in Bioreaktoren beim mikrobiellen Kohlenwasserstoffabbau," Diss. No. 4900. Eidg. Techn. Hochschule, Zürich.
Einsele, A., Fiechter, A., and Knöpfel, H.-P. (1972). *Arch. Mikrobiol.* **82**, 247.
Einsele, A., Blanch, H. W., and Fiechter, A. (1973). *Biotechnol. Bioeng. Symp.* **4**, 455.
Einsele, A., Schneider, H., and Fiechter, A. (1974). *Proc. Int. Symp. Yeasts, 4th.* (in press).
Fencl, Z. (1966). In "Theoretical and Methodological Basis of Continuous Culture of Microorganisms" (I. Málek and Z. Fencl, eds.), p. 69. Publ. House Czech. Acad. Sci., Prague.
Fiechter, A. (1965). *Biotechnol. Bioeng.* **1**, 101.
Fiechter, A., and Schatzmann, H. (1972). In "Fermentation Technology Today" (G. Terui, ed.). Soc. Ferment. Technol., Kyoto, Japan.
Fiechter, A., and von Meyenburg, K. (1968). *Biotechnol. Bioeng.* **10**, 535.
Finn, R. K. (1954). *Bacteriol. Rev.* **18**, 255.
Flury, U. (1973). "Isoenzyme der Malat-Dehydrogenase in *Schizosaccharomyces pombe*," Diss. No. 5129. Eidg. Techn. Hochschule, Zürich.
Herbert, D. (1961). *SCI (Soc. Chem. Ind., London) Monogr.* **12**, 21.
Herbert, D., Elsworth, R., and Telling, R. C. (1956). *J. Gen. Microbiol.* **14**, 601.
Hug, H. (1973). "Die Bedeutung der Lipide beim mikrobeillen Kohlenwasserstoffabbau," Diss. No. 5068. Eidg. Techn. Hochschule, Zürich.
Hug, H., Blanch, H. W., and Fiechter, A. (1974). *Biotechnol. Bioeng.* (in press).
Knöpfel, H.-P. (1972). "Zum Crabtree-Effekt bei *Saccharomyces cerevisiae* und *Candida tropicalis*," Diss. No. 4906. Eidg. Techn. Hochschule, Zürich.
Krebs, A. H. (1972). *Essays Biochem.* **8**, 1.
Kubitschek, H. E. (1970). "Introduction Research with Continuous Culture." Prentice-Hall, Englewood Cliffs, New Jersey.
Küenzi, M. T. (1970). "Ueber den Reservekohlenhydratstoffwechsel von *Saccharomyces cerevisiae*," Diss. No. 4544. Eidg. Techn. Hochschule, Zürich.
Küenzi, M. T., and Fiechter, A. (1969). *Arch. Mikrobiol.* **64**, 396.
Küenzi, M. T., and Fiechter, A. (1972). *Arch. Mikrobiol.* **84**, 254.
Lowdon, M. J., Gordon, P. A., and Stewart, P. R. (1972). *Arch. Mikrobiol.* **85**, 355.
Matile, P., and Bahr, G. F. (1968). *Exp. Cell Res.* **52**, 301.
Maag, M. (1973). In preparation.
Meyer, C. (1974). In preparation.
Mian, F. A., Prokop, A., and Fencl, Z. (1971). *Folia Microbiol. (Prague)* **16**, 249.
Monod, J. (1942). "Recherches sur la croissance des cultures bactériennes." Hermann, Paris.
Monod, J. (1950). *Ann. Inst. Pasteur, Paris* **79**, 390.
Mor, J.-R., and Fiechter, A. (1968). *Biotechnol. Bioeng.* **10**, 787.
Mor, J.-R., Zimmerli, A., and Fiechter, A. (1973). *Anal. Biochem.* **52**, 614.

Moser, H. (1958). *Carnegie Inst. Wash. Publ.* **614**.

Nagley, P., and Linnane, A. W. (1970). *Biochem. Biophys. Res. Commun.* **39**, 989.

Novick, A., and Szilard, L. (1950). *Proc. Nat. Acad. Sci. U.S.* **36**, 708.

Oldshue, J. Y. (1966). *Biotechnol. Bioeng.* **8**, 3.

Pfenning, N., and Jannasch, H. W. (1962). *Ergeb. Biol.* **25**, 93.

Řičica, J. (1973). *Folia Mikrobiol. (Prague)* **17**, 418.

Robinson, C. W. (1971). Ph.D. Thesis, University of California, Berkeley.

Rogers, P. J., and Stewart, P. R. (1973). *J. Gen. Microbiol.* **79**, 205.

Schatzmann, H. (1974). *Pathol. Microbiol.* **41**, 186.

Schatzmann, H., and Fiechter, A. (1974). *Chem. Ing. Tech.* **46**, 703.

Sols, A., Gancedo, G., and De la Fuente, G. (1971). *In* "The Yeasts" (A. H. Rose and J. S. Harrison, eds.), Vol. 2, p. 271. Academic Press, New York.

von Meyenburg, K. (1969). "Katabolit-Repression und der Sprossungszyklus von *Saccharomyces cerevisiae*," Diss. No. 4279. Eidg. Techn. Hochschule, Zürich.

Weibel, K. E., Mor, J.-R., and Fiechter, A. (1974). *Anal. Biochem.* **58**, 208.

Chapter 7

Methods for Monitoring the Growth of Yeast Cultures and for Dealing with the Clumping Problem

JOHN R. PRINGLE AND JUAN-R. MOR

Institute of Microbiology,
Swiss Federal Institute of Technology,
Zürich, Switzerland

I. Introduction

A. General Remarks

Most biochemical, physiological, cytological, and developmental studies of yeasts require monitoring the growth of yeast cultures. In some cases determinations of the rate or extent of growth provide important experimental results; in many other cases it is necessary to know the amount of cellular material analyzed, or the physiological state of the cells, or both. The methods used for determining amounts of yeast, and for monitoring the growth of yeast cultures, are the same as those used with other microorganisms, and are widely known. Unfortunately, they are not always applied with sufficient attention either to the limitations of the methods themselves, or to the limitations imposed by the properties and peculiarities of yeast cells. This has led to a significant amount of confusion in the interpretation and in the communication of experimental results.

In this chapter we have two goals. The first is to contribute to a lessening of confusion by calling attention to the virtues and limitations of, and the relationships between, various methods of monitoring growth. This discussion pays particular attention to the complications imposed by the budding mode of reproduction and by the clumping problem (Section IV). Previous useful discussions of these issues, partially overlapping with ours, have been given by Williamson (1964) and Mitchison (1970); the latter reference focuses on the problems encountered in work with fission yeasts, a subject with which we do not deal directly here. (Although much of the discussion is more generally applicable, this chapter is based explicitly on our experiences with *Saccharomyces cerevisiae*, and reflects this bias in several places.) Our second goal is to supply tidbits of practical information, derived from our own experience, which may be useful to others. We have tried to avoid writing about either theory or practice that is generally known, available in instruction manuals, or described in detail elsewhere (particularly useful general references are Mallette, 1969; Kubitschek, 1969; Postgate, 1969; Herbert *et al.*, 1971; Koch, 1970; others are given below). We do not deal, except incidentally, with the growth of *individual* yeast cells (for information on this topic, see Williamson, 1964; Johnson, 1965; Mitchison, 1970, 1971; Sebastian *et al.*, 1971; Wells and James, 1972; Hayashibe *et al.*, 1973; Vraná *et al.*, 1973; Hartwell, 1974).

In keeping with common usage, we have used "precision" to refer to the reproducibility of repeated measurements on a single sample (i.e., as a measure of random error), and "accuracy" to refer to the closeness of the mean of many measurements to the true value of the measured parameter (i.e., as a measure of systematic error).

B. Different Measures Measure Different Things

Several different types of measurements can be used to determine amounts of yeast and to monitor the growth of yeast cultures. The most commonly used are measurements of the turbidity of cell suspensions (as absorbance), of the total number of cells, of the number of viable cells, of the total biomass (as dry weight), and of certain components of biomass, such as total protein. Each of these types of measurements is discussed in more detail in subsequent sections, but we comment here on the relationships among them.

In using these various methods it is essential to remember that they measure different parameters of the cell population, and that the relative values of these different parameters can vary significantly. Some data illustrating this point are given in Table I. (These data serve also to give an idea of the kinds of absolute values that can be anticipated for the several parameters, as well as of the time courses of such cultures. In these connec-

TABLE I

RELATIONSHIPS AMONG VARIOUS GROWTH PARAMETERS

Sample no.[a]	Time of sampling (hours)	Budded cells[b]	Absorbance, (A)[c]	Dry weight, (D)[d]	Cell number, (N)[e]	A/D[f]	A/N[f]	D/N[f]
1	0	78	1.92	0.30	1.29	6.40	1.49	0.233
2	10	18	18.4	2.95	16.2	6.24	1.13	0.182
3	20	7	28.3	4.62	24.9	6.13	1.14	0.185
4	58	3	47.0	8.84	32.7	5.32	1.44	0.270
5	0	67	0.922	0.16	0.752	5.76	1.23	0.213
6	3	43	1.76	0.245	1.53	7.18	1.15	0.160
7	6	6.7	2.34	0.33	2.33	7.09	1.00	0.141
8	11	0.7	2.71	0.44	2.55	6.16	1.06	0.172
9	16	0.3	3.17	0.535	2.59	5.92	1.22	0.207

[a]Samples 1 to 4 represent various points in the history of a culture of the prototrophic diploid strain C276 (probably equivalent to strain X2180; see Wilkinson and Pringle, 1974) which was allowed to grow to stationary phase in YM-1 medium. (YM-1 contains 7 gm Difco yeast nitrogen base, 5 gm Difco yeast extract, 10 gm Difco Bacto peptone, 10 gm succinic acid, 6 gm NaOH, and 20 gm glucose, per liter.) The times of sampling are indicated; it should be noted that the culture had already been growing exponentially for about five generations when sample 1 was taken. Samples 5 to 9 represent various points in the history of a culture of C276 growing exponentially in minimal medium (salts and vitamins, plus 10 gm succinic acid, 6 gm NaOH, and 20 gm glucose, per liter) until it was shifted abruptly (at time 0) to a medium identical except in that no nitrogen source was included. All cultures were grown in flasks with rotary shaking, at 23 °C.

[b]As proportion (in percent) of the total population; determined after sonication.

[c]Read at 660 nm with 1-cm cuvettes in a Beckman DB-G spectrophotometer. Each sample was diluted until $0.1 < A < 0.25$ (in this range absorbance is proportional to the amount of cell material), and the reading obtained was then multiplied by the dilution factor.

[d]As milligrams dry weight per milliliter of culture. The filtration method described in section III was used.

[e]Given as cells per milliliter $\times 10^{-7}$. The Celloscope 302, with a 48-μm aperture, was used (Section V).

[f]The values given are simply the ratios of those in the other columns.

tions, however, it must be remembered that the culture of samples 5 to 9 was shifted to nitrogen-free medium at an arbitrary point.) It is clear that the relationships among absorbance, dry weight, and cell number do not remain constant as the physiological state of the cells varies. It is important to realize that the varying ratios shown in Table I are only the tip of an iceberg. For one thing, even when these particular ratios do not vary, the populations involved can be very different with respect to other parameters. Samples 5 and 9 in Table I provide an example. In contrast to sample 5, sample 9 consists almost entirely of unbudded cells, moreover these unbudded cells are much more heterogeneous in size than the unbudded cells of sample 5 (J. Pringle, unpublished observations). In addition, the two populations differ drastically in chemical composition; reserve carbohydrates (glycogen and trehalose) account for about 40% of the dry weight of the cells of sample 9, but less than 1% of the dry weight of the cells of sample 5. It should also be noted that the data of Table I were obtained with a single nonclumpy (see Section IV) diploid strain; the use of a haploid strain, or of a clumpy strain, would have significantly altered some of the ratios. Finally, much greater variations in these ratios occur following certain types of experimental manipulations. For example, dry weight and adsorbance can increase extensively while increases in cell number are completely prevented by mating factor (Throm and Duntze, 1970), by various inhibitors (Mitchison and Creanor, 1971; Slater, 1973), or by cell division cycle mutations (Hartwell, 1974).

Thus the value of growth parameter y cannot always simply be inferred from the value of growth parameter x. It is important to realize also that it may be much more *informative* to know the value of y than to know the value of x. Which parameters are the most relevant and interesting depends of course on the goals of the particular experiment; each of the parameters mentioned (with the exception of turbidity) is preeminently useful in at least some types of experiments. (The amount of light scattered by cells is not, in its own right, a property of much biological interest. As far as we know, measurements of turbidity always derive their usefulness from what can be inferred from them about *other* parameters of the cell population. See also Section VI.) Some examples will illustrate this point.

1. In the experiment represented by samples 5 to 9 (Table I), our primary interest was in the question how cell division is coordinated with the availability of nitrogen supplies. Thus the most informative data seemed to be those on total cell number (taken together with the data on budded cells). We might have done the same experiment with the intention of learning about the survival of cells undergoing nitrogen starvation; in this case measurements of viable cells clearly would have been essential. The dry-weight data obtained in this experiment do not seem very informative, because the composition of the cells changes so much. It would be more

interesting to be able to compare the cell number data to data on total protein; this would reveal how much protein synthesis is associated with the last round of cell division in cells exhausting their nitrogen supplies.

2. In many studies of nutrition, metabolism, and cell composition, the total biomass seems to be the most reasonable point of reference. Typical questions are: How many grams of cellular material are produced from 1 gm of substrate w? How rapidly does biomass increase when substrate w is present at concentration c? What proportion of the total biomass is reserve carbohydrate? How much respiration is there per gram of cellular material? There are, however, complications. For example, the significance of a growth yield of 1 gm of total biomass may depend on the composition of that biomass. Also, it may be more enlightening to relate composition, or even metabolic activity, to the number of cells rather than to their total mass (e.g., it is frequently useful to know the DNA content per cell), or to relate an enzyme activity to the total protein rather than to the total mass (e.g., in cases in which total mass is changing rapidly but total protein is not). Yet another example is that the meaning of a metabolic measurement can depend on whether the sample analyzed contains $2m$ cells, each of mass n, or m cells, each of mass $2n$.

3. In many cultures in which cell division is highly synchronized, so that the number of cells increases in a sharply stepwise fashion, the total biomass increases steadily (Scopes and Williamson, 1964; Williamson, 1964; Mitchison, 1970); on the basis of biomass measurements alone, these cultures cannot readily be distinguished from asynchronous cultures. The biomass measurements have contributed some information, but it is clear that for most experiments involving synchronous cultures it is cell number measurements that are crucial.

It is also important to note that the types of measurements originally listed are not the only possibilities, and are sometimes not the most meaningful or convenient ones. For example, in a study of reserve carbohydrate metabolism during the course of glucose-to-acetate diauxic growth, it is much more useful to know when the extracellular glucose is exhausted than it is to know the precise cell number (or biomass) at which this occurs. To generalize, it is not infrequently more useful to monitor the properties of the medium than it is to monitor the cells themselves. A second example is that it is sometimes useful to know the total volume of cells per unit volume of culture; such measurements can be made with some electronic particle counters (see Section V). Still a third example is that the relative numbers of budded and unbudded cells can be a useful indicator of the status of a culture (see Section VII).

To summarize, then, considerable care is required in the selection of growth parameters to be measured and reported. Frequently it is useful to measure several different parameters; for example, it is very often helpful to know both the total biomass and the total cell number. When this is not done, it is important to select the parameter that will be most revealing in the context of the experiments to be performed. Under some well-defined conditions the relationships between various growth parameters are sufficiently stable that it is profitable to construct and use calibrations for interconverting values of different growth parameters. For example, in performing a series of diauxic growth experiments with a group of related strains, we found that growth curves obtained by measuring absorbance, and converting the data to cell number data using a proper calibration (see Section VI), were exactly parallel to the growth curves obtained by measuring cell numbers directly with a Coulter Counter, during each of the periods of exponential growth in each culture. In using such calibrations, however, it is important to remember that they may break down, and often at the most interesting times. For example, in the experiments just cited, the two methods did *not* give parallel results during the diauxic lag period. Because of these limitations, it seems to us to be in general prudent to measure directly the parameter of most interest, particularly when accurate results are desired.

II.   Stopping Growth Rapidly

In measuring any growth parameter it is of course desirable that the measurement reflect the status of the culture at the instant of sampling. This problem is particularly acute, since it is usually convenient, and often essential, to be able to make the measurement at a time significantly after the actual time of sampling. Thus it is desirable to be able to inactivate growth processes rapidly, and to do this in a way that neither alters the value of the parameter to be measured nor interferes with the measuring technique. There exist several methods for stopping growth rapidly; which of these is most appropriate depends on the measurements to be made.

A.   Fixation

For many purposes the use of fixatives is a convenient and effective method of stopping growth. We have used formaldehyde-fixed samples extensively for measurements of dry weight, total cell number, absorbance, cell volumes, and proportion of cells that carry buds. Fixation has been conducted in two ways:

1. One part of concentrated (37–40%) formaldehyde solution is mixed directly with 9 parts of culture. This protocol is convenient when dry weight or absorbance is to be read, and also for other purposes when the concentration of cells is low.

2. One part of concentrated formaldehyde is mixed with 9 parts of saline (0.9% NaCl in water). For sampling, 0.5-ml portions of culture are mixed with 4.5-ml portions of the formaldehyde–saline solution. This procedure is often convenient, because only small samples of culture are taken and because these are given a first dilution which is often convenient for subsequent electronic cell counting (Section V). In this protocol a sudden dilution of the medium accompanies the exposure to formaldehyde. It is possible that this dilution contributes to stopping growth rapidly (see below), or allows more effective formaldehyde action on the cells by reducing the concentration of other reactive substances, but no evidence for such effects has been observed. Protocol 2 has been used successfully to fix samples prior to nuclear staining (Hartwell, 1970), but protocol 1 may work as well.

It is of course important to know whether fixation with formaldehyde stops growth processes rapidly and effectively, and whether the values obtained after fixation are identical to those that would have been obtained with an instantaneous measurement on an unfixed sample. The following observations are relevant.

1. We have seen little or no systematic difference between dry weights measured by immediate filtration of unfixed samples and dry weights measured by adding formaldehyde (protocol 1) and leaving the samples 24–48 hours at room temperature before filtration (Table II, experiments 1 and 2). It is conceivable that some increase in mass occurs after formaldehyde addition but is balanced by some loss of pool constituents (Mitchison, 1971, p. 10); however, it seems more likely that growth stops quickly and that mass (or at least most of it) is preserved.

2. Small but significant changes in absorbance do occur after formaldehyde treatment (Table II, experiment 3). It is conceivable that some residual growth is involved, but this seems unlikely in view of the observations on dry weight, the fact that part of the absorbance change occurs almost immediately on formaldehyde addition, and the fact that the effect seems somewhat greater at higher formaldehyde concentrations. Probably the absorbance changes are due to the cells shrinking somewhat on fixation (Mitchison, 1971, p. 10); this would lead to increased absorbance (Koch, 1961, 1970; Mallette, 1969; Bussey, 1974). It may still be convenient to measure absorbances on fixed samples, but it seems that more accurate results will come from measurements on unfixed samples; this is of course not a great hardship, given the ease of making the measurements and their usual uses (see Section VI). The possibility of fixation-induced volume changes should of

TABLE II

RELIABILITY OF DRY-WEIGHT AND ABSORBANCE MEASUREMENTS ON FIXED SAMPLES

Experiment 1: Dry-weight measurements[a] on replicate samples from a growing culture
  Unfixed: 1.07, 1.07, 1.07, 1.06; fixed: 1.05, 1.05, 1.03, 1.03

Experiment 2: Dry-weight measurements[a] on replicate samples from a stationary culture
  Unfixed: 5.46, 4.96, 5.38, 5.43; fixed: 5.33, 5.27, 5.35, 5.33

Experiment 3: Successive absorbance measurements[b] made on samples fixed with various
  concentrations of formaldehyde
  a. 1.8% formaldehyde, storage at 23 °C: 1.00, 1.05, 1.07, 1.13
  b. 3.6% formaldehyde, storage at 23 °C: 1.00, 1.08, 1.11, 1.17
  c. 3.6% formaldehyde, storage at 4 °C: 1.00, 1.07, 1.09, 1.12
  d. 6.0% formaldehyde, storage at 23 °C: 1.00, 1.11, 1.13, 1.21

[a] Dry-weight measurements were made by the filtration method described in Section III, and are expressed as milligrams dry weight per milliliter of culture.

[b] The measurements were made in a Beckman DB-G spectrophotometer, at 660 nm, in 1-cm cuvettes. The values given (reading from left to right) were measured immediately before fixation, immediately after fixation, 1 hour later, and 12 hours later. All values have been corrected for dilution and normalized to the value obtained with the unfixed sample, which was in each case between 0.660 and 0.700 (the cultures were in exponential growth with a generation time of about 100 minutes).

course also be kept in mind when volumes are to be measured, although it seems likely that measurements of *relative* volumes would not be affected.

3. It seems very likely that reliable results are obtained when fixed cells are used for measurements of the total cell number and of the proportion of budded cells. Attempts to compare budded cell counts made with fixed and unfixed samples are plagued by an obvious inaccuracy affecting the counts of unfixed cells. Unless the environment of the cells is changed radically (which may have effects of its own), new buds appear, while old buds do not effectively separate from their mother cells, during the interval between sonication (Section IV) and the end of the count. However, there are certainly no gross differences between a count obtained on a fixed sample and one made quickly on an unfixed sample. This fact taken together with the dry-weight data (Table II) makes it very unlikely that total cell number can change appreciably through growth after the addition of formaldehyde. It might be the case that fixation affects the precise stage at which mother and daughter cells become separable by sonication; however, it should be remembered that this is a somewhat arbitrarily defined point in any case (see Section IV). Thus the best argument for the reliability of working with fixed cells is that the results obtained seem internally consistent. Sonication of fixed cells does seem to separate pairs that have progressed to a rather definite stage of the division process (Section IV). Growth curves obtained

by counting cells in fixed and sonicated samples seem reasonable; where synchronous division is expected, synchronous division is seen, and the exponential rates of increase observed with asynchronous cultures are identical to those obtained by absorbance readings (properly calibrated; see Section VI) on unfixed samples, or by dry-weight readings on either fixed or unfixed samples. No obvious changes either in total count or in the proportion of budded cells occur if samples are scored the same day as fixation and again several weeks later. (For prolonged storage it helps to keep samples in the cold and well wrapped.)

It is said that much better preservation of internal structure is obtained when fixation is performed with the glutaraldehyde and acrolein fixative of Hess (1966). We have not evaluated this procedure for the rapidity with which it stops growth.

B.   Other Methods

For some purposes, such as measurement of various cell components and medium constituents, the best approach to stopping growth may be to chill the cells as rapidly as possible. Sometimes it is practical simply to add ice to the cell suspension. Another approach we have used is to surround a condenser coil with ice water. Samples are poured in at the top of the coil and collected as they emerge at the bottom. In the roughly 15 seconds of passing through the coil, the samples can be reduced in temperature by $15°–20°C$. It is important to note that merely cooling is not sufficient to stop metabolic activity. For example, strain C276 grows exponentially, with a generation time of about 24 hours, at $8.4°C$ (J. Pringle, unpublished observations).

Another approach to stopping growth that is sometimes useful is an abrupt deprivation of nutrients. With most nutrients some further growth seems to be possible at the expense of intracellular reserves (see, for example, Table I) but, in all cases tested so far, abruptly depriving yeast cells of an external carbon source leads to an almost immediate and almost complete cessation of growth (certainly less than a 10% increase in cell number within 12 hours after the shift in medium; J. Pringle, unpublished observations). Although the point has not been directly tested, this almost certainly means that diluting a growing culture 100-fold with water (as in preparation for a viable cell count by plating) is in itself sufficient to prevent further changes in cell number.

Rapid collection and inactivation of cell samples is especially crucial when it is necessary to determine the levels of low-molecular-weight metabolites whose turnover rates *in vivo* are high. Weibel *et al.* (1974) compared several methods and described what seems to be an optimal procedure, using ATP concentration as a point of reference.

III.   Measuring Biomass and Its Components

When it is useful to know the total mass of cellular material present in a particular sample (see the discussion in Section I), a direct measurement of the dry weight (the amount of cellular material remaining after drying to constant weight) is almost always the method of choice. Several exceptional situations may be briefly noted:

1. If only an approximate value for the total biomass is required, or if a biomass value is needed immediately for guiding an experiment in progress, an absorbance measurement may be preferable. Such measurements must of course be used with their limitations in mind (see Sections I and VI).

2. If the amount of material available is limiting, an indirect determination of the total biomass may be necessary. Absorbance measurements, cell counts, and measurements of various cellular components (perhaps after radioactive labeling) may all be converted to biomass data. The reliability of such data is in each case limited by the reliability of the conversion factor; except in certain well-defined situations, this is a rather serious problem (see Section I).

3. If the amount of intracellular water is itself a parameter of interest, or if the total cellular volume must be known (for example, in interpreting permeability data), it may be desirable to make measurements of wet weight or of packed cell volume. Ordinarily, however, such measurements should be eschewed in favor of dry-weight (for accuracy) or absorbance (for speed) measurements, since the difficulties in controlling and/or assessing the amounts of intracellular and extracellular (interstitial) water are simply too great (Mallette, 1969; Herbert et al., 1971).

Dry weights can be measured with several different methods, which differ only in detail. All are simple and offer good precision, although the absolute values obtained may vary somewhat with the details of washing and drying procedures (Mallette, 1969). The errors are probably small, and the procedures are in any case readily standardized.

We have employed two different methods for determining dry weights. In the first of these, cell suspension is pipetted (generally with a mark-to-mark volumetric pipette) into a previously weighed glass centrifuge tube. The sample is centrifuged, the supernatant removed by decantation or aspiration, and the cells resuspended in a volume of distilled water equal to the original volume of cell suspension. Centrifugation is repeated, the supernatant is again removed, and the tube is placed in an oven at 105°C for 12–16 hours. The tube is then allowed to cool to room temperature in a desiccator, and is weighed within a few minutes of removal from the desiccator. This method has been used mostly with relatively dense cell suspensions (1–6 mg dry weight per milliliter, such as are encountered in many chemostat populations; see Mor and Fiechter, 1968b; Fiechter, this volume), where it is convenient and precise. The precision was tested by measuring the dry matter in 30 replicate 10-ml samples from a single cell suspension; a mean of 45.4 mg per sample, with a standard deviation of 0.3 mg, was obtained. Reproducible results were also obtained after drying *in vacuo* at lower temperatures (down to 40 °C), but drying under an infrared lamp was unsatisfactory.

The second method involves collecting and washing the cells by filtration, rather than by centrifugation. We have used Millipore filters (HAWP, 25-mm diameter, 0.45-μm pore size), but similar filters would probably work as well. The individual filters are weighed (without preliminary washing or drying), the cell suspension is filtered, the collected cells are washed with 15 ml of distilled water, and the filters are put in an oven. After drying, the filters are removed from the oven and weighed within 15 minutes. It is convenient to handle the filters in a shallow aluminum pan; a cover of foil wards off dust and prevents filters and dried cells from blowing around during transport. The following points should be noted.

1. Preliminary washing and drying of the filters (Table III, experiments

TABLE III

DRY-WEIGHT MEASUREMENTS BY THE FILTRATION METHOD[a]

Net weights of various individual filters after (a) drying for 10 hours at 65 °C; (b) rinsing with 15 ml of water and drying for 10 hours at 65 °C; (c) rinsing with 15 ml of water and drying for 20 hours at 105 °C; (d) rinsing with 20 ml of YM-1 (see Table I), and then rinsing with 15 ml of water and drying for 5 hours at 105 °C:

(a) −0.1, −0.1, −0.2, 0;   (b) −0.7, −0.8, −0.6, −0.6;   (c) 0, +0.1, 0;   (d) 0, 0

Net weights of several cell samples were measured (e) after drying for 2.5 hours at 65 °C (first value of each pair) and again after an additional 11 hours at 65 °C (second value of each pair); and (f) after drying for 4 hours at 105 °C (first value of each pair) and again after an additional 60 hours at 105 °C (second value of each pair):

(e) 11.1, 11.0; 9.0, 8.8; 11.6, 11.4; 18.5, 18.1;   (f) 3.7, 3.6; 3.3, 3.2

Net weights of two cell samples (g and h) were measured 0, 30, 70, and 840 minutes, respectively, after removal from the oven after drying for 20 hours at 105 °C:

(g) 8.9, 9.0, 9.3, 9.6;   (h) 16.9, 16.9, 17.2, 17.7

Net weights of pairs of identical samples were measured after drying for 12 hours at 105 °C:

(i) 5.3, 5.4; 7.2, 7.2; 12.3, 12.2; 26.6, 26.5; 41.4, 41.8; 80.3, 78.9

---

[a]Some measurements (experiments a, b, and e) were made in Seattle using filters punched by hand from a large sheet, house distilled water for washing, a 65°C oven for drying, and a balance with a proclivity for zero-point drift of as much as 0.7 mg. The other measurements were made in Zürich using commercially prepared 25-mm filters, house deionized water for washing, a 105°C oven for drying, and a balance whose zero was generally stable to within 0.1 mg. All values are net weights, in milligrams per sample.

a to d) makes either no difference (Zürich series), or a small, reproducible difference (Seattle series). The reason for the discrepancy between the Zürich and Seattle results is not known, but it is recommended that this control be performed from time to time.

2. Neither temperature nor time of drying seems critical (Table III, experiments e and f). At either 65° or 105 °C samples come to constant weight within a few hours, and hold this weight for extended periods.

3. Samples dried at 105 °C slowly regain some weight as they sit at room temperature (Table III, experiments g and h). It is perhaps significant that our laboratory is not air-conditioned or dehumidified.

4. With samples of 5 mg or more, measurements on duplicate samples seldom differ by more than 1 or 2% (Table II; Table III, experiment i). With smaller samples errors are somewhat greater, because of the small uncertainty introduced by handling, washing, and reweighing the filters. This uncertainty has consistently been less in Zürich than in Seattle, apparently because of the superior stability of the balance.

5. Filtering speed becomes annoyingly slow when too many cells are present. It is thus desirable to adjust sample volumes so that each sample contains roughly 10–15 mg dry weight; this represents a good compromise between precision of the measurement and speed of filtering.

6. In comparison to the centrifugation method, the filtration method is faster (both in terms of manipulation time and in the time required to reach constant weight), and is more convenient and precise when cell suspensions are dilute (i.e., when relatively large volumes of sample must be taken, when the danger of loss of some cells during decantation is relatively great, and when it may be necessary to weigh a small difference between two large numbers). However, the centrifugation method is cheaper, and also conveniently provides a sample of the cell-free supernatant which can be saved for subsequent analysis.

It is worth noting that special difficulties in making biomass measurements arise when yeasts are grown on certain substrates, notably alkanes. Various approaches to these problems have been described and compared by Yoshida et al. (1971) and Hug and Fiechter (1973).

It is also worth noting that there are situations in which it is more interesting and relevant, or more practical, to measure the total amount of some cellular component than to measure the total biomass. For example, the desirability of knowing the total cellular protein in cultures undergoing nitrogen starvation has already been noted in Section I,B. Other examples are the frequent desirability of knowing the total DNA content of a culture, and the interesting data obtained by measuring total nitrogen contents of synchronized cultures (summarized by Williamson, 1964). Analytical methods for most cellular components have been given by Herbert et al. (1971).

IV.   Dealing with the Clumping Problem and Measuring Cytokinesis

Throughout the early stages of the cell division cycle, the cytoplasm of the mother cell is continuous with that of the growing bud. At some point subsequent to nuclear division there comes a moment, in principle quite sharply defined, at which the formerly continuous cytoplasms become completely separated by a barrier of cell membrane (perhaps with associated cell wall). This moment marks the completion of cytokinesis, and it is clear that from this moment mother and daughter are in a real sense two separate cells. If the world had been designed for the convenience of biologists, the following two conditions would be met.

1. Mother and daughter cells would separate immediately after cytokinesis.

2. Cell division cycle $n + 1$ would never begin before completion of the cytokinesis of cell division cycle n. (For the purposes of the following discussion, the beginning of a new cell cycle is taken to be the time at which the new bud appears. See comments in Section VII.)

Thus microscopic examination of a population of growing cells would reveal only a forms and b forms (Fig. 1), and in each b form the mother and daughter cytoplasms would still be continuous through the connecting neck. Condition 1, however, is clearly somewhat unrealistic in view of the thick

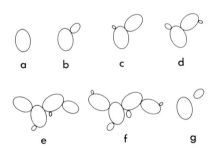

FIG. 1.   Diagrammatic representation of unbudded cells (a), budded cells and mother plus daughter pairs (b), and clumps of various sizes (c to f). Each form is intended to symbolize the whole class of forms composed of the same number of entities (single cells, mother cells, or buds). For example, form b represents all such "doublets," regardless of the relative sizes of mother and daughter, and regardless of whether they have or have not become physiologically separate as a result of the occurrence of cytokinesis. In addition, form f represents all clumps of more than eight entities. It should be noted that these diagrams are provided only for the purpose of facilitating discussion; any resemblance between the detailed appearance of these cells and clumps, and the cells and clumps of any actual yeast strain, living or dead, is purely coincidental. This disclaimer applies in particular to the detailed spatial arrangement of the various mothers, daughters, and buds.

yeast cell wall, and one should thus hope only for the validity of a modified condition:

1'. A well-defined, and relatively short, interval of cell wall development would intervene between cytokinesis and the actual separation of mother and daughter cells. (In the simplest case, a growing population would still contain only a and b forms, but a definite fraction of b forms would actually be pairs in which cytokinesis was complete.)

Real life, however, is much more complicated. In most actual cultures (even well-agitated liquid cultures), one sees not only a and b forms but also c and d forms, and not infrequently e and f forms as well. The occurrence of c to f forms (known as clumps) is what we refer to as the clumping problem; clumping is a problem because it greatly complicates the making and interpretation of counts of total cells, of viable cells, and of the relative numbers of budded and unbudded cells (see further comments in Sections V and VII). The degree of clumping (i.e., the relative numbers of a to f forms) varies greatly both with the strain and with the culture conditions. [Various strains in common use differ greatly in their clumping behavior. In addition, Hartwell found it easy to isolate mutants that were very much clumpier than the parent strain (see Hartwell *et al.*, 1974); the mutants isolated by Gilmore and Murphy (1972; see also Farris and Gilmore, 1974), and by Francis and Hansche (1972), are apparently also of this type.] In some cases only a and b forms, with a few c and d forms, are observed, while in other cases the appearance of the culture is dominated by f forms. Clumping is generally more pronounced in haploids than in diploids, and is generally more pronounced in actively growing than in stationary cultures (although some stationary cultures do contain many clumps). The clumps of a clumpy strain are not in general very uniform in size; i.e., a and b forms, as well as f forms, are usually observed at any given time.

Clumping might, in principle, arise from a departure from condition 2, above. It seems, however, that this is rarely if ever the case. The behavior of some mutants in which cytokinesis is specifically blocked indicates that the beginning of division cycle $n + 1$ is not necessarily dependent on the successful completion of cytokinesis n (Hartwell, 1971; Hartwell *et al.*, 1974). However, the arguments that follow (based on observations of a considerable variety of normal strains under a great variety of physiological conditions) suggest that this sort of overlap of division cycles rarely, if ever, occurs with normal strains. Burns (1956) examined several strains, using a micromanipulator, and found that the elements of a b form could almost never be pulled apart. In contrast, all c and d forms could easily be separated into a and b forms. This manipulation did not damage the cells in any detectable way. Williamson and Scopes (see Williamson, 1964) extended these observations; with their strain several manipulations, including exposure to ultrasound,

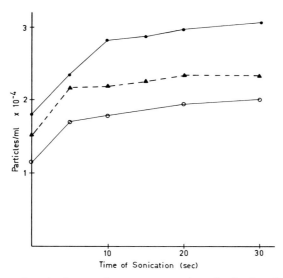

FIG 2. Effect of sonication on the apparent number of cells. Samples from a culture of diploid strain A364A D5 (Hartwell, 1970) growing exponentially in YM-1 medium (Table I) were fixed with formaldehyde according to protocol 2 (Section II,A). Several identical dilutions were then made of each original sample, using 0.9% NaCl as diluent. The various diluted samples were then subjected to various periods of sonication, using the microtip of a Branson S75 sonifier (Branson Co., Danbury, Connecticut), at a power setting of 3. The concentration of particles was then determined for each sonicated sample using a Model B Coulter Counter (Section V). The number of particles versus the amount of sonication is shown for three groups of samples. The differences between the groups in time course and in the final extent of the changes in particle number may reflect small differences in the initial state of clumping, or in the effectiveness of sonication (neither depth of insertion of the sonicator probe, nor tuning, was closely controlled). They may, however, also simply be random, since the experiment was not performed under optimal conditions of reproducibility of dilutions, avoidance of extraneous particles, and so on (see Section V). The counts for each group of course also reflect the cell concentration in the original sample.

were found to disperse essentially all *c* and *d* forms, although essentially no *b* forms were split. Our own experience is similar. Figure 2 and Table IV show results typical of what we have observed with several strains differing significantly in their initial degrees of clumping. Exposure to ultrasound for short periods gives a sharp increase in the number of countable particles, but longer periods of sonication lead to very little further change (Fig. 2). [The details of the sonication procedure are indicated in the figure legend. It should be noted that the amount of sonication necessary to disperse clumps depends on the strain and on the precise conditions of sonication (instrument, power setting, tuning, types of samples treated, etc.). Thus it is recommended that control experiments like those of Fig. 2 or Table IV be done

TABLE IV

EFFECT OF SONICATION ON THE CLUMPING PATTERN AND
ON THE APPARENT PROPORTIONS OF BUDDED AND
UNBUDDED CELLS[a]

Amount of sonication (seconds)	Proportion of clumps (forms c and d)	Proportion of a forms[c]	Proportion of b forms[c]
0	26	21	79
5	1	31	69
10	1	35	65
25	0.5	36	64

[a] Samples similar to those of Fig. 2 were sonicated for varying periods (as described in the legend to Fig. 2), and were then examined by phase-contrast microscopy. Forms a to d are defined in Fig. 1. No e or f forms were observed in this experiment.

[b] As percent of the total number of units (i.e., cells or clumps; i.e., a or b or c or d forms) observed.

[c] As percents of the total apparent numbers of cells (i.e., a or b forms). For the purpose of this calculation each c form has been assumed to be one a form plus one b form; each d form has been assumed to be two b forms.

when necessary to show what is sufficient in a particular situation. It may be worth noting that under our conditions the temperature of our samples rises only about 5°C during sonication.] This suggests that only clumps and/or a discrete set of b forms are affected by the treatment, an impression that is confirmed by microscopic examination (Table IV). Essentially all clumps are dispersed within a few seconds, and about 15% of the b forms are separated to give a forms; in all cases the b forms that are split are ones in which the buds (or daughter cells) are relatively large. More prolonged sonication leads to very little further change in the appearance of the population. The implication is clearly that all cells that have passed a certain stage in the cell division cycle, and only such cells, are split by sonication. In these non-clumpy and mildly clumpy strains, the stage of separability by sonication precedes the time at which the next bud appears by a small but significant amount. These experiments were performed with fixed cells, but comparable doses of sonication also separate clumps of living cells without detectably damaging their viability. (This has been tested both by viable cell platings and by time-lapse observations on individual cells.) It is important to note that clumps, once dispersed by sonication, show no tendency to re-form; this presumably reflects the fact that they are due to incompleted division processes rather than to any generalized stickiness of the cells. [Under some conditions (notably during mating; see Sakai and Yanagishima, 1971, or Campbell, 1973) yeast cells undergo "agglutination"; this is a phenomenon distinct from the clumping being considered here, as is the Ca^{2+}-dependent flocculation which some strains undergo at late stages in the growth of cultures (Lewis and Johnston, 1974; Stewart et al., 1974).] The clonal nature of the clumps is helpful in some situations, such as mutagenesis. Provided that a few generations elapse between mutagenesis and plating, the colonies that appear will be clones even if the clumps have not been successfully dispersed.

Thus with many strains rather mild procedures, which do no detectable harm to the cells, eliminate essentially all clumps. This strongly suggests that these strains, at least, all meet condition 2 but differ in the times required to proceed from cytokinesis to the natural (i.e., in the absence of sonication) separation of mother and daughter cells. The times of cell separability when sonication is applied, however, seem rather similar. Unfortunately, some populations persist in containing many c to f forms even after extensive sonication. It is conceivable that in some cases condition 2 is not met, but in general it seems that the differences are in the times that must elapse between cytokinesis and the stage at which the cells become separable by sonication. The strongest evidence for this point of view comes from the application of a technique developed by Hartwell (1971) in an attempt to monitor directly the occurrence of cytokinesis. He fixed cells with formaldehyde (protocol 2, Section II,A), washed them with water, and treated them with a snail enzyme preparation to digest the cell walls. [The preparation used was Glusulase (Endo Laboratories, Garden City, N.Y.); an apparently comparable preparation, Helicase, is available from L'Industrie Biologique Française, Gennevilliers, France. The cells were incubated for 2 hours at 23°C in Glusulase (diluted 30-fold with water and filtered to remove particles). In other experiments we used Glusulase diluted 20-fold, and overnight incubations, with no obvious changes in the results. If subsequent scoring is to be visual only (e.g., a count of budded versus unbudded cells), the filtration step can be omitted.] After treatment the suspensions were mixed vigorously with a vortex mixer and diluted with 0.9% NaCl; particles were counted with a Coulter Counter. When this procedure was applied to samples of a synchronously dividing population of strain A364A D5 (Fig. 3), it was found that mother and daughter cells became separable at a point in the cell cycle that followed nuclear division but preceded the point at which they could be separated by sonication (cf. Fig. 2 and Table IV). Morphological observations confirmed that small buds remained attached to their mother cells, while some large buds were separated. It thus seems reasonable to suppose that this procedure splits b forms in which cytokinesis has occurred, and does not split b forms in which cytokinesis has not been completed.

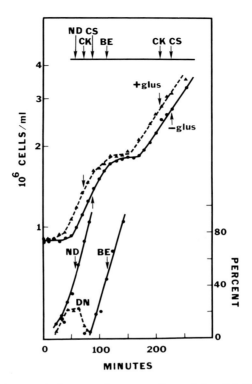

FIG. 3. Monitoring events in a synchronous culture of the diploid strain A364A D5. Shown are the proportion of cells that have completed bud emergence (BE), the proportion of cells with nuclei in the process of division (DN), the proportion of cells that have completed nuclear division (ND), and the total numbers of cells as determined by two methods. In one method (−glus), fixed samples were sonicated before counting with the Coulter Counter. In the second method (+glus), fixed cells were digested with snail enzyme prior to counting with the Coulter Counter. The first method was considered by Hartwell to separate cells that have completed "cell separation" (CS), the second method to separate cells that have completed cytokinesis (CK). See text for further discussion. Arrows indicate the times at which 50% of the cells had completed a particular event. (Figure from Hartwell, 1971, q.v. for further details.)

Support for this concept has come from two sources:

1. Observations were made on temperature-sensitive mutants in which nuclear division is completed normally but cell separation (monitored after sonication) does not occur. The Glusulase procedure also did not lead to separation of mothers from daughters (Hartwell, 1971), and observation of the digested cells by phase-contrast microscopy with an oil-immersion lens revealed what seemed to be a narrow cytoplasmic connection between each

mother cell and its bud (J. Pringle, unpublished observations). This detail could not be seen before the removal of the wall. That such a seemingly tenuous connection suffices to hold mothers and daughters together encourages the view that cytokinesis really must be complete before mother and daughter can be separated by this procedure.

2. When nonclumpy or mildly clumpy strains are subjected to various kinds of nutritional limitations, it is often the case that nearly all the cells in the resulting stationary-phase populations are arrested as unbudded cells, at least if scoring is performed after sonication (Pringle et al., 1974). In some experiments, however, a significant number of b forms remains. The Glusulase method has been used to try to ask whether these b forms have really been arrested prior to completing cytokinesis, or whether they have merely been arrested between cytokinesis and cell separation. Populations 2 to 5 (Table V) show the results obtained with several such populations; population 1, a more typical energy-limited population (see also populations 7 and 8, discussed below), is shown for comparison. In each case there is a definite number of b forms that are not split by the Glusulase treatment. Moreover, few of these are split even when the Glusulase-treated samples are gently sonicated. That the b forms remaining after Glusulase treatment can be resistant to subsequent sonication is further illustrated by the growing population 6. It seems reasonable to suppose that the b forms that are not split are those that have not reached a certain stage of the cell cycle, and it seems reasonable to suppose that this stage is the end of cytokinesis.

With these particular nonclumpy populations, the number of b forms resistant to Glusulase is not greatly different from the number resistant to sonication. (The larger difference in the case of the growing population 6 is compatible with that shown in Fig. 3.) However, this was not the invariable result. In at least two cases (Milne, 1972; M. Unger and L. H. Hartwell, personal communication), nonclumpy stationary-phase populations displayed many b forms after sonication and very few after Glusulase treatment. The case studied by Milne is particularly informative. After a shift of exponentially growing cells to sporulation medium (in which continued growth was not possible), the arrested population contained many b forms in which the bud was quite small in relation to the mother cell. However, nuclear staining revealed that even the small daughters had received nuclei. Sonication did not separate these b forms effectively, but the Glusulase procedure gave partial separation. That is, although the two members of a pair were still associated in some way (since they floated as a unit through the microscope field), they now had the appearance of g forms (Fig. 1) rather than of b forms. No cytoplasmic connection could be seen either by oil-immersion phase-contrast or by staining mother and daughter cytoplasms intensely. Gentle pressure on the cover slip caused most of these pairs to

TABLE V

PROPORTIONS OF BUDDED CELLS AFTER VARIOUS DISPERSION REGIMENS[a]

Population[b]	Regimen			
	No treatment	Sonication only[c]	Glusulase plus vortex[d]	Glusulase plus sonication[e]
1	(1)	(0) 3	(0) 2.5	(0) 2.5
2	(1)	(0) 18	(0) 17	(0) 16
3	(1)	(0) 22	(0) 18	(0) 18
4	(1)	(0) 26	(0) 26	(0) 22
5	(2)	(0) 29	n. e.	(0) 24
6	(2)	(0) 57	(0) 44	(0) 44
7	(2)	(0) 3	(1) 10	(0) 2
8	(4)	(3) —	(2) 12	(0) 4
9	(4)	(2) 46	n. e.	(0) 27
10	(4)	(4) —	(4) —	(1) 12.5
11	(4)	(4) —	(2) 59	(0) 46

[a]Values given are budded cells (b forms) as percent of the total population (a forms plus b forms). For these experiments g forms were scored as b forms. The numbers in parentheses are subjective evaluations of the degree of clumping; the scale runs from 0 (no clumps observed) to 4 (population exists primarily as f forms). A count of budded versus unbudded cells was not possible when the clumping rating was 3 or 4.

[b]Populations listed are: 1, an energy-limited stationary-phase population of a haploid strain; 2 and 3, energy-limited stationary-phase populations of two mutant haploid strains defective in glycogen accumulation; 4, a Mg^{2+}-limited stationary-phase population of a diploid strain; 5, an isoleucine-limited stationary-phase population of an isoleucine auxotroph; 6, an exponentially growing population of the same strain as population 1; 7 and 8, energy-limited stationary-phase populations of two other haploid strains; 9, population of a mutant strain temperature-sensitive for protein synthesis which has arrested development at the restrictive temperature; 10, energy-limited stationary-phase population of an unusually clumpy haploid strain; 11, exponentially growing population of the same strain as population 8.

[c]Sonication was for 15 seconds at power setting 3 on a Branson S75 sonifier. In some cases the populations were reexamined after an additional 15 seconds of sonication, but no significant differences were observed.

[d]Glusulase treatment was as described in the text. n.e., Not Examined.

[e]Glusulase treatment as in footnote d was followed by 10 seconds of sonication at a power setting of 2.

become fully separated, and it was concluded that only remaining traces of cell wall had held them together.

Populations 4 and 7 (Table V) also contained g forms after Glusulase treatment. In the case of population 7, the likelihood of there being cytoplasmic connections between mothers and daughters seemed especially small in view (α) of the lower percent of budded cells obtained by sonication alone, and (β) of the residual presence of some clumps (most of which were actually analagous to g forms in that the constituent entities were not in direct contact). In light of the probable validity of condition 2, the residual clumps suggested that the separation of postcytokinesis pairs had not been complete. As expected, in both populations it was primarily or exclusively the g forms that were separated by gentle sonication after Glusulase treatment.

Thus a variety of arguments supports the view that the Glusulase method, even with subsequent gentle sonication, *does not* separate b forms in which cytokinesis is not complete, and *does* separate b forms in which cytokinesis is complete, although this separation may not be total unless gentle sonication or other mechanical stress is applied after the Glusulase treatment. (The proportion of b forms that is between cytokinesis and cell separation, and which is thus separable by Glusulase but not by sonication, varies considerably with the strain and the growth conditions.) Thus, if condition 2 is met by clumpy strains, Glusulase treatment should always reduce clumps to a mixture of a and b forms, thus allowing meaningful counts both of the total numbers of individual cells and of the relative numbers of cells between bud emergence and cytokinesis (b forms) and between cytokinesis and bud emergence (a forms). (It should be emphasized that such counts do have a somewhat different meaning than comparable counts made on sonicated populations of nonclumpy strains.) In our experience these expectations have always been met; populations 8 to 11 (Table V) provide examples. Population 8 remained a confusion of clumps after sonication alone. Even after Glusulase treatment substantial numbers both of clumps and of g forms remained. A subsequent sonication, however, eliminated both clumps and g forms, and gave a percent of budded cells similar to that obtained with related, but less clumpy, strains under identical physiological conditions (e.g., populations 1 and 7). In population 9 a percent of budded cells was obtained after sonication, but it was substantially different from the value obtained after Glusulase and sonication. The presence of residual clumps in the sample that had only been sonicated had suggested that such a discrepancy would exist. Population 10 was so clumpy that scoring was not even possible until both Glusulase and sonication had been applied. Even then, however, some clumps remained; this casts some doubt on the significance of the observation that the percent of budded cells here was higher than in the similar populations 1, 7, and 8. Finally, a growing culture (population 11) also attained, after Glusulase and sonication, a percent of budded cells similar to that attained with Glusulase alone with the similar, but less clumpy, population 6.

To recapitulate, then, there are three approaches to the clumping problem. The first is to apply no special measures and to wait for the natural

separation of the cells. Unless one is specifically interested in the processes
occurring between cytokinesis and cell separation, this approach should be
avoided. It creates difficulties in comparing results obtained with different
strains and, even in studies of a single strain, it is clear that it is usually more
useful to score an *e* form as four cells than it is to score it as one clump. The
second approach is the application of mild doses of sonication. With most
strains this eliminates most or all clumps, and separates what seems to be a
small and well-defined fraction of *b* forms into two *a* forms. The precise
time (in relation to the time of cytokinesis and to the time of appearance of
the next bud) at which mother and daughter can be separated by sonication
is undoubtedly dependent on strain and growth conditions, and is not clearly
defined with respect to the morphological stages of septation so elegantly
elucidated by Cabib and his colleagues (1974). It is, however, clear that
results with various strains are more nearly comparable, and more nearly
approach the information of greatest interest, when sonication is applied
than when natural separation is allowed.

Sometimes, however, sonication is not sufficient. [It should be emphasized that it is a great convenience to work with a strain that is sufficiently
nonclumpy that clumps are readily dispersed by sonication. It is thus well
worth taking this factor into account in choosing a strain to work with.
Fortunately, the widely used S288C family (including X2180 and C276)
meets this criterion reasonably well (although different stocks may differ
somewhat). A strain of interest which is clumpy can of course also be
crossed with a nonclumpy strain to generate a new strain of satisfactory
properties.] Normal doses of sonication can leave appreciable numbers
of clumps. These may be so numerous that they completely prevent reasonable counting of cells or scoring of budded cells. Even when a count can be
obtained, its meaning is obscured by the fact that the stage at which mothers
and daughters become separable by sonication does not seem well defined
in such populations. (Recall that a mixture of *a* to *f* forms is generally
present.) Moreover, it seems clear that even with nonclumpy strains some
cell wall development must occur after cytokinesis if the cells are to be
separable by sonication; sometimes one would like to eliminate this factor
and determine more directly where the *b* forms stand in relation to cytokinesis. The arguments presented in this section suggest that the procedure
of fixation, treatment with snail enzyme, and (if necessary) gentle sonication,
is a satisfactory approach to these situations. The point at which cells become separable by this procedure is again not really well defined with
respect to known morphological stages, but the existing evidence suggests
that if the point of separability is not actually the end of cytokinesis it is
at least a point quite close to that.

V.   Measuring the Number of Cells

A.   Hemocytometer Counts

Total numbers of cells can be determined by visual counts, using a hemocytometer. It seems to us, however, that this method should be used only when superior alternatives are unavailable. For a quick assessment of the status of a culture, an absorbance measurement is much faster and (when properly used; see Section VI) at least as reliable; for accurate measurements of cell numbers, hemocytometer counts cannot compete with electronic particle counts. An electronic counter has a much lower inherent error, simply because it counts a much larger number of cells. [The standard deviation of a count, due solely to statistical effects during counting, is equal to the square root of the number of cells counted (Berkson *et al.*, 1940). Thus, if 100 cells are counted, the standard deviation *due to this source* is 10, or 10% of the total count. If 10,000 cells are counted, the standard deviation is 100, or 1% of the total count.] In addition, hemocytometer counts are subject to a variety of random and systematic errors, including unconscious human error, which are difficult to control (Berkson *et al.*, 1940; Brecher *et al.*, 1956; Mattern *et al.*, 1957; Magath and Berkson, 1960; Turpin, 1963; Mallette, 1969). Finally, electronic counting is faster and less tedious, particularly when many samples must be handled.

When hemocytometer counts must be used, the following point should be noted. A cell with a bud is a single cell until cytokinesis occurs and mother and daughter are physiologically separate. Counting should attempt to reflect this biological fact, since the practice of scoring a bud as a new cell from the moment it appears, or from the moment it reaches a certain arbitrarily defined (and usually subjectively evaluated) size, leads to much confusion. If the methods of Section IV are successful in eliminating clumps [i.e., if no *c* to *f* forms (Fig. 1) are present], then each remaining unit (i.e., an *a* form or a *b* form) should be scored as one cell. This approach is still somewhat arbitrary (see Section IV), but it can be objectively applied, and it corresponds relatively closely to biological reality. The persistence of mild clumping (i.e., of *c* and *d* forms, Fig. 1) does not prevent the application of this approach; one simply scores each such form as two cells. This procedure should be explicitly noted, however, since a cell number, as so determined, has a somewhat different meaning than an identical cell number obtained with a less clumpy strain. If large clumps persist, there may be no choice but to score each entity (single cell, mother cell, or bud) as one cell; this procedure cannot, of course, be carried out very accurately, and the situation must again be explicitly noted.

In denigrating the usefulness of hemocytometer *counts*, we certainly do *not* mean to denigrate the usefulness of microscopic observations, which are in fact valuable in a variety of ways. See further comments in Sections IV, V,B, VII, and VIII.

B.   Electronic Particle Counts

The theory and practice of electronic particle counting are covered in detail elsewhere (Mattern *et al.*, 1957; Kubitschek, 1969; Pruden and Winstead, 1964; Magath and Berkson, 1960; Mansberg and Ohringer, 1969; Mor *et al.*, 1973; detailed instructions and extensive bibliographies supplied by the manufacturers). We offer here only a few possibly useful comments derived from our personal experience. It is perhaps worth noting that another advantage (see also Section V,A) of electronic particle counters is that they can also be used to obtain cell volume distributions (Harvey, 1968; Clark, 1968; Mor and Fiechter, 1968a; Kubitschek, 1969, 1971; Sebastian *et al.*, 1971; Shuler *et al.*, 1972).

1.   GENERAL COMMENTS

One must remember that these machines count any particles of suitable size that happen along. Thus a yeast cell, a clump of yeast cells, a bacterium, a fleck of dust, a small air bubble, etc., can each give rise to one count. This fact has several corollaries:

1. It is essential to attempt to eliminate clumps before counting, and to supplement these attempts with microscopic evaluations of their effectiveness. (The evaluation step is of course less crucial when a familiar strain is being studied under familiar growth conditions. Nevertheless, periodic microscopic observations are highly recommended, since they can reveal a variety of abnormalities, such as abnormal clumping behavior, contamination, the presence of abnormal cells, etc. Table VII, discussed below, provides an illustration. Besides, the cells are cute.)

2. The electrolyte used as the diluting and counting fluid must be reasonably free of particles. We have done most of our work with commercially available, sterile, filtered, bottled, 0.9% saline for irrigation (a standard hospital item). We have also made satisfactory solutions by filtering saline twice through membrane filters of 0.15-μm pore size. The bottled products (once opened), as well as homemade solutions, show an extraordinary tendency to acquire countable particles on standing; this can introduce significant errors if it is not guarded against. Thus it is desirable to establish, at the beginning of any set of measurements, that the system has a low background when electrolyte alone is counted. This also guards against the

possibility of a high background from electronic noise. Any irreducible background can of course be subtracted from each subsequent count. Electrolytes other than simple 0.9% saline can be used; this may alter the instrument settings (aperture current, amplification, threshold settings) necessary for optimal counting.

3. Particles must not be introduced from other sources. For example, we have noticed that ordinary clean test tubes may contain as many as 2000 counts/ml. It does not introduce much error to put a sample (with perhaps 10^6 cells/ml) into such a tube, but significant errors may arise if such tubes are used for precounting dilutions. The tubes can be specially rinsed and dried with precautions against picking up dust, but it is probably better to make the final 10- to 50-fold dilution of each sample directly in the beaker into which the counting probe will be put. This beaker can be rinsed carefully before beginning the series of measurements. We have also observed that sonication of a tube containing nearly count-free saline may lead to the acquisition of countable particles. The origin of these is not known, but it is probably a good idea to perform this test from time to time. Bottles of concentrated formaldehyde generally contain much particulate matter; this can be avoided by letting a large bottle stand quietly in a corner, and carefully decanting small amounts when needed. If snail enzyme preparations are used on samples to be counted (Section IV), they must be filtered before use.

4. A short delay should be introduced between mixing the final dilution and the beginning of counting, to give air bubbles a chance to disappear. The settling of yeast cells is not rapid enough to be a problem during ordinary counting operations.

As already mentioned, we have found it convenient to make each final dilution in the beaker into which the counting probe will be put. We have used Eppendorf pipettes for the cell suspension and a L/I (Labindustries) Repipet for the diluting saline; both speed and reproducibility are excellent. No doubt other similar systems work as well. Proper handling of the Repipet gives adequate mixing without splashing out cell suspension. When two samples differing greatly in number of cells are to be handled in succession, it helps to rinse Eppendorf tip, beaker, *and probe*, with the new sample (appropriately diluted), and then to make a second dilution for the actual count.

One must be alert to signs of machine malfunction, such as excessive electronic noise, unusually slow or unusually fast movement of the mercury column, or irregularities in counting cadence or oscilloscope pattern. Counting instruments have a high inherent accuracy and precision, but these are realized only when they are functioning properly (which is not invariably the case).

2. Specific Counting Systems

We have had experience with three particle-counting systems. Two of these, the Model B Coulter Counter (Coulter Electronics Inc., Hialeah, Florida), and the Celloscope, Models 202 and 302 (AB Lars Ljungberg and Co., Stockholm, Sweden), are identical in working principle, and very similar in the mechanics of operation. A Coulter Counter is substantially more expensive than a Celloscope of comparable nominal capacity. (The price ratio in Switzerland is approximately 2:1. This might be significantly different elsewhere.) However, our experiences with a Coulter Counter (acquired in Seattle) were rather good. The instrument was rarely out of order and, when so, was promptly and effectively serviced. Experience with the Celloscope (acquired in Zürich) was less satisfactory. Our Model 202 did not function well, and repeated attempts to have it serviced successfully were ineffective. We do not know whether these experiences are typical or not, and it is only fair to note that our Model 302 has now been in service for several months and appears to be functioning properly. The counting times of the two instruments are comparable, but the Celloscope is much slower in returning to zero pressure after a count. This prevents rapid processing of successive samples, and is a major nuisance when many samples must be counted. Possibly there is a way to avoid this problem that we have not discovered. The Coulter Counter is fast enough that operations (counting, recording, mixing and diluting of the next sample) can be effectively integrated with very little waiting time; a single efficient operator, working carefully, can handle a sample (duplicate counts on each sample) every 2–3 minutes for extended periods.

Table VI shows the instrument settings we have routinely used for counting

TABLE VI
SUITABLE INSTRUMENT SETTINGS FOR ELECTRONIC COUNTING OF YEAST CELLS

Instrument	Aperture diameter (μm)	Cell type	Amplification[a]	Aperture current[b]	Lower threshold	Upper threshold
Coulter B	100	Haploids	2	4	5	Infinity
Coulter B	100	Diploids	1	4	5	Infinity
Celloscope 302	48	Diploids	1/1	8	12	Infinity
Celloscope 302	80	Diploids	1/1	8	11	Infinity

[a]The Coulter Counter dial shows 1/amplification; thus the numbers given here are the reciprocals of the actual dial settings. For the Celloscope, the gain (F) setting is shown directly as it appears on the dial.

[b]The Coulter Counter dial shows 1/aperture current; thus the numbers given here are the reciprocals of the actual dial settings. For the Celloscope, the aperture current (P) setting is shown directly as it appears on the dial.

yeasts. With these settings background problems are normally not severe, and in most cases all yeast cells present are counted. It is important to note, however, that with these settings very small cells, such as those that occur in stationary phase (Williamson, 1964; Milne, 1972; J. R. Pringle, unpublished observations) and in very slowly growing (Williamson, 1964; Polakis and Bartley, 1966; Mor and Fiechter, 1968a; Vraná et al., 1973; M. Küenzi, personal communication) populations, are not counted. This was discovered when we studied a diploid population undergoing nitrogen starvation (Table VII). The obvious discrepancy between the budded cell data and the cell number data obtained when counting with a threshold of 5 led us to repeat the cell number counts using a threshold of 2. The results obtained agreed quite well with the budded cell counts. Evidently, the daughter cells produced during the approach to stationary phase were initially too small to register on the counter when it was set at threshold 5, but subsequently grew into the countable range. It should be noted that working at the lower threshold setting demanded special attention to the problem of reducing background counts. This case seems to us to be a good illustration of the value of regular microscopic observations.

The other cell counting system we have used (Technicon Instruments Corp., Tarrytown, N.Y.) employs a different working principle; pulses of

TABLE VII

DISCREPANCY BETWEEN CELL COUNTS AND
PERCENT BUDDED CELLS DATA CAUSED BY
POOR CHOICE OF COULTER COUNTER SETTINGS

Time[a]	Budded cells[b]	Cell number[c]	
		Threshold 5	Threshold 2
0	58	2.90	3.05
5	30	4.18 (1)	5.92 (1)
18	1	4.32 (1.03)	7.64 (1.29)
40	0	7.84 (1.87)	8.04 (1.35)

[a] In hours since the shift of an exponentially growing diploid population (the prototrophic strain C276) to nitrogen-free medium (see also Table I).

[b] As proportion (in percent) of the total population.

[c] As cells/ml $\times 10^{-6}$. The numbers in parenthesis are the cell counts normalized to the 5-hour count in each series. The thresholds indicated are the lower threshold settings on the Coulter Counter; amplification was set at 1, aperture current at 4, and the upper threshold at infinity, in each case.

scattered light, rather than momentary impedance changes, are counted (Mansberg and Ohringer, 1969). The instrument has lower reproducibility than the counters discussed above, although it is still appreciably better in this regard than hemocytometer counting. Moreover, the data provided by the Technicon counter can be converted to absolute counts only after a calibration has been established for the cells of interest. Establishment of the calibration of course demands that accurate counts be made by some other method. The major advantage of the Technicon counter is that it is designed for automatic or continuous-flow operation. A disadvantage, for most laboratories, is that it is designed as an integral part of the whole Technicon AutoAnalyzer system. A further disadvantage is that this system cannot be used to provide size distributions. More information about the use of this system is given by Mor *et al.* (1973).

C. Viable Cell Counts

The number of viable cells per unit culture volume is generally assessed by plating suitably diluted aliquots of culture and counting the colonies that appear. The inaccuracies inherent in this procedure, with suggestions on how to avoid them, have been discussed at length by Postgate (1969); he also comments on alternative approaches to obtaining the same information. The special difficulty in obtaining meaningful viable counts with yeast is of course the clumping problem. For example, if a particular mutagenesis procedure kills 60% of the individual cells, but a typical cell is in a clump of six cells (not so rare in exponential-phase cultures of haploid strains), less than 5% of the clumps will contain no viable cells. Plating such a population without disrupting the clumps gives the false impression that essentially no killing has occurred. Sonication to break up clumps, coupled with microscopic assessment of the success of this procedure, is highly recommended when accurate viable cell counts are desired. Sonication sufficient to break up clumps in most strains (see Section IV) results in no loss of viability. For most purposes the sonicator probe is adequately sterilized by rinsing well with 70% ethanol. With at least some sonifiers it is possible to disrupt clumps sterilely by dipping samples, in closed, sterile, disposable, plastic tubes, into a small "bath" of sonicated water (E. Cabib, personal communication). In our experience diluting and sonicating the cells in sterile water, rather than in medium or in buffer, results in no loss in viability. A trick that saves time, and possibly even increases accuracy, is to spread the cells on the agar surface using the tip of the pipette, rather than a special spreading rod; one must be careful, however, to remove the agar that occasionally plugs the pipettes.

To learn the *proportion* of viable cells in a population, one must compare a viable cell count to a count of total cells. This introduces additional possibilities for error, and for some purposes it may be preferable to determine the proportion directly by making time-lapse observations on single cells (a simple procedure for such time-lapse observations has been described by Hartwell *et al.*, 1970). This approach can be very little trouble to do (note that, in principle, only two observations are needed), has the potential for high accuracy, allows the viabilities of budded and of unbudded cells to be assessed independently (Beam *et al.*, 1954), and has the additional advantage of revealing any residual clumps for what they are. There is, however, one major disadvantage. In many populations for which viability determinations are of interest, individual viable cells may vary greatly in the time they require before beginning discernible growth and division. Thus it may be difficult to know how long to observe a cell before pronouncing it dead, and to keep early-starting cells from obscuring, by extensive growth, the behavior of late-starting (or dead) cells.

VI.   Measuring Turbidity

A.   Absorbance Measurements

Yeast cells scatter visible light. The resulting turbidity is usually measured and expressed as optical density, or absorbance (i.e., as $\log_{10} I_0/I$, where I_0 and I are the intensities of the incident and transmitted light beams, respectively). Since the absorbance of a suspension of light-scattering material depends on the concentration of suspended material, and since absorbance measurements are quickly, easily, and precisely made, such measurements are widely used for determining cell concentrations. Unfortunately, in much of this wide use insufficient attention is paid to the real complications and limitations of such measurements. These complications and limitations arise from the following facts (Koch, 1961, 1970; Mallette, 1969; Mitchison, 1970; Bussey, 1974).

1. Ordinary spectrophotometers are not ideal turbidimeters, since much of the light that is scattered reaches the detector along with the light that was not scattered. The proportion of scattered light reaching the detector is a function of the geometry of the particular instrument; different commercially available spectrophotometers differ appreciably in this regard.

2. A suspension of yeast cells scatters light in a way that is a complex function of the number, sizes, shapes, refractive indexes (in comparison to

that of the medium), and state of aggregation of the cells, as well as of the wavelength of light used.

These facts have the following consequences.

1. If absorbance measurements are made on various dilutions of *a single cell suspension*, in *a single, properly functioning, spectrophotometer*, it is found that absorbance is proportional to cell concentration only at rather low cell densities. With most instruments this lack of proportionality is significant at absorbances greater than 0.4 (Fig. 4). It should be noted that Fig. 4 shows only cell concentrations less than 4×10^7 cells/ml; actual cultures may grow to nearly 10 times this cell density (Table I), with correspondingly greater deviations from proportionality.

2. A single cell suspension gives different absorbance values in different spectrophotometers (Fig. 4). Moreover, the rate of curvature of the plot of absorbance against cell concentration also varies from instrument to instrument. The spectrophotometers tested for Fig. 4 do not represent extreme cases; a 4-fold difference in the absorbance at 660 nm was noted when a single cell suspension was measured in Coleman Jr. and Gilford spectrophotometers (E. Cabib, personal communication).

3. The dry-weight value, or the cell number value, corresponding to a particular absorbance value is not fixed, but is a function of various properties (ploidy, clumpiness, physiological state, etc.) of the cells (see Table I and the discussion in Section I,B).

It is clear that points 1 to 3 must be attended to if absorbance measurements are to be used properly for determining amounts of yeast and for monitoring the growth of cultures. Even when all measurements are made on a single instrument, and when no attempt is made to relate the absorbance data to more basic parameters such as cell number or total biomass (i.e., even when the amounts of yeast are to be expressed only in relative terms, as is sufficient for some purposes such as the determination of generation times), it is necessary to attend to point 1. This can be done by diluting all samples (ideally with additional amounts of the medium in which the cells are growing) until their absorbances are within the range in which absorbance is proportional to cell concentration. Alternatively, a calibration curve of absorbance versus relative cell concentration can be constructed (ideally using cells similar to those subsequently to be measured) and used to correct absorbance values out of the range of proportionality. Ordinarily, it is convenient to use a combination of the two approaches.

When absorbance data are to be converted to cell number data, biomass data, etc. (see the next paragraph), it is necessary to attend to points 2 and 3. That is, a calibration must be made using the spectrophotometer that will subsequently be used, and the strain of interest (or a closely related one) under the physiological conditions of interest.

Because of the complications discussed, it is impossible for a reader to interpret a published absorbance value unless he has himself made similar measurements with the same strain, under the same physiological conditions, on the same type of spectrophotometer, and at the same wavelength.

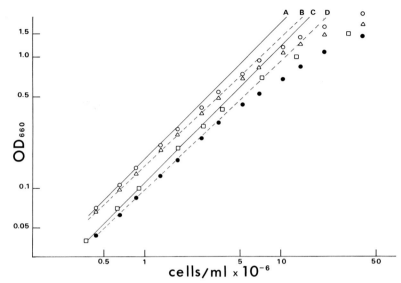

FIG. 4. Absorbance (OD) at 660 nm as a function of cell concentration. A culture of strain C276 was allowed to grow to approximately 4×10^7 cells/ml in YM-1 medium (Table I); the culture is still in exponential phase at this cell density. The cells were then fixed with formaldehyde (using protocol 1, Section II,A), and a sample was taken for the determination of cell number (using the Celloscope 302). A series of independent dilutions was then made, using water as diluent, and the absorbance of each dilution was measured on each of three spectrophotometers: a Zeiss PMQ II (open circles and line A), a Beckman DB-G (triangles and line B), and a Beckman Acta III (solid circles and line D). A similar experiment was done on a separate occasion using a Gilford 300N spectrophotometer (squares and line C). The data are shown in a logarithmic plot (a linear plot of similar data is shown by Mitchison, 1970). Thus, if absorbance were proportional to cell concentration (OD = kC), the experimental points for each instrument should fall on a straight line of slope 45°, whose height above the abscissa would be a measure of the value of k for that instrument. As can be seen, this is essentially the case at lower cell densities; the straight lines A to D, which were drawn by eye through as many of the experimental points as seemed reasonable, have slopes of about 45°. At higher cell concentrations the experimental points fall significantly below these lines, reflecting the fact that absorbance fails to increase in proportion to cell concentration at higher cell concentrations. It should be noted that, since a single cell suspension was used to generate the data, there is a fixed relationship between cell number and dry weight. Thus an identical plot would have been obtained if absorbance had been plotted against dry weight. It should also be noted that both fixing the cells and using water rather than medium as diluent introduce errors; these are, however, small with respect to the effects the figure is designed to show.

Thus it seems to us that absorbance values per se should never be published, but should always first be converted to values of one of the more basic parameters.

In practice, we have found it extremely convenient to use calibrations that relate the absorbance observed on our spectrophotometer directly to the cell number or dry weight of exponential-phase cultures of typical strains from our collection. Such a calibration is readily constructed by taking a relatively dense population of exponentially growing cells, measuring the dry weight and cell number, and then making various dilutions and reading the absorbances. This procedure generates the kind of calibration data shown in Fig. 4. It is somewhat more convenient to use a table constructed from the graph than to use the graph itself. Growth curves obtained with such a calibration parallel those obtained by direct measurement of dry weight and cell number throughout exponential phase although, as expected, deviations occur as the physiological status of the cells begins to change later in the course of growth. If exponential cells are not the only ones of interest, it may be profitable to make other, similar, calibrations (e.g., it might be useful to know that, if a population has been in sporulation medium for 5 hours and has an absorbance of x, it contains z cells per milliliter). Another approach that can be useful is to relate absorbance to cell number or dry weight throughout the growth history of a particular type of culture that is frequently used (e.g., it might be useful to know that when strain C276, growing in YM-1 medium, attains a corrected absorbance of 18 it has attained a biomass of 2.9 mg/ml and a cell number of 1.6×10^8 cells/ml; see Table I).

To summarize, absorbance measurements can be very useful, but must be used with discretion. We use them in almost every experiment for guiding the experiment as it is in progress. We use them as data when a highly accurate measurement of cell concentration does not seem to be required. We use them, with precautions, to determine growth rates. [A procedure that avoids some uncertainties is to obtain a population in exponential growth at a rather low cell concentration, to determine the absorbance, and then to dilute the culture and time the return to the original absorbance. It seems likely that the relation between absorbance and cell number is the same each time the population is at that particular absorbance value, and that a valid generation time therefore results.] We do not think that absorbance measurements should substitute for direct measurements of more basic parameters, such as cell number and dry weight, when accuracy is desired, and we do not think that absorbance values should ever be reported without first being converted properly to values of one of the more basic parameters.

B. Nephelometry; Continuous Monitoring of Turbidity

In absorbance measurements the diminution in intensity of a transmitted light beam is measured. It is also possible to measure directly the amount of light scattered at particular angles relative to the direction of the incident light beam. Such measurements, known as nephelometric, have limitations similar to those of absorbance measurements (Koch, 1961; Mallette, 1969; Wyatt, 1973), but have the advantage that they can be used over a significantly greater range of cell concentrations. Nephelometry has been relatively little used in microbiology, probably because the instruments are not so universally available as spectrophotometers.

Several schemes have been described for the continuous monitoring of the turbidity of growing cultures. Under some conditions "cultivation cuvettes" (cuvettes that serve also as cultivation chambers) can be used to good advantage (Řičica, 1966). More generally useful are flow cuvettes (available for most spectrophotometers). With a continuous culture the effluent can simply pass through the cuvette on its way to waste or collection vessels. With batch cultures in appropriate cultivation chambers, culture can be continuously pumped through the cuvette and back into the cultivation chamber. A new development of great promise is the use of fiber optic probes. Light is conducted by means of a glass fiber bundle from the monochromator of a spectrophotometer to the end of the probe, which is inserted directly in the growing culture. There the light must traverse a gap, at which time it is in part scattered by the cells. The remaining light is conducted by a second fiber bundle back to the photomultiplier of the spectrophotometer, and an absorbance value is obtained. The use of such systems with yeast cultures has been described by Robrish et al. (1971) and by Blanch and Fiechter (1974).

VII. Taking Advantage of the Budding Mode of Reproduction

The budding mode of reproduction is in certain respects a great experimental convenience. It allows the differentiation of mother cells from daughter cells, even after they have separated (Williamson, 1964; Beran, 1968; Hayashibe and Sando, 1970; Lieblová and Beran, 1971; Hayashibe et al., 1973; Vraná et al., 1973). Also, since the bud grows progressively in size during the cycle, the size of the bud provides an approximate visual marker for the position of the cell in the cycle. This fact has proved useful in a variety of contexts (Johnson, 1965; Williamson, 1965; Wiemken et *al.*, 1970; Hartwell *et al.*, 1970; Shulman *et al.*, 1973; Hartwell, 1974). It must be remembered, however, that the precise relationship between bud size and cell cycle stage is a function of the culture conditions and of the properties of the mother cell (Williamson, 1964; Vraná *et al.*, 1973; Hayashibe *et al.*, 1973).

Another great convenience of the budding mode of reproduction derives from the fact that, in all physiological situations examined to date, the first appearance of the bud coincides quite closely in time with the occurrence of the other known early events of the cell division cycle (Hartwell, 1974; Hartwell *et al.*, 1974). Thus, if care is taken to separate mature daughter cells from their mothers (Section IV), visual scoring of the relative numbers of budded and unbudded cells provides a measure of the relative numbers of cells within (i.e., engaged in) and between cell cycles. (A cell between cell cycles is one that has completed division cycle *n* but has not begun division cycle *n* + 1.) Such scoring of the proportion of budded cells has proved useful in evaluating synchrony (see, for example, Ogur *et al.*, 1953; Williamson, 1964, 1965; Esposito, 1968; Küenzi and Fiechter, 1969; Wiemken *et al.*, 1970; Hayashibe *et al.*, 1973), and in assessing the effects of various inhibitors (Slater, 1973; Bücking-Throm *et al.*, 1973; Wilkinson and Pringle, 1974), mutations (Reid and Hartwell, 1974), and nutritional regimens (Beam *et al.*, 1954; von Meyenburg, 1968; Pringle *et al.*, 1974; Hartwell *et al.*, 1974; Hartwell, 1974).

Another example of the usefulness of scoring budded cells was given in Table VII; to generalize somewhat, we note that in the history of such a culture the proportion of budded cells is closely correlated with the position of the culture on the path leading from exponential to stationary phase (see also Table I). Once the correlation is established, a budded cell count can serve to locate a culture on that path without the necessity of making a complete set of cell number counts. Most notably, it is clear that, if all budded cells have disappeared, the culture *must* be in stationary phase; clearly, no further increase in cell number is occurring. This criterion has frequently proved convenient. It should be noted, however, that *some* stationary-phase cultures (i.e., some cultures in which no further increase in cell number will occur) do contain appreciable numbers of budded cells (Pringle *et al.*, 1974), so that this criterion cannot always be applied.

It should be obvious that the successful use of budded cell counts depends on the use of a successful, or at least consistent, approach to the clumping problem (Section IV). It is also true that a practiced and patient eye is required if fully accurate budded cell counts are to be obtained. The detection of small buds demands focusing up and down, waiting for the cells to roll, and some experience in recognizing small buds. (This experience is probably best gained by following closely the emergence of the first generation of new buds after a stationary-phase population, such as those of Tables I and VII, is put in fresh medium.)

VIII. Monitoring for Contamination

Yeasts grow rapidly enough that contamination is not usually a problem. However, serious contamination by bacteria and by other fungi (including other yeasts) does occasionally occur. Such problems are of course much less likely in short-term experiments involving large inocula than in longer-term operations such as continuous culture, repeated subculturing, or the growth of large batches of cells for biochemical work. Contamination can be detected in a variety of ways; each of the following approaches has proved useful to us on one or more occasions:

1. Observation of the gross properties of the culture, such as color, odor, apparent state of cellular aggregation, presence or absence of mycelia, settling rate of cells in a standing culture, and rate of growth.

2. Direct observation of the culture by phase-contrast microscopy.

3. Plating aliquots of the culture, with subsequent observations of colony morphologies and microscopic observations of the cells in at least some colonies.

None of these approaches, taken alone, is foolproof. Probably the best strategy is to attend in all cases to approaches 1 and 2, and to take the trouble to attend to approach 3 only when the risk is relatively high (see above), or when the other approaches have provided grounds for suspicion. With this strategy it is not likely that contamination by bacteria or mycelial fungi will be overlooked.

Contamination by other yeasts, however, is a trickier problem. One of us (J. R. Pringle) has on four separate occasions, in a 6-year period, had serious contamination by wild yeasts. The contaminating cells in each case were about the same size as *Saccharomyces cerevisiae* cells, reproduced by budding, and produced colonies not very different from those of *S. cerevisiae*. The generation times were also not grossly different. Closer microscopic observation revealed some differences: the contaminating cells were more elongated, budded more nearly at the apices of the mother cells, and had a somewhat darker appearance under phase-contrast microscopy. It is not clear whether these differences would have been noticed if the color and odor of the contaminated cultures had not already provided grounds for suspicion. The contaminated cultures developed a red color which was somewhat different in hue and in time course of appearance from the red color of *ade 1* and *ade 2* strains of *S. cerevisiae*. (On the first two occasions when this contaminant was encountered, an *ade 2* strain of *S. cerevisiae* was actually being studied; this made the color clue at first something of a red herring.) Moreover, the contaminated cultures did not have the pleasant, wholesome odor that comparable cultures of *S. cerevisiae* have.

Contamination by other strains of the yeast species being studied is of course also a potentially tricky problem. This can only be dealt with by occasional checks of the ploidy, mating type, and known genetic markers of the strain under investigation.

In conclusion, a word of caution may be in order. Although both visual and olfactory clues can be very useful in detecting contamination, gustatory clues should be utilized only with circumspection. Taste tests of yeast culture supernatants can indeed provide important insights, but such tests should in general follow contamination checks by other methods.

ACKNOWLEDGMENT

We thank Armin Fiechter and Lee Hartwell, in whose laboratories this work was done, for their advice and encouragement. We also thank L. Hartwell for permission to use Fig. 3, R. Maddox and S. Hochmann for their help with various experiments, and B. S. Mitchell for unlimited good advice.

Portions of this work were supported by Swiss National Science Foundation Grant No. 3.628.71 (to A. Fiechter), by U.S. Public Health Service Grant GM-17709 (to L. Hartwell), by a U.S. Public Health Service Training Grant to the Department of Genetics, University of Washington, and by U.S. Public Health Service postdoctoral fellowship FO2-GM41910 (to J.R.P.).

REFERENCES

Beam, C. A., Mortimer, R. K., Wolfe, R. G., and Tobias, C. A. (1954). *Arch. Biochem. Biophys.* **49**, 110–122.

Beran, K. (1968). *Advan. Microbial Physiol.* **2**, 143–171.

Berkson, J., Magath, T. B., and Hurn, M. (1940). *Amer. J. Physiol.* **128**, 309–323.

Blanch, H. W., and Fiechter, A. (1974). *Biotechnol. Bioeng.* **16**, 539–543.

Brecher, G., Schneidermann, M., and Williams, G. Z. (1956). *Amer. J. Clin. Pathol.* **26**, 1439–1449.

Bücking-Throm, E., Duntze, W., Hartwell, L. H., and Manney, T. R. (1973). *Exp. Cell Res.* **76**, 99–110.

Burns, V. W. (1956). *J. Cell. Comp. Physiol.* **47**, 357–375.

Bussey, H. (1974). *J. Gen. Microbiol.* **82**, 171–179.

Cabib, E., Ulane, R., and Bowers, B. (1974). *Curr. Topi. Cell. Regul.* **8**, 1–32.

Campbell, D. A. (1973). *J. Bacteriol.* **116**, 323–330.

Clark, D. J. (1968). *J. Bacteriol.* **96**, 1214–1224.

Esposito, R. E. (1968). *Genetics* **59**, 191–210.

Farris, J. S., and Gilmore, R. A. (1974). *Genetics* **77**, s21–s22.

Francis, J. C., and Hansche, P. E. (1972). *Genetics* **70**, 59–73.
Gilmore, R. A., and Murphy, J. (1972). *Genetics* **71**, s19–s20.
Hartwell, L. H. (1970). *J. Bacteriol.* **104**, 1280–1285.
Hartwell, L. H. (1971). *Exp. Cell Res.* **69**, 265–276.
Hartwell, L. H. (1974). *Bacteriol. Rev.* **38**, 164–198.
Hartwell, L. H., Culotti, J., and Reid, B. (1970). *Proc. Nat. Acad. Sci. U.S.* **66**, 352–359.
Hartwell, L. H., Culotti, J., Pringle, J. R., and Reid, B. J. (1974). *Science*, **183**, 46–51.
Harvey, R. J. (1968). *In* "Methods in Cell Physiology" (D. M. Prescott, ed.), Vol. 3, pp. 1–23. Academic Press, New York.
Hayashibe, M., and Sando, N. (1970). *J. Gen. Appl. Microbiol.* **16**, 15–27.
Hayashibe, M., Sando, N., and Abe, N. (1973). *J. Gen. Appl. Microbiol.* **19**, 287–303.
Herbert, D., Phipps, P. J., and Strange, R. E. (1971). *In* "Methods in Microbiology" (J. R. Norris and D. W. Ribbons, eds.), Vol. 5B, pp. 209–304. Academic Press, New York.
Hess, W. M. (1966). *Stain Technol.* **41**, 27–35.
Hug, H., and Fiechter, A. (1973). *Arch. Mikrobiol.* **88**, 77–86.
Johnson, B. F. (1965). *Exp. Cell Res.* **39**, 577–583.
Koch, A. L. (1961). *Biochim. Biophys. Acta* **51**, 429–441.
Koch, A. L. (1970). *Anal. Biochem.* **38**, 252–259.
Kubitschek, H. E. (1969). *In* "Methods in Microbiology" (J. R. Norris and D. W. Ribbons, eds.), Vol. 1, pp. 593–610. Academic Press, New York.
Kubitschek, H. E. (1971). *J. Bacteriol.* **105**, 472–476.
Küenzi, M. T., and Fiechter, A. (1969). *Arch. Mikrobiol.* **64**, 396–407.
Lewis, C. W., and Johnston, J. R. (1974). *Proc. Soc. Gen. Microbiol.* **1**, 73.
Lieblová, J., and Beran, K. (1971). *Folia Microbiol. (Prague)* **16**, 241–248.
Magath, T. B., and Berkson, J. (1960). *Amer. J. Clin. Pathol.* **34**, 203–213.
Mallette, M. F. (1969). *In* "Methods in Microbiology" (J. R. Norris and D. W. Ribbons, eds.), Vol. 1, pp. 521–566. Academic Press, New York.
Mansberg, H. P., and Ohringer, P. (1969). *Ann. N.Y. Acad. Sci.* **157**, 5–12.
Mattern, C. F. T., Brackett, F. S., and Olson, B. J. (1957). *Appl. Physiol.* **10**, 56–70.
Milne, C. P. (1972). Master's Thesis, Dept. of Genetics, University of Washington, Seattle (see also Hartwell, 1974).
Mitchison, J. M. (1970). *In* "Methods in Cell Physiology" (D. M. Prescott, ed.), Vol. 4, pp. 131–165. Academic Press, New York.
Mitchison, J. M. (1971). "The Biology of the Cell Cycle." Cambridge Univ. Press, London and New York.
Mitchison, J. M., and Creanor, J. (1971). *Exp. Cell Res.* **67**, 368–374.
Mor, J.-R., and Fiechter, A. (1968a). *Biotechnol. Bioeng.* **10**, 159–176.
Mor, J.-R., and Fiechter, A. (1968b). *Biotechnol. Bioeng.* **10**, 787–803.
Mor, J.-R., Zimmerli, A., and Fiechter, A. (1973). *Anal. Biochem.* **52**, 614–624.
Ogur, M., Minckler, S., and McClary, D. O. (1953). *J. Bacteriol.* **66**, 642–645.
Polakis, E. S., and Bartley, W. (1966). *Biochem. J.* **98**, 883–887.
Postgate, J. R. (1969). *In* "Methods in Microbiology" (J. R. Norris and D. W. Ribbons, eds.), Vol. 1, pp. 611–621. Academic Press, New York.
Pringle, J. R., Maddox, R., and Hartwell, L. H. (1974). In preparation.
Pruden, E. L., and Winstead, M. E. (1964). *Amer. J. Med. Technol.* **30**, 1–35.
Reid, B. J., and Hartwell, L. H. (1974). In preparation.
Řičica, J. (1966). *In* "Theoretical and Methodological Basis of Continuous Culture of Microorganisms" (I. Málek and Z. Fencl, eds.), pp. 155–313. Publ. House Czech. Acad. Sci., Prague.

Robrish, S. A., LeRoy, A. F., Chassy, B. M., Wilson, J. J., and Krichevski, M. I. (1971). *Appl. Microbiol.* **21**, 278–287.

Sakai, K., and Yanagishima, N. (1971). *Arch. Mikrobiol.* **75**, 260–265.

Scopes, A. W., and Williamson, D. H. (1964). *Exp. Cell Res.* **35**, 361–371.

Sebastian, J., Carter, B. L. A., and Halvorson, H. O. (1971). *J. Bacteriol.* **108**, 1045–1050.

Shuler, M. L., Aris, R., and Tsuchiya, H. M. (1972). *Appl. Microbiol.* **24**, 384–388.

Shulman, R. W., Hartwell, L. H., and Warner, J. R. (1973). *J. Mol. Biol.* **73**, 513–525.

Slater, M. L. (1973). *J. Bacteriol.* **113**, 263–270.

Stewart, G. G., Russell, I., and Garrison, I. F. (1974). *Proc. 4th Int. Symp. Yeasts (Vienna)*, Part I, 139–140.

Throm, E., and Duntze, W. (1970). *J. Bacteriol.* **104**, 1388–1390.

Turpin, B. C. (1963). *Amer. J. Med. Technol.* **29**, 45–51.

von Meyenburg, H. K. (1968). *Pathol. Microbiol.* **31**, 117–127.

Vraná, D., Lieblová, J., and Beran, K. (1973). *Proc. Int. Spec. Symp. Yeasts, 3rd, 1973* Part 2, pp. 285–296.

Weibel, K. E., Mor, J.-R., and Fiechter, A. (1974). *Anal. Biochem.* **58**, 208–216.

Wells, J. R., and James, T. W. (1972). *Exp. Cell Res.* **75**, 465–474.

Wiemken, A., Matile, P., and Moor, H. (1970). *Arch. Mikrobiol.* **70**, 89–103.

Wilkinson, L. E., and Pringle, J. R. (1974). *Exp. Cell Res.* **89**, 175–187.

Williamson, D. H. (1964). *In* "Synchrony in Cell Division and Growth" (E. Zeuthen, ed.), pp. 351–379. Wiley (Interscience), New York.

Williamson, D. H. (1965). *J. Cell Biol.* **25**, 517–528.

Wyatt, P. J. (1973). *In* "Methods in Microbiology" (J. R. Norris and D. W. Ribbons, eds.), Vol. 8, pp. 183–263. Academic Press, New York.

Yoshida, F., Yamine, T., and Yagi, H. (1971). *Biotechnol. Bioeng.* **13**, 215–228.

Note added in proof: Our comments (p. 156) on the Celloscope now appear to us to be too harsh. For one thing, we have spoken to one completely satisfied user of a Celloscope obtained from Particle Data, Inc. (Elmhurst, Illinois). In addition, we have continued to enjoy trouble-free operation of our Celloscope 302, and have also discovered that appropriate manipulation of the stopcock allows a nearly air-bubble-free change of samples.

Chapter 8

Preparation and Growth of Yeast Protoplasts

S.-C. KUO[1] AND S. YAMAMOTO[2]

Waksman Institute of Microbiology, Rutgers University,
The State University of New Jersey,
New Brunswick, New Jersey

I. Introduction

Eddy and Williamson (1957) first described the formation of protoplasts from *Saccharomyces carlsbergensis* using lytic enzymes present in gut juice of the snail *Helix pomatia*. [The term protoplast as used in this chapter simply denotes osmotically sensitive spherical bodies formed from yeasts, without regard to completeness of removal of the wall material (Brunner *et al.*, 1958).] Since then many investigators have studied the formation

---

[1] *Present address*: Department of Biological Research, Lederle Laboratories, Pearl River, New York.

[2] *Permanent address*: Department of Agricultural and Biological Chemistry, Kochi University, Nankoku, Kochi, Japan.

of yeast protoplasts by a variety of methods. Many enzymes from micro-organisms have also been shown to produce protoplasts. Unfortunately, most of these lytic enzymes are not commercially available, and their use is therefore limited to a few laboratories. Recently, Yamamoto isolated and crystallized in relatively large quantity two lytic enzymes from species of *Rhizopus* and fungi imperfecti. The lytic action of these enzymes has provided useful information on the structure of the yeast cell wall, and these enzymes are of great value in the preparation of protoplasts.

Many metabolic activities of yeast protoplasts, prepared with either snail gut or microbial enzymes, are comparable to those of intact cells when the protoplasts are incubated in appropriate medium. These protoplasts are able to synthesize protein and nucleic acid, as well as cell wall materials (Eddy and Williamson, 1959; Shockman and Lampen, 1962; Tabata *et al.*, 1965; Hutchson and Hartwell, 1967). In solid medium containing gelatin or agar, protoplasts grow and eventually develop into normal yeast cells (Nečas, 1961, 1971; Ota, 1972). Yeast protoplasts have recently proved to be particularly useful for study of the regulation of macromolecular synthesis, especially of enzymes and cell wall materials, since protoplasts, unlike intact yeasts, can be lysed readily by osmotic shock or detergents. This chapter describes methods adopted by several investigators for the preparation of yeast protoplasts and for the use of protoplasts to study growth and macromolecular synthesis.

II.   Preparation of Yeast Protoplasts

Procedures for the preparation of yeast protoplasts can be classified in three major groups: (1) mechanical or autolytic methods (Holden and Tracey, 1950; Nečas, 1956); (2) specific inhibition of cell wall synthesis (Berliner and Reca, 1970); and (3) enzymatic dissolution of the wall. We discuss only the use of enzymes from snail gut juice (snail enzyme) and microorganisms, because the first two methods are not practical.

A.   Factors Affecting Protoplast Formation

1.   STRAIN AND GROWTH PHASE

Variations in susceptibility to attack by snail and microbial enzymes have been observed between one yeast strain and another of the same species and between species. Even within a single strain, susceptibility is highly dependent on the physiological state of the yeast. Yeasts used for the preparation of protoplasts are mainly *Saccharomyces* and *Candida*. Therefore these two genera are primarily discussed.

Generally, exponential-phase yeasts are susceptible to protoplast formation, while stationary-phase yeasts are relatively resistant (Eddy and Williamson, 1957; Holter and Ottolenghi, 1960; Rost and Venner, 1965a,b). During the transition from exponential to stationary phase, there is a rapid increase in resistance to protoplast formation (Russell *et al.*, 1973; Deutch and Perry, 1974). Such differences may be due to variations in the structure of the cell wall. As will be described later, yeasts can be rendered more susceptible to protoplast formation by treatment with a mercapto compound or even protease (Russell *et al.*, 1973).

2. OSMOTIC STABILIZERS

Since protoplasts are osmotically fragile, an osmotic stabilizer is required during their formation from yeasts. Mannitol, sorbitol, and potassium chloride are most frequently used (Table I). Magnesium sulfate is also recommended as an excellent stabilizer by Gascon and Villanueva (1965). The compounds (0.6–1.0 osmolal) are usually added to a buffer solution, such as tris–HCl, sodium acetate, potassium phosphate, and sodium citrate, at a pH between 5.8 and 6.8.

3. MERCAPTO COMPOUNDS

Burger *et al.* (1961) first reported that cysteine accelerated the preparation of protoplasts from yeasts by snail enzyme. Since then, pretreatment or concurrent treatment of cells with a mercapto compound such as 2-mercaptoethylamine (Duell *et al.*, 1964), 2-mercaptoethanol, (Davis and Elvin, 1964), thioglycollate (Kovac *et al.*, 1968), or dithiothreitol (Sommer and Lewis, 1971) has been used to render the cells susceptible to the protoplast-forming enzyme (Table I). For some strains, however, the addition of a mercapto compound does not have an effect (Deutch and Perry, 1974).

Davis and Elvin (1964) suggested that the site of action of mercapto compounds is the cell wall and not the snail enzyme. Acceleration by a mercapto compound may result from ruputre of disulfide bonds of cell wall proteins (Falcone and Nickerson, 1956).

B. Procedure

1. THE USE OF SNAIL ENZYME

Snail enzyme can be obtained as Glusulase from Endo Laboratories, Inc., Garden City, N.Y., or as *Suc-digestif d'Helix pomatia* or Helicase from L'industrie Biologique Francaise, 35–49 Quai du Moulin de Cage, Genevilliers, Paris, France.

TABLE I

OSMOTIC STABILIZERS AND MERCAPTO COMPOUNDS

Osmotic stabilizer	Mercapto compound	Yeast species	Reference
Sorbitol	—	*S. cerevisiae*	Ottolenghi (1966); Hutchison and Hartwell (1967)
	2-Mercaptoethylamine	*S. cerevisiae*	Duell *et al.* (1964)
	2-Mercaptoethylamine	*S. carlbergensis*	Charalampous and Chen (1974)
Mannitol	—	*S. cerevisiae*	Eddy and Williamson (1959); Longley *et al.* (1968)
	—	*C. utilis*	Svihla *et al.* (1961); Mendoza and Villanueva (1962, 1964)
	2-Mercaptoethanol	*S. cerevisiae*	Bauer *et al.* (1972)
		W. fluorescens	
	Dithiothreitol	*S. cerevisiae*	Russell *et al.* (1973); Sommer and Lewis (1971)
Rhamnose	—	*S. carlsbergensis*	Eddy and Williamson (1957)
Maltose	2-Mercaptoethanol	*S. fragilis*	Davis and Elvin (1964)
Sorbitol and mannitol	Thioglycollate	*S. cerevisiae*	Kovac *et al.* (1968)
Lactose	Cysteine	*S. cerevisiae*	Burger *et al.* (1961)
KCl	—	*C. utilis*	Svihla *et al.* (1961)
	—	*S. carlsbergensis*	Holter and Ottolenghi (1960)
		Schiz. pombe	
KCl	2-Mercaptoethanol	*Saccharomyces* sp.	Kuo and Lampen (1971)
		S. cerevisiae	Yamamoto and Nagasaki (1971a,b); Yamamoto *et al.* (1972), (1974)
MgSO$_4$	—	*C. utilis*	Gascon and Villanueva (1965)
(NH$_4$)$_2$SO$_4$	—	*S. fragilis,*	Rost and Venner (1965a,b)
		S. cerevisiae,	
		Schiz. pombe, S. willanus,	
		S. ludwigii	
NaCl	—	*S. cerevisiae*	Tabata *et al.* (1965)

Procedure (Kuo and Lampen, 1971). Exponential-phase cells of *Saccharomyces* 1016 from modified Vogel's medium N (Vogel, 1956) are washed twice with distilled water and resuspended at a concentration of 4×10^8 cells/ml (8 mg dry-weight equivalent). To 5 ml of yeast suspension, the following additions are made: 0.05 M tris–HCl (pH 7.5), 1 ml; 1.2 M KCl containing 0.02 M MgSO$_4$, 6 ml; snail enzyme (Glusulase), 0.5 ml; and 1 M 2-mercaptoethanol, 0.2 ml. The mixtures are incubated at 30 °C on a reciprocal shaker (75 cpm with a 2.5-cm stroke); after 1 hour more than 95% of the cells are converted into protoplasts. The protoplasts are centrifuged at 1000 g for 3 minutes and washed twice with 5 mM MgSO$_4$ and 0.02 M potassium phosphate buffer (pH 6.0) containing 0.6 M KCl or 0.8 M sorbitol.

2. THE USE OF MICROBIAL ENZYMES

Since Salton (1955) reported the isolation of several actinomycetes and myxobacteria that showed lytic activity toward the cell wall of yeasts, enzymes from various microbial sources have been utilized for the preparation of protoplasts from yeasts, and some of them have been purified and crystallized (Table II). The formation of protoplasts using microbial enzymes appears to be dependent on the operation of enzymes similar to those present in snail enzyme.

Recently, two lytic endo-$\beta = (1 \rightarrow 3)$-glucanases have become commercially available. Zymolase 5000 can be obtained from Kirin Brewery Co., Ltd., Takasaki, Gunma, Japan, and yeast glucanase from Dr. S. Nagasaki, Department of Agricultural and Biological Chemistry, Kochi University, 200B Monobe, Nankoku, Kochi, Japan. These enzyme preparations readily produce protoplasts from the following species: *Saccharomyces* (six species), *Candida* (eight species), *Pichia* (three species), *Torulopsis* (two species), *Brettanomyces anomalus*, *Debaryomyces subglobus*, *Endomyces capsularis*, *Hansenula capsulata*, *Nematospora coryli*, and *Saccharomycodes ludwigii* (Kitamura *et al.*, 1971, 1972; Kaneko *et al.*, 1973, Yamamoto *et al.*, 1974).

Procedure. (Yamamoto *et al.*, 1974). *Saccharomyces cerevisiae* LK2G12 is grown overnight in 100 ml of Wickerham's medium with 2% glucose at 30 °C on a reciprocal shaker. This culture medium (10 ml) is transferred into 100 ml of fresh Wickerham medium with 2% glucose. Cells harvested in the exponential phase of growth (approximately 4 hours' growth) are washed once with water and twice with 0.02 M potassium phosphate buffer (pH 6.0) containing 0.6 M KCl (buffer). The washed cells are suspended in buffer at a concentration of 10 mg (dry-weight equivalent) per milliliter. The protoplast-forming mixture consisting of 10 ml of buffer, 5 ml of the cell suspension, 3 ml of water, and 2 ml of enzyme solution (100 μg of crystalline

TABLE II

LYTIC ENZYMES PRODUCED BY MICROORGANISMS

Microorganism	Active enzyme identified[a]	Reference
Rhizopus 69	I, C	Yamamoto *et al.* (1972)
Fungi imperfecti species	I, C	Yamamoto *et al.* (1974)
Arthrobacter luteus	I, C	Kitamura *et al.* (1971, 1972); Kaneko *et al.* (1973)
Arthrobacter YCWD-3	I	Doi *et al.* (1971)
Flavobacterium dormitator var. *glucanolyticae*	I, II	Yamamoto and Nagasaki (1971a,b)
Rhizopus sp.	I, III, VI	Kuroda *et al.* (1968a,b)
Basidomyces sp.	I	Mada *et al.* (1970, 1971a,b)
Basidomycetes QM 806	II, VII	Bauer *et al.* (1972)
Chaetomium thermophilum	II, VI, VII	Okazaki and Iizuka (1970, 1971)
Myriococcum albomyces	II, VI, VII	Okazaki and Iizuka (1970, 1971)
Thermomyces lanuginosus	II, VI, VII	Okazaki and Iizuka (1970, 1971)
Micropolyspora 434	?	
Bacillus circulans AKU0211	II, IV	Horikoshi and Sakaguchi (1958); Nagasaki *et al.* (1966); Nagasaki and Yamamoto (1968); McLellan and Lampen (1968)
Bacillus circulans WL-12	I, II, VI	Tanaka and Phaff (1965)
Oerskovia xanthineolytica	I, V, VII	Mann *et al.* (1972)
Streptomyces albidoflavus	G, VI, VIII	Tabata *et al.* (1965)
Streptomyces sp.	G, VI	Furuya and Ikeda (1960a–d)
Streptomyces GM	?	Mendoza and Villanueva (1962)
Streptomyces WL-6	U, III, VI	Tanaka and Phaff (1965); Schwencke *et al.* (1969).
Rhizopus arrhizus	?	Villanueva *et al.* (1966)
Physarium sp.	?	Kawakami and Kawakami (1971)
Micromonospora RA	G	Ochoa *et al.* (1963)
Sclerotinia libertiana	G, VIII	Satomura *et al.* (1960)
Corticium centrifugum	?	Sugimori *et al.* (1972)

[a] I, Endo-β-(1→3)-glucanase; II, exo-β-(1→3)-glucanase; III, β-(1→6)-glucanase; IV, phosphomannanase; V, α-mannanase; VI, protease; VII, chitinase; VIII, lipase and phospholipase; G, uncharacterized glucanase; C, enzyme being crystallized.

yeast glucanase from a fungi imperfecti species) is incubated at 30°C with gentle shaking. After 1 hour more than 95% of the cells have been converted to protoplasts. The protoplasts are centrifuged at 1000 *g* for 3 minutes, washed twice with buffer, and resuspended in the same buffer. The addition of 2-mercaptoethanol or dithiothreitol (0.01 *M*) to the incubation medium is required, particularly for the preparation of protoplasts from stationary-phase yeasts.

III.   Factors Affecting the Growth and Metabolic Activities of Protoplasts

A.   Osmotic Stabilizers

As in the preparation of protoplasts (see Section II,A, 2), a wide variety of inorganic salts and nonmetabolizable sugars has been employed as osmotic stabilizer by different investigators for the study of macromolecular synthesis by yeast protoplasts (see Tables III and IV). Hutchison and Hartwell (1967) established the conditions for the synthesis of various macromolecules by protoplasts. Essentially no RNA was synthesized in medium lacking an osmotic stabilizer. With 1 M sorbitol as the osmotic support, protoplasts were able to increase their net content of protein, RNA, and DNA approximately 5- to 6-fold on incubation for several hours. Kuo and Lampen (1971) reported that the synthesis and secretion of invertase by protoplasts of *Saccharomyces* are remarkably sensitive to the osmolarity of the supporting medium. As shown in Fig. 1, the rate of invertase (exo-enzyme) synthesis and the leakage of an intracellular enzyme, α-glucosidase,

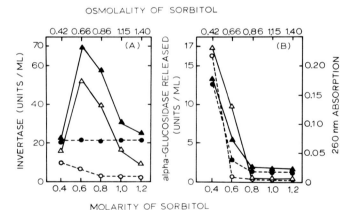

FIG. 1.   Formation and secretion of invertase, and release of α-glucosidase and 260 nm-absorbing materials from protoplasts suspended in various concentrations of sorbitol. Protoplasts were suspended at 5×10^7 per milliliter in the indicated concentrations of sorbitol with 10 mM fructose as the energy source at 30°C. (A) Invertase activity. Solid circles, zero time, total activity; solid triangles, 2 hours, total activity; open circles, zero time, supernatant fluid; open triangles, 2 hours, supernatant fluid. (B) α-Glucosidase and 260 nm-absorbing materials in the supernatant fluids. Open circles, zero time, α-glucosidase; open triangles, 2 hours α-glucosidase; solid circles, zero time, absorbancy at 260 nm; solid triangles, 2 hours absorbancy at 260 nm (Kuo and Lampen, 1971).

and of 260 nm-absorbing materials are both functions of the osmolarity of the medium. Synthesis of the enzyme was optimal at 0.65–0.75 osmolal and was severely reduced by increasing osmolarity. Essentially identical results were obtained with KCl or $MgSO_4$ as osmotic stabilizer. The reduction in invertase formation and secretion at high osmolarity was eliminated promptly when protoplasts were transferred to a medium or lower osmolarity. The inhibitory effect of high osmolarity may be the result of decreased penetration of sugar into the cells, since the reduction in invertase formation could be partially reversed by increasing the level of sugar supplied as an energy source. Thus changes in the permeability of the plasma membrane are important in the response of protoplasts to high osmolarity.

Ota (1972) reported that the regeneration of protoplasts from *Candida albicans* in solid agar medium was also markedly influenced by the nature and concentration of the stabilizer. Maximal regeneration of walled cells was obtained when the protoplasts were incubated with 0.68 M $MgSO_4$ and 0.5 M KCl, $MnSO_4$, sucrose, or NaCl.

B.   Nutrient Medium

The nutritional requirements for the growth or metabolic activities of protoplasts are generally the same as those for intact cells. However, freshly prepared protoplasts have suffered some membrane damage and are in a relatively "energy-poor" state. Incubation with an energy source such as glucose, in the presence of additional growth factors, permits membrane repair and the regeneration of metabolic energy stores to facilitate subsequent macromolecular synthesis. For the growth of protoplasts from *S. cerevisiae*, Shockman and Lampen (1962) employed a medium containing 5 mM $MgCl_2$, 20 mM ammonium phosphate buffer (pH 6.0), 0.1 M glucose, and (per liter): 2.0 mg of inositol, 0.2 mg of calcium pantothenate, 0.2 mg of pyridoxine · HCl, 0.2 mg of thiamine · HCl, and 5 μg of biotin. This level of vitamins was essential for good growth of protoplasts, growth being strikingly reduced if only one-tenth as much was used (M. S. Pollack and J. O. Lampen, unpublished results). Various metabolizable sugars, usually monosaccharides or disaccharides, have been used as the carbon and energy source for macromolecular synthesis by protoplasts. The formation of certain enzymes is sensitive to glucose repression as in intact cells, for instance, invertase synthesis and secretion by some strains of yeasts (Kuo and Lampen, 1971). For these experiments the sugar concentration should be kept as low as possible; alternatively, a nonrepressive carbon source such as raffinose, maltose, or mannose can be used. The phosphatases (acid and alkaline) of yeast protoplasts are actively synthesized only when the protoplasts are incubated in a high-glucose (100 mM), low-phosphate medium (Kuo and Lampen, 1972, 1974; Van Rijn *et al.*, 1972).

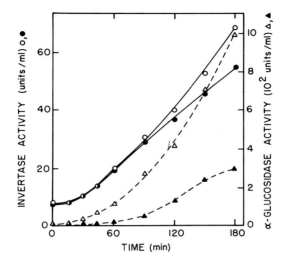

FIG. 2. Effect of amino acids on the formation of α-glucosidase and invertase by protoplasts. Protoplasts from cells grown in 0.2 M glucose medium were suspended at 5 × 10⁷ per milliliter in phosphate buffer containing 0.8 M sorbitol, 10 mM fructose, and with or without 20 mM maltose (for induction of α-glucosidase). Open symbols, 0.5% Casamino acids (Difco) added; solid symbols, no Casamino acids. After incubation at 30°C for various time intervals, 0.2-ml samples were removed and assayed for enzyme activity.

The induction of α-glucosidase by yeast protoplasts is unusual in that synthesis of this enzyme, induced by maltose, is greatly enhanced in the presence of an amino acid mixture. However, synthesis and secretion of invertase was not stimulated by amino acids during the first 2 hours of incubation (Fig. 2).

IV.   Growth and Macromolecular Synthesis

A.   Liquid Medium

Eddy and Williamson (1959) first described the growth and formation of aberrant cell walls by yeast protoplasts in nutrient media (1 M mannitol was the osmotic support). Protein and insoluble anthrone-positive carbohydrate increased severalfold during the first few hours. Most of the nitrogen of the aberrant walls was present as N-acetylglucosamine. In addition, cell wall materials, proteins (enzymes), and nucleic acids are actively formed by protoplasts under appropriate conditions (Table III). Recent studies in our laboratory on the synthesis of cell wall materials by

TABLE III

MACROMOLECULAR SYNTHESIS BY YEAST PROTOPLASTS IN LIQUID MEDIUM

Process studied	Stabilizing medium	Organism	Reference
Morphology, wall material synthesis	1.0 M mannitol (complex malt extract or synthetic medium)	*S. carlsbergensis*	Eddy and Williamson (1959)
Morphology, growth (turbidity)	0.6 M KCl, 0.8 M mannitol, or 1.0 M sorbitol (synthetic medium)	*S. cerevisiae*	Shockman and Lampen (1962)
Glycolysis, invertase formation and secretion	0.6 M KCl (synthetic medium)	*S. cerevisiae*	Islam and Lampen (1962)
DNA and protein synthesis	0.7 M mannitol, (complex medium)	*S. cerevisiae, S. carlsbergensis*	Millbank (1964)
Respiratory and fermentative activities, DNA and RNA synthesis	0.5 M NaCl, (synthetic medium)	*S. cerevisiae*	Tabata *et al.* (1965)
DNA, RNA, and protein synthesis	1.0 M sorbitol (complex medium)	*S. cerevisiae*	Hutchison and Hartwell (1967)
α-Glucosidase synthesis	0.6 M KCl, (synthetic medium)	*S. cerevisiae*	Van Wijk *et al.* (1969)
Acid and alkaline phosphatase synthesis	12% mannitol (synthetic medium)	*S. cerevisiae*	Van Rijn *et al.* (1972)
Cytochrome oxidase synthesis	0.67 M sorbitol (synthetic medium)	*S. carlsbergensis*	Charalampous and Chen (1974)
Synthesis of RNA and protein (including invertase, α-glucosidase, acid and alkaline phosphatase), cell wall materials, mannan, glucan, and chitin	0.8 M sorbitol (synthetic medium)	*Saccharomyces* sp.	Kuo and Lampen (1971, 1972, 1974; Kuo *et al.* (1973)

TABLE IV

INCORPORATION OF MANNOSE-^{14}C OR GLUCOSAMINE-^{3}H
INTO MANNAN, GLUCAN, AND CHITIN BY PROTOPLASTS[a]

Precursor	Medium	Protoplasts		
	Mannan	Mannan	Glucan	Chitin
Mannose-^{14}C	3442	1446	1649	351
Glucosamine-^{3}H	1077	408	107	2983

[a] Protoplasts (5 × 10^7/ml) in modified Vogel's medium N were preincubated in 40 mM mannose for 60 minutes. The protoplast suspension was then divided into two portions; one received mannose-^{14}C (0.2 μCi/ml), and the other 1mM glucosamine-^{3}H (2 μCi/ml). After 60 minutes samples were removed, and incorporation of radioactivity into the wall polymers was measured as described by Kuo and Lampen (1972, 1974). Essentially no glucan and chitin was detected in the medium. Values given are in counts per minute per milliliter.

yeast protoplasts revealed that glucan and chitin are formed and retained by the protoplasts, but most of the mannan is secreted into the medium (Kuo and Lampen, 1972, 1974) (Table IV). Glycoprotein enzymes such as invertase of acid phosphatase are synthesized and liberated into the medium, whereas nonglycoprotein enzymes, α-glucosidase, and alkaline phosphatase are retained by the protoplasts (Kuo and Lampen, 1972).

The procedures used in our laboratory for protoplasts from *Saccharomyces* sp. 1016 are as follows. Protoplasts prepared with snail enzyme (see Section II,B,1) are centrifuged at 1000 g for 3 minutes and washed twice at 2 × 10^8 per milliliter with either 0.02 M phosphate buffer (pH 6.0) and 5 mM MgSO$_4$ containing 0.8 M sorbitol, or Vogel's medium containing 0.8 M sorbitol. Washed protoplasts are suspended at a concentration of 6 × 10^6 per milliliter (for growth experiments), or 5 × 10^7 per milliliter (for macromolecular synthesis), in 0.8 M sorbitol solution containing an appropriate energy source. The suspensions are incubated at 30°C on a reciprocal shaker. Growth is evaluated principally by the measurement of increases in protein (Shockman and Lampen, 1962); turbidity measurements are less valid because of the tendency of growing protoplasts to aggregate. Growth can also be evaluated microscopically for the qualitative presence of enlarged protoplast forms. For measurement of the enzyme that is synthesized and released into the medium, 1 ml of the incubation mixture is withdrawn at intervals and centrifuged at 1000 g for 3 minutes; 0.2 ml of the supernatant fluid is pipetted into 0.8 ml of ice-cold water and assayed for enzyme activity. For determination of total enzyme synthesis, samples are periodically removed from the incubation mixture and transferred to chilled tubes containing ice-cold water to lyse the protoplasts. The resulting suspensions are assayed for enzyme activity (see Figs. 1 and 2).

B. Solid Medium

Nečas and co-workers (Nečas, 1961; Svoboda and Nečas, 1966) first demonstrated that yeast protoplasts have not only the ability to grow but also the capacity to regenerate walled cells when they are embedded in gelatin medium containing an osmotic stabilizer (Table V). Nearly all the protoplasts regenerate into normal cells after a 24-hour incubation. No formation of microcolonies is found when the concentration of gelatin is less than 15% (at 22°C), or if the protoplasts are merely growing on the surface of the gelatin.

Regeneration of walled cells from protoplasts can also be obtained in agar gels. According to Svoboda (1966), protoplasts were embedded at 10^4 to 10^6 per milliliter in agar gel (1.3–2.0%) in the presence of nutrient medium of 0.8 M mannitol. After 24 hours microcolonies of normal yeast cells were found. The highest frequency of regeneration, 50–70%, was found in 2% agar; with 1.3% agar only 5% regeneration occurred.

Fukui et al. (1969) established the so-called thin-layer agar method for the regeneration of protoplasts from Geotrichum candidium. Ota (1972), from Fukui's laboratory, extended the method for the study of protoplasts from C. albicans. Protoplasts were suspended at 1.5×10^4 per milliliter in citrate–phosphate buffer (5 mM, pH 6.0) containing 0.75 M MgSO$_4$. 0.1 ml

TABLE V

GROWTH AND CELL WALL REGENERATION OF PROTOPLASTS IN SOLID MEDIUM

Osmotic stabilizer	Solid support	Organism	Reference
0.8 M mannitol (synthetic medium)	15–30% gelatin	S. cerevisiae	Nečas (1961); Svoboda and Nečas (1966)
0.8 M mannitol (synthetic medium)	2% agar	S. cerevisiae, S. carlsbergensis, C. utilis	Svoboda (1966)
0.6 M KCl (complex medium)	30% gelatin	Schiz. pombe	Havelkova (1969)
0.75 M NaCl (synthetic medium)	2% agar	G. candidium	Fukui et al. (1969)
0.75 M MgSO$_4$ (synthetic medium)	2% agar	C. albicans	Ota (1972)

of the protoplast suspension was then put in the center of a petri dish, and 0.5 ml of citrate–phosphate buffer containing 2% agar and 0.75 M NaCl or $MgSO_4$ as osmotic stabilizer (at 52 °C) was placed just beside it. The two droplets were quickly mixed and spread out as a thin-layer agar using a thin bent glass rod at 42 °C. The protoplasts embedded in the agar were then overlayed with Sabouraud's medium containing glucose and yeast extract (basal YEP medium), and with 0.75 M $MgSO_4$ as osmotic stabilizer. The plates were incubated at 30 °C for 7 hours and then gently washed twice for 15 minutes by pouring sterilized water onto the agar layer. This procedure bursts osmotically sensitive cells that have not resynthesized an osmobarrier. The residual cells were then overlayed with nonstabilizing basal YEP medium and kept at 30 °C for 48 hours until colonies developed. Ota (1972) has reported that with this procedure approximately 70–90% of the protoplasts regenerated and developed into colonies within 48 hours.

ACKNOWLEDGMENTS

We thank Dr. J. Oliver Lampen for his critical reading of the manuscript and his stimulating comments. Part of the work (S.-C. K) was supported by Public Health Service Grant AI-04572 from the National Institute of Allergy and Infectious Diseases.

REFERENCES

Bauer, H., Bush, D. A., and Horisberger, M. (1972). *Experientia* **28**, 11.
Berliner, M. D., and Reca, M. E. (1970). *Science* **167**, 1255.
Brenner, S., Dark, F. A., Gerhardt, P., Joynes, M. H., Klienberger-Nobel, E., McQuillen, K., Roio-Hucetos, M., Salton, M. R. J., Strange, R. E., Tomcsik, J., and Weibull, C., (1958). *Nature (London)* **181**, 1713.
Burger, M., Bacon, E. E., and Bacon, J. S. D. (1961). *Biochem. J.* **78**, 504.
Charalampous, F. C., and Chen, W. L. (1974). *J. Biol. Chem.* **249**, 1007.
Davis, R., and Elvin, P. A. (1964). *Biochem. J.* **93**, 8P.
Deutch, C. E., and Perry, J. M. (1974). *J. Gen. Microbiol.* **80**, 259.
Doi, K., Doi, A., and Fukui, T. (1971). *J. Biochem. (Tokyo)* **70**, 711.
Duell, E. A., Inoue, S., and Utter, M. F. (1964). *J. Bacteriol.* **88**, 1762.
Eddy, A. A., and Williamson, D. H. (1957). *Nature (London)* **179**, 1252.
Eddy, A. A., and Williamson, D. H. (1959). *Nature (London)* **183**, 1101.
Falcone, G., and Nickerson, W. J. (1956). *Science* **124**, 272.
Fukui, K., Sagara, Y., Yoshida, N., and Matsuoka, T. (1969). *J. Bacteriol.* **98**, 256.
Furuya, A., and Ikeda, Y. (1960a). *Nippon Nogei Kagaku Kaishi* **34**, 33.
Furuya, A., and Ikeda, Y. (1960b). *Nippon Nogei Kagaku Kaishi* **34**, 38.
Furuya, A., and Ikeda, Y. (1960c). *J. Gen. Appl. Microbiol.* **6**, 40.
Furuya, A., and Ikeda, Y. (1960d). *J. Gen. Appl. Microbiol.* **6**, 49.
Gascon, G., and Villanueva, J. R. (1965). *Nature (London)* **205**, 822.
Havelkova, M. (1969). *Folia Microbiol. (Prague)* **14**, 155.
Holden, M., and Tracey, M. V. (1950). *Biochem. J.* **47**, 407.
Holter, H., and Ottolenghi, P. (1960). *C. R. Trav. Lab. Carlsberg* **31**, 409.
Horikoshi, K., and Sakaguchi, K. (1958). *J. Gen. Appl. Microbiol.* **1**, 1.
Hutchison, H. T., and Hartwell, L. H. (1967). *J. Bacteriol.* **94**, 1697.

Islam, M. F., and Lampen, J. O. (1962). *Biochim. Biophys. Acta* **58**, 294.
Kaneko, T., Kitamura, K., and Yamamoto, Y. (1973). *Agr. Biol. Chem.* **37**, 2295.
Kawakami, N., and Kawakami, H. (1971). *J. Ferment. Technol.* **49**, 479.
Kitamura, K., Kaneko, T., and Yamamoto, Y. (1971). *Arch. Biochem. Biophys.* **145**, 402.
Kitamura, K., Kaneko, T., and Yamamoto, Y. (1972). *J. Gen. Appl. Microbiol.* **18**, 57.
Kovac, L., Bednarova, H., and Greksak, M. (1968). *Biochim. Biophys. Acta* **153**, 32.
Kuo, S.-C., and Lampen, J. O. (1971). *J. Bacteriol.* **106**, 183.
Kuo, S.-C., and Lampen, J. O. (1972). *J. Bacteriol.* **111**, 419.
Kuo, S.-C., and Lampen, J. O. (1974). *Biochem. Biophys. Res. Commun.* **58**, 287.
Kuo, S.-C., Cano, F. R., and Lampen, J. O. (1973). *Antimicrob. Ag. Chemother.* **3**, 716.
Kuroda, A., Tokumara, A., and Tawada, T. (1968a). *J. Ferment. Technol.* **46**, 926.
Kuroda, A., Tawada, T., and Tokumaru, A. (1968b). *J. Ferment. Technol.* **46**, 930.
Lillenhoj, E. B., and Ottolenghi, P. (1967). *In* "Symposium on Yeast Protoplasts" (R. Muller, ed.), p. 145. Akademie-Verlag, Berlin.
Longley, R. P., Rose, A. H., and Knithts, B. A. (1968). *Biochem. J.* **108**, 401.
McLellan, W. L., and Lampen, J. O. (1968). *J. Bacteriol.* **95**, 967.
McLellan, W. L., McDaniel, L. E., and Lampen, J. O. (1970). *J. Bacteriol.* **102**, 261.
Mada, M., Hirao, K., Kimura, Y., and Noda, K. (1970). *Nippon Nogei Kagaku Kaishi* **44**, 393.
Mada, M., Hirao, K., Kimura, Y., and Noda, K. (1971a). *Nippon Nogei Kagaku Kaishi* **45**, 260.
Mada, M., Hirao, K., Kimura, Y., and Noda, K. (1971b). *Nippon Nogei Kagaku Kaishi* **45**, 269.
Mann, J. W., Heintz, C. E., and MacMillan, J. D. (1972). *J. Bacteriol.* **111**, 821.
Mendoza, C. G., and Ledien, M. N. (1968). *Nature (London)* **220**, 1035.
Mendoza, C. G., and Villanueva, J. R. (1962). *Nature (London)* **195**, 1326.
Mendoza, C. G., and Villanueva, J. R. (1964). *Nature (London)* **202**, 1241.
Millbank, J. W. (1964). *Exp. Cell Res.* **35**, 77.
Monreal, J., De Uruburu, F., and Villanueva, J. R. (1967). *J. Bacteriol.* **94**, 241.
Nagasaki, S., and Yamamoto, S. (1968). *Res. Rep. Kochi Univ.* **17**, 93.
Nagasaki, S., Neumann, N. P., Arnow, P., Schnable, L. D., and Lampen, J. O. (1966). *Biochem. Biophys. Res. Commun.* **25**, 158.
Nečas, O. (1956). *Nature (London)* **177**, 898.
Nečas, O. (1961). *Nature (London)* **192**, 580.
Nečas, O. (1971). *Bacteriol. Rev.* **35**, 149.
Ochoa, A. G., Acha, G. I., Gascon, G., and Villanueva, J. R. (1963). *Experientia* **19**, 581.
Okazaki, H., and Iizuka, H. (1970). *J. Ferment. Technol.* **50**, 228.
Okazaki, H., and Iizuka, H. (1971). *Nippon Nogei Kagaku Kaishi* **45**, 461.
Ota, F. (1972). *Jap. J. Microbiol.* **16**, 359.
Ottolenghi, P. (1966). *C. R. Trav. Lab. Carlsberg* **35**, 363.
Rost, K., and Venner, H. (1965a). *Arch. Mikrobiol.* **51**, 122.
Rost, K., and Venner, H. (1965b). *Arch. Mikrobiol.* **51**, 130.
Russell, I., Garrison, F., and Stewart, G. G. (1973). *J. Inst. Brew., London* **79**, 48.
Salton, M. R. J. (1955). *J. Gen. Microbiol.* **12**, 25.
Satomura, Y., Ono, M., and Fukumoto, J. (1960). *Bull. Agr. Chem. Soc. Jap.* **24**, 313.
Schwencke, J., Gonzales, G., and Farias, G. (1969). *J. Inst. Brew., London* **75**, 15.
Shockman, G. D., and Lampen, J. P. (1962). *J. Bacteriol.* **84**, 508.
Sommer, A., and Lewis, M. J. (1971). *J. Gen. Microbiol.* **68**, 327.
Sugimori, T., Uchida, Y., and Tsukada, Y. (1972). *Agr. Biol. Chem.* **36**, 669.
Svihla, G., Schlenk, F., and Dainko, J. L. (1961). *J. Bacteriol.* **82**, 808.

Svoboda, A. (1966). *Exp. Cell Res.* **44**, 640.
Svoboda, A., and Nečăs, O. (1966). *Nature (London)* **210**, 845.
Tabata, S., Imai, T., and Terui, G. (1965). *J. Ferment. Technol.* **43**, 221.
Tanaka, H., and Phaff, H. J. (1965). *J. Bacteriol.* **89**, 1570.
Van Rijn, H. J. M., Boer, P., and Steyn-Parve, E. P. (1972). *Biochim. Biophys. Acta* **268**, 431.
Van Wijk, R., Duwehand, J., Van Den Bos, T., and Koningsberger, V. V. (1969). *Biochim. Biophys. Acta* **186**, 178.
Villanueva, J. R., Elorza, M. V., Monreal, J., and Uruburu, F. (1966). *Proc. Symp. Yeasts, 2nd, 1965*, p. 203.
Vogel, H. J. (1956). *Microbiol. Genet. Bull.* **13**, 42.
Yamamoto, S., and Nagasaki, S. (1971a). *J. Ferment. Technol.* **50**, 117.
Yamamoto, S., and Nagasaki, S. (1971b). *J. Ferment. Technol.* **50**, 127.
Yamamoto, S., Shiraishi, T., and Nagasaki, S. (1972). *Biochem. Biophys. Res. Commun.* **46**, 1802.
Yamamoto, S., Fukuyama, J., and Nagasaki, S. (1974). *Agr. Biol. Chem.* **38**, 329.

Chapter 9

Dissecting Yeast Asci without a Micromanipulator

P. MUNZ

Institute of General Microbiology,
University of Bern, Bern,
Switzerland

The genetic characterization of the progeny of individual meioses may be interesting for various reasons. In organisms such as the yeasts *Saccharomyces cerevisiae* and *Schizosaccharomyces pombe*, the task of separating the four meiotic spores is usually accomplished with the aid of a micromanipulator. An inexpensive micromanipulator has been designed by Sherman (1973).

Yet it is also possible to do tetrad analysis by free hand. This method may be appreciated by workers having no access to a micromanipulator; it is also helpful in solving additional experimental problems in yeast genetics.

This free-hand method has been described for *S. cerevisiae* (Munz, 1973). Here attention is given to *Schiz. pombe*; the procedure is practically the same for both organisms.

Yeast extract agar plates are allowed to solidify in an exactly horizontal position. This minimizes refocusing of the microscope during the actual dissection work. Plastic petri dishes are preferred. The constituents of the medium should be particle-free; Difco products (yeast extract and agar) have proved satisfactory.

In order to place the spores of individual tetrads at specific positions, we use a device consisting of eight blades with four gaps, each at regular intervals (gap width 1 mm; Fig. 1). In the corresponding pattern produced by slightly pressing the agar surface with this instrument, the spores are placed at the gaps. Alternatively, plates may be prepared individually by inscribing four vertical and eight intersecting horizontal lines on the agar. Spores are then placed at the intersections.

Sporulated cell material is suspended in water, diluted if necessary, and streaked on the left part of the plate. A gradient in cell density will arise if the plate is slanted immediately after streaking, such that the suspension will tend to flow to the far left. However, the material may also be placed directly and distributed with a loop, so as to create some areas of reasonably low cell density.

Glass tubes (25 cm long, 6 mm diameter) are heated in the middle and pulled out to give two pipettes resembling Pasteur pipettes. Next, the capillary section is carefully heated on a small flame and pulled out. The resulting microcapillary is broken at a suitable point to give an orifice approximately 0.06 mm in diameter. A good check is to blow air forcefully through the pipette, holding the tip in ethanol; a *single* file of small air bubbles should be observed. Not every tip will prove satisfactory, but since the starting material is inexpensive and little time is required to prepare pipettes, this presents no problem.

A glass rod with a rounded tip is prepared having overall dimensions similar to the pipette (Fig. 1).

The agar plate is placed under the microscope (e.g., objective $10 \times$, eyepieces $10 \times$; distance from surface to objective 5 mm), and an isolated ascus brought into the field. The pipette is connected with a piece of plastic tubing, the free end of which is held in the mouth. The arm holding the pipette is supported by an armrest. The tip of the pipette is brought into the microscope field, gently lowered into a cell-free area and filled with liquid by capillary action. Then the ascus is touched and picked up, and the tip slightly lifted. The tip *normally* more or less moves around under the microscope. The other hand moves the plate so as to bring the first agar mark into view. Then the tip is lowered again, and the ascus blown out *without* dis-

Fig. 1. Instruments used for free-hand ascus dissection. Device for inscribing marks on agar surface (lower left; see also text). Microrod (left) and microneedle (right) lying on empty petri dish and blackened for better visualization, and piece of plastic tubing.

pensing the entire capillary liquid. If the entire fluid is blown out, the transferred unit (ascus or spore) is often lost as the result of an explosionlike
discharge. Additional asci are then placed at other predetermined positions.

The plates are incubated 4–6 hours at 30°C to allow spontaneous disintegration of ascus walls. Then the glass rod is used to separate the four
spores, and the pipette once more to transfer three of them to additional
preassigned places on the plate. After incubation for 3–5 days at 30°C, single
spores give rise to clones. Individual colonies are well spaced, so that
material can be easily picked for subculturing (Fig. 2).

No sterility problems are encountered with these microinstruments. The
tip of the pipette is occasionally washed in ethanol, and the tip of the rod
is cleaned by stabbing repeatedly in a sterile agar layer.

Beginners should invest a small amount of time per day in acquiring the
technique, and try first simply to touch a cell with a glass rod, then to pick
up a cell and place it back at the same position, and only last to dissect
tetrads. Some workers will perhaps be deterred by the seemingly insurmountable task of holding instruments reasonably still. The first trials are
most disappointing indeed, yet it is only a matter of practice.

Micromanipulators are by no means obsolete; anyone well trained

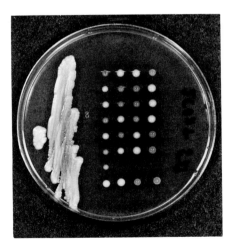

FIG. 2. Result of free-hand tetrad dissections. This *Schiz. pombe* cross involved the wild-type
strain (mating type *h⁺*) and an adenine-requiring strain (*ade6*, mating type *h⁻*) which accumulates a red pigment on media with suboptimal adenine concentrations. Regular 2:2
segregations are observed in seven asci (light, wild type; dark, adenine-dependent). The wall
of the seventh ascus had not dissolved after an appropriate incubation time, and dissection
work was discontinued with this ascus.

in this free-hand technique will be amazed at the ease with which asci can be dissected and, as a consequence, any single vegetative cell or random spore manipulated.

REFERENCES

Munz, P. (1973). *Experientia* **29**, 251–252.
Sherman, F. (1973). *Appl. Microbiol.* **26**, 829.

Chapter 10

Use of Micromanipulators in Yeast Studies

FRED SHERMAN

Department of Radiation Biology and Biophysics,
University of Rochester School of Medicine and Dentistry,
Rochester, New York

I. Introduction

Micromanipulators are used in yeast genetic studies not only for the separation of ascospores from individual asci (see Mortimer and Hawthorne, Chapter 12, this volume), but also for the separation of zygotes from mass-mating mixtures and in special cases for positioning of vegetative cells and spores for mating purposes and for single-cell analyses. The dissection of asci by free hand, not requiring micromanipulators, is dealt with separately in this volume (Munz, Chapter 9).

II. Apparatus

A. Micromanipulators

Over 200 different types of micromanipulators have been described since the middle of the last century, but only a relatively few are commercially available; several of these are discussed in detail by El-Badry (1963). Unfortunately, most of the commercially available models have not been systematically evaluated for the specific requirements for studies with yeasts, and only several types are in common use.

Micromanipulators usually can be subdivided into two major types: those having three separate controls for positioning the microtool, and those having a single lever or joystick which freely controls the microtool either in three dimensions of space or in one plane. Micromanipulators having three separate controls depend on rack-and-pinion or feed-screw movements, and they lack maneuverability since one has to change constantly from one control to another during the manipulation. Simultaneous rotation of three knobs is required to produce diagonal movements not in the principal axes of the instrument. While these micromanipulators are low in cost, they are not recommended for studies with yeasts, although they have been used by some investigators for dissection of asci.

Micromanipulators that operate with control levers or joysticks can translate hand movements into synchronously reduced movements of microtools. Most of the instruments were designed so that the movement in the horizontal plane is directly related to the movement of the control handle, while the vertical movement is determined by rotating a knob located either on or near the horizontal control handle. Transmission of the motions is based on pneumatic principles (de Fonbrune, 1949), hydraulic principles (Cailloux, 1943; May, 1953), piezoelectric principles (Ellis, 1962), thermal expansion of electrically heated wires (Bush et al., 1953; the B-D-H micromanipulator previously marketed by the American Optical Co.), or on several ingenious mechanical principles involving direct coupling to sliding components. Mechanical micromanipulators include several models designed by Emerson (1931), the Singer single-control sliding micromanipulator (Barer and Sauders-Singer, 1948), the Leitz lever-activated micromanipulator (see El-Badry, 1963), and the Zeiss sliding micromanipulator (see El-Badry, 1963). Most of the commercially available micromanipulators having single control levers are listed in Table I, along with distributors and prices in the United States. Of these, only the de Fonbrune Series A and the Singer MKIII micromanipulators are commonly used for yeast genetics studies.

The de Fonbrune Series A micromanipulator, shown in Fig. 1, pneumatically transmits a fine degree of motion from a single joystick. The micromanipulator consists of two basic parts, a control unit with a joystick operating three piston pumps, and a receiver with a lever holding the microtool which is actuated by the pneumatic pressure of the pumps by way of

TABLE I

SELECTIVE LIST OF COMMERCIALLY AVAILABLE MICROMANIPULATORS WITH SINGLE-LEVER CONTROLS

Micromanipulator	Distributor	Catalog number	Price[a] ($)
de Fonbrune Series A micromanipulator	Beaudouin (Paris, France)	—	—
	Orion Research Incorp. (Cambridge, Mass.)	50001	3575
	Curtin Matheson Scientific Inc. (Wayne, N.J.)	266–320	2018
Singer MKIII micromanipulator	Singer Instrument Co. Ltd. (Reading, Berkshire, England)	—	2530
	Eric Sobotka Co., Inc. (Sarasota, Fla.)	—	1050
Cailloux micromanipulator	C. H. Stoelting Co. (Chicago, Ill.)		
Emerson Model B micromanipulator	A. H. Thomas Co. (Philadelphia, Pa.)	6592-M10	825
Leitz lever-activated micromanipulator	E. Leitz (Rockleigh, N.J.)	520137	1957
Zeiss sliding micromanipulator	International Micro Optics (Fairfield, N.J.)	300017:020.20/A	1250
de Fonbrune Series B micromanipulator	Beaudouin (Paris, France)	—	—
	Orion Research, Inc. (Cambridge, Mass.)	54001	1495
de Fonbrune Series C micromanipulator	Beaudouin (Paris, France)	—	—
	Orion Research, Inc. (Cambridge, Mass.)	51001	3150
Abacus Model Amp-1 micropositioners	Alessi Industries (Santa Ana, Calif.)	—	225
Motorized micromanipulator	Eric Sobotka Co., Inc. (Sarasota, Fla.)	MM 33 M + CM-2	720

[a] Current prices in the United States (June 1974).

Fig. 1. The de Fonbrune Series A micromanipulator. The control unit (right) and the receiver (left) are interconnected by flexible tubing on opposite sides of a Leitz Laborlux II microscope. The complete assembly is on a vibration eliminator (Vibration Damping Mount, Cat. No. 9705/601, The Laboratory Apparatus Co., Cleveland, Ohio).

three diaphragms or aneroids. A sliding adjustment ring changes the length of the levers controlling the horizontal movements, making it possible to change the ratio of movement between the joystick and the microtool from approximately 1:40 to 1:2000. The instrument is manufactured in France by Beaudoiun and is distributed in the United States by Orion Research, Inc. A similar, but not completely identical instrument, the Sensaur micromanipulator, is manufactured in the United States and is distributed by Curtin Matheson Scientific, Inc. at a lower price (Table I).

The Singer MKIII micromanipulator, shown in Fig. 2, is a robust mechanical micromanipulator used for yeast genetic studies in several laboratories in England. In the Singer MKIII micromanipulator, as in the de Fonbrune micromanipulator, lateral hand movements control the horizontal movements of the microtool, while rotation of the same lever controls vertical movements. Also as in the de Fonbrune micromanipulator, the ratio of movement is variable.

Other commercially available micromanipulators with single-lever controls have capabilities similar to those of the de Fonbrune Series A and the

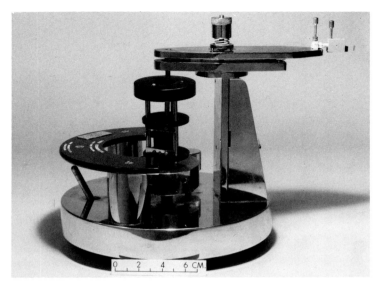

FIG. 2. The Singer MKIII micromanipulator. The joystick, seen in the center of the figure, transmits corresponding movements in the horizontal plane to the microtool, while rotation of the same control produces vertical movements. A semicircular hand rest can be seen at the left of the figure above the coarse-control lever. The micromanipulator has been modified to accommodate the microneedle holder (C in Fig. 4) seen at the right of the figure.

Singer MKIII micromanipulators, and there is no reason to doubt that they could be employed in studies of yeast. The Cailloux micromanipulator has pneumatic controls similar to those of the de Fonbrune Series A micromanipulator but has a fixed ratio of movement of approximately 1:90. The de Fonbrune Series B and Series C are smaller than the Series A units, and they can be mounted directly on a microscope stage, which results in less vibration. Several models of the Emerson mechanical micromanipulator are commercially available for use in biological reserch (Model B) and in probe testing and inspection of electronic circuits (Abacus Model Amp-1 micropositioner). The Leitz lever-activated micromanipulator and the Zeiss sliding micromanipulator have been used extensively for micrurgical studies on living cells.

In addition, there are two mechanical micromanipulators, shown in Figs. 3 and 4, that were specifically designed for dissection of asci. One designed by Dr. R. K. Mortimer approximately 20 years ago (see Burns, 1956) consists of two components: a needle holder or micromanipulator (Fig. 3), and a very fine mechanical stage or "two-axes" stage used to move the chamber during the dissection. The micromanipulator is basically like Barber's single pipette holder which was described at the beginning of the century (see Chambers, 1918). While other fine-microscope stages have been used with the micromanipulator, the two-axes stage is an important feature of the unit. One complete turn of the micrometer wheels causes a movement of 1 mm of the chamber. During the operation the cells are positioned in the horizontal plane by the two controls on the stage, while the vertical position is controlled by a screw (d in Fig. 3) on the micromanipulator. Vibration is at a minimum, since the micromanipulator is clamped directly to the stage. The unit is used extensively in the United States, and the micromanipulator and the two-axes stage can be purchased from Lawrence Precision Machine (1416 Winton Avenue, Hayward, Calif.) for, respectively, $250 and $700.

Recently, an inexpensive micromanipulator (Fig. 4), primarily intended for dissection of asci, was described by Sherman (1973). Its components can be purchased for approximately $125, and it requires little custom machining. The joystick controls the fine Y and Z motion, and a knurled knob controls the X motion. Coarse adjustments are made with a microclamp (C) and a dovetail slide (B). Micromanipulators are either directly fastened to the base of the microscope with a custom-made C-clamp (D), or attached to the microscope stage with a bracket (H). The mechanical stage of the microscope should have left-handed controls. The direct coupling of the micromanipulator to the microscope minimizes vibration. The low price makes

FIG. 3. The micromanipulator manufactured by Lawrence Precision Machine. The micromanipulator is attached with a knurled screw (a) to either a special two-axes microscope stage or to a suitable graduated microscope stage with fine controls. The microneedle (b) is secured with a screw (c). Horizontal movements are controlled by knobs e and f, while the vertical movement is controlled by knob d.

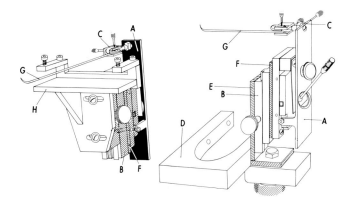

FIG. 4. Two inexpensive mechanical micromanipulators (from Sherman, 1973). These manipulators were assembled from the following commercially available components: A, micromanipulator (Model MPM-0-100), Affiliated Manufacturers, Inc., Whitehouse, N.J.; B, dovetail slide (Cat. No. SA-808), C. H. Stoelting Co., Chicago, Ill.; C, microclamp (Cat. No. V58096A), Curtin Matheson Scientific, Inc., Wayne, N. J. Also shown are the simple brass plate (F) and the glass microneedle (G). The top edge of the manipulator (A) base was cut off, and the microclamp (C) was mounted on a 6-32 machine screw which had had some of its threads removed. The micromanipulators were designed to be attached to either a microscope stage (left) by a custom-made bracket (H), or to the base of a Leitz Laborlux II microscope (right), which has a fixed object stage, by a custom-made C-clamp (D) and a bracket (E).

this model ideal for class instruction when large numbers are required.

A simple and inexpensive hydraulic micromanipulator was designed by Tomlinson (1967). Three large syringes arranged at right angles to each other are connected to three smaller syringes in the same configuration. Hand movement of the plungers of the small syringes is transmitted to the plungers of the large syringes, the ratio of movement being the same as the ratio of the volumes of the two sizes of syringes. While the capability and precision are not as high as those of the instruments described above, performance tests suggests that it may be adequate for yeast studies.

It is difficult to recommend objectively one specific micromanipulator, since most workers strongly prefer the model with which they have had the most experience. The de Fonbrune Series A, the Singer MKIII, and other micromanipulators not attached directly to the microscope stage require more space, and in some instances heavy base plates or vibration eliminators (see Fig. 1). It has been our experience that the skill of asci dissection can be taught more quickly with the de Fonbrune micromanipulator. Although the cost may be prohibitive, the de Fonbrune Series A micromanipulator, with the two-axes stage (Lawrence Precision Machine), provides a convenient and rapid means for dissection of asci and for other micromanipulation of yeast cells.

B. Microscopes

It should be stressed that microscope requirements depend on the type of micromanipulator. Micromanipulators not directly attached to the stage should be used in conjunction with microscopes having fixed stages and tube focusing, such as the Leitz Laborlux II microscope. Fine mechanical stages with graduations are essential with all micromanipulators. It is convenient to have long working distance objectives for magnifications in the range of 150–300×. Some workers have used stereomicroscopes which at magnifications of 250 × have working distances of 28 mm.

C. Microneedles

The separation of ascospores, zygotes, and vegetative yeast cells has been carried out with glass microneedles, microloops, and micropipettes. Although the transport of vegetative cells is more easily performed with microloops and micropipettes, simple microneedles are sufficient for most purposes.

While some workers rely on a microforge (see Thaysen and Morris, 1947; de Fonbrune, 1949; El-Badry, 1963) for constructing microneedles, a simple microneedle for separating ascospores and cells can be easily made by hand, using the small flame from the pilot light of an ordinary bunsen burner. A useful microburner, specifically designed for preparing glass microtools by hand is commercially available from Zeiss (Jena, Germany). A 2-mm-diameter glass rod is drawn out to a fine tip with the bunsen burner; by using the pilot burner, an even finer tip is drawn out at a right angle with an auxiliary piece of glass rod. The end is cut off so that the tip has a diameter of 10–100 μm and a length of a few millimeters. The exact size is not important, and various investigators have different preferences. Manipulation in higher densities of cells is more conveniently carried out with microneedles having tips of smaller diameter, while spores are more readily picked up and transferred with microneedles having tips of larger diameter. The length of the perpendicular end should be compatible with the height of the chamber; too short an end may result in optical distortions from the main shank of the microneedle. Several trials are usually required to produce an end with the correct degree of taper. The drawn-out tip can be cut with microscissors, or broken between a glass slide and the edge of a cover slip. Belkin (1928) has described a "microguillotine" for breaking off the tips of microtools in a reliable and precise manner.

D.   Chambers

Micromanipulation is usually carried out on thin slabs of agar mounted within a chamber. However, the same procedures can be implemented directly on the surface of ordinary petri plates filled with nutrient medium. A few investigators have constructed special rigs for holding petri plates, face down, on microscope stages. Several workers in England dissect asci on petri plates, face up, with microloops, using the Singer MKIII micro-manipulator. Nevertheless, chambers can be positioned with all commercially available microscope stages, and their use lessens the chance of contamination.

A simple but adequate chamber for dissection of asci and micromanipulation of yeast cells can be built from a U-shaped metal frame having outside dimensions of 3 × 1.5 inches and a thickness of 3/8 inch. The bottom of the frame is permanently closed by a 3 × 1.5 inch glass slide attached with epoxy glue, and the top is covered by a glass slide carrying the agar slab and held in position with petroleum jelly. The dissecting needle enters the chamber through the open end of the U-frame. A chamber can be seen on the microscope stage in Fig. 1. If extended periods of micromanipulation are anticipated, dessication of the agar can be prevented by lining the sides of the chamber with moist filter paper or wicks. Agar slabs are simply prepared by cutting approximately 2.5 × 6 cm rectangular areas from thin agar plates; the agar slabs are placed in the center of the glass slides, making sure that no air bubbles are trapped under the slabs. An apparatus for preparing agar slabs of uniform thickness has been described by Haefner (1967).

III.   Micromanipulation

A.   Dissection of Asci

Since the dissection of asci is required for tetrad analysis, it is a necessary skill needed by all serious yeast geneticists. The initial method for the separation of ascospores from individual asci, first described by Winge (see Winge and Laustsen, 1937; Lindegren, 1949) requires rupture of the ascus wall with a microneedle and transferral of the four spores to different locations on a nutrient medium. The individual asci are transferred from liquid drops or agar surfaces to dry areas of the slide where the actual dissection is carried out by carefully applying momentary pressure to the ascus. This older method must be performed with considerable precision, since the spores can easily collapse or be injured if they are compressed between the microneedle and glass slide.

A less demanding procedure (Johnston and Mortimer, 1959), currently in use in most if not all laboratories, relies on digestion of the ascus sac with snail juice or mushroom extract (Bevan and Costello, 1964) without dissociating the four spores from the ascus. Snail juice can be obtained commercially as Glusulase (Endo Laboratories, Garden City, N.Y.) and *Suc d'Helix pomatia* (L'Industrie Biologique Française, Génévilliers, France). A water suspension of a sporulated culture is usually treated for 10–30 minutes, depending on the particular yeast strain and on the strength of the Glusulase or other digestive agent. The sample is ready for dissection when at least three spores are out of the ascus sac. Extensive treatment sometimes can decrease the viability and dissociate the clusters of four spores. The culture is suspended by gently rotating the tube, and the suspension is streaked along the edge of the thin agar slab. The agar slab on the glass slide is inverted over a chamber, and individual clusters of four spores are picked up with the microneedle; each of the spores is placed at a regular interval across the slab. About 20 tetrads can be dissected on a single slab which is then placed on the surface of a nutrient plate with the spores facing upward.

Considerable patience is required in attaining the ability to pick up a cluster of four spores on a microneedle and place each one at a different location on the agar slab. The difficulties of ascus dissection are often exaggerated, however, and most workers can master the technique after a few days of practice.

B.   Isolation of Cells

In addition to ascus dissection, micromanipulation is occasionally required for separating zygotes from mating mixtures, for pairing vegetative cells and spores for mating, and for separating mother cells and daughter cells during vegetative growth. For unknown reasons microneedles cannot be readily used to pick up vegetative cells, although these cells can be transported with microloops or micropipettes, or simply by dragging them across the agar surface with microneedles. The use of microneedles is rather effective, since the cells usually follow closely in the wake of the microneedle as it is moved along the liquid surface film of the agar.

ACKNOWLEDGMENTS

The writing of this chapter was supported by Public Health Service research grant GM-12702 from the National Institute of General Medical Sciences and by the Atomic Energy Commission at the University of Rochester Atomic Energy Project, Rochester, N.Y. It has been designated U.S. Atomic Energy Commission report no. UR-3490-557.

REFERENCES

Barer, R., and Saunders-Singer, A. E. (1948). *Quart. J. Microsc. Sci.* **89**, 439.
Belkin, M. (1928). *Science* **68**, 137.
Bevan, E. A., and Costello, W. P. (1964). *Microbial Genet. Bull.* **21**, 5.
Burns, V. W. (1956). *J. Cell. Comp. Physiol.* **47**, 357.
Bush, V., Duryee, W. R., and Hastings, J. A. (1953). *Rev. Sci. Instrum.* **24**, 487.
Cailloux, M. (1943). *Rev. Can. Biol.* **2**, 528.
Chambers, R. (1918). *Biol. Bull.* **34**, 121.
de Fonbrune, P. (1949). "Technique de Micromanipulation." Masson, Paris.
El-Badry, H. M. (1963). "Micromanipulators and Micromanipulation." Academic Press, New York.
Ellis, G. W. (1962). *Science* **138**, 84.
Emerson, J. H. (1931). U. S. Patent 1,828,460.
Haefner, K. (1967). *Z. Allg. Mikrobiol.* **7**, 229.
Johnston, J. R., and Mortimer, R. K. (1959). *J. Bacteriol.* **78**, 292.
Lindegren, C. C. (1949). "The Yeast Cell. Its Genetics and Cytology." Educational Publishers, St. Louis, Missouri.
May, K. R. (1953). *J. Roy. Microsc. Soc.* [3] **73**, 140.
Sherman, F. (1973). *Appl. Microbiol.* **26**, 829.
Thaysen, A. C., and Morris, A. R. (1947). *J. Gen. Microbiol.* **1**, 221.
Tomlinson, J. (1967). *Turtox News* **45**, 209.
Winge, Ö., and Lausten, O. (1937). *C. R. Trav. Lab Carlsberg, Ser. Physiol.* **22**, 99.

Chapter 11

Cell Cycle Analysis

J. M. MITCHISON AND B. L. A. CARTER

*Department of Zoology, University of Edinburgh,
Edinburgh, Scotland*

I. Introduction

Yeasts have been widely used in recent years for studying the events of the cell cycle, and many methods of cell cycle analysis are available. For this reason we provide here what is mainly a guide and a commentary on the methods that have been described, rather than detailed laboratory instructions.

There are two recent short reviews concerned specifically with the yeast cell cycle. One is by Duffus (1971) on methods and results in various yeasts. The other describes the work by Hartwell and his colleagues on the genetic analysis of the cycle in *Saccharomyces cerevisiae* (Hartwell *et al.*, 1974). Two more general reviews on the cell cycle by Mitchison (1971) and by Halvorson *et al.* (1971a) contain much information about yeasts. In addition, there are articles about yeasts in the multiauthor books on the cell cycle edited by Zeuthen (1964), Cameron and Padilla (1966), Padilla, Cameron and Whitson (1969), and Padilla, Cameron, and Zimmerman (1975).

It may be helpful to show in Fig. 1 some of the main events of the cell cycle in two well-studied yeast species, the budding yeast *S. cerevisiae* and the fission yeast *Schizosaccharomyces pombe*. In each case the end of the cycle is taken as the time at which the daughter cells become physically separate, since this is the normal criterion for other cells. But it should be remembered that the daughter cells are physiologically separated (by a membrane) for some time before the final physical cleavage. Nuclear division has to precede physiological separation, and this leads to a conspicuous difference between the cell cycle of yeasts and that of higher eukaryotes—the much earlier occurrence of nuclear division in the cycle.

The cycle in budding yeasts starts with a single unbudded cell. Bud enlargement continues through the cycle, although there is some evidence of a slowing down of the volume increase toward the end of the cycle. DNA synthesis takes place in about the first quarter of the cycle. Nuclear division and the migration of a nucleus into the bud happen relatively slowly in midcycle. The G_2 period is short, but there is an appreciable G_1 period at the end of the cycle. G_2 precedes G_1 in the cell cycle because the G_1-S-G_2 sequence must start and finish at nuclear division and not at cell division.

Fission yeasts grow only in length, and a cell nearly doubles its length during the first three-quarters of the cycle. Thereafter the length and volume stay constant, while a cell plate (or septum) forms across the middle of the cell. Nuclear division happens at the end of the third quarter of the cycle, and DNA is replicated at the time of cell division. In contrast with budding yeasts, fission yeasts have a different mode of growth, faster nuclear division, a shorter G_1, and a longer G_2.

II.   Single-Cell Techniques

Most of the methods for analyzing the cell cycle fall into two broad classes. One of them relies on microscopic examination of single cells, and the other relies on the production of synchronous cultures. These cultures are far more versatile in that they can be used for biochemical analysis by standard methods, but they have the disadvantage that the synchrony is never perfect.

A.   Living Cells

A simple and precise method of following volume growth and division timing is to observe (and photograph) single yeast cells growing under a

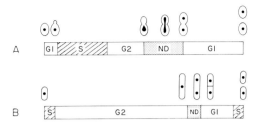

FIG. 1.   Cell cycles of *S. cerevisiae* (A) and *Schiz. pombe* (B). S, Period of DNA synthesis; G₁, G₂, pre- and postreplicative gaps; ND, nuclear division. A, slightly modified from Williamson (1966). B, from data in Mitchison (1970) and Mitchison and Creanor (1971b).

microscope. This has a long history (e.g., Bayne-Jones and Adolph, 1932; Knaysi, 1940; Lindegren and Haddad, 1954; Burns, 1956; Mitchison, 1957, 1958; Swann, 1962; Hartwell *et al.*, 1970; Streiblová and Wolf, 1972). There are no great technical problems involved. A photomicroscope is desirable, and it is convenient to have an automatic timer for photography. The slide must be kept warm by means of a warm chamber around the slide or the microscope, or by using the microscope in a warm room. A flat-field object-ive is useful if several cells are being followed, and we have found that dark-ground illumination gives good definition of the cell wall and cell plate of *Schiz. pombe*. The exact way of preparing a slide is often not described in articles. The method we use with *Schiz. pombe* is to make a small pad of medium with agar (about 1 mm thick and 5 mm square) on a slide, streak it with a thin suspension of cells from a growing culture, cover it with a cover slip (22 mm square) and then seal the edges of the cover slip with petroleum jelly or paraffin wax. In addition, we insert a thin glass capillary tube (ca. 20 mm long and 0.5 mm in diameter) under the cover slip before sealing. This equalizes pressure between the inside and the outside and prevents the cover slip from being forced up (and therefore out of focus) by the evolution of carbon dioxide during growth. It may be necessary to blow a slow current of air through the chamber in order to maintain growth, possibly because the carbon dioxide concentration becomes inhibitory. This can be done with two capillary tubes, one of which is connected to an air pump.

   There are various ways of using this technique to observe the effect of agents on the cell cycle. Growth and division can be followed in the presence of a chemical agent incorporated in the agar medium. Alternatively, pulse treatment with an agent can be applied to a culture before setting up the slide culture. The first microphotograph gives the stage in the cycle of each individual cell when the agent was applied. In a few situations (e.g., change of a temperature-sensitive mutant from the permissive to the restrictive temperature), the treatment can be applied while the slide culture is under observation. The cycle stage of each cell can be determined by the morphological criteria shown in Fig. 1 (for *Schiz. pombe*, see also Mitchison, 1970).

The power of this technique is that it can give a precise cell cycle analysis without the use of synchronous cultures. The limitation is that only a few cycle parameters can be measured. Cell division can be followed easily in *Schiz. pombe* and with somewhat greater difficulty in *S. cerevisiae* in which the final cleavage may not be apparent. Nuclear division can be detected with phase-contrast microscopy and a high-refractive-index mounting medium (e.g., 21% gelatin; McCully and Robinow, 1971). Absolute cell volume can be calculated from measurements of cell length and diameter, with the assumption that the cell is a solid of revolution. Relative cell volume can be estimated by a single length measurement with *Schiz. pombe* and by length and diameter measurements of elliptical buds and mother cells of budding yeasts. Because of the uniaxial mode of growth in *Schiz. pombe*, length measurements give estimates of volume increase more accurate than the equivalent estimates for budding yeasts. Total dry mass can also be measured on single growing cells by interferometry, but the technique is complicated (Mitchison, 1957, 1958).

B. Dead Cells

The measurements that can be made on single living cells can also be made on dead fixed cells. Volume can be calculated from linear dimensions, although there is often severe shrinkage on fixation. For example, fixed cells of *Schiz. pombe* are only 45% of the volume of living cells (Mitchison and Cummins, 1964). Total dry mass (less pool components) can also be measured by interferometry (Mitchison and Cummins, 1964). Nuclear division and cell plates can be observed by staining (for *Schiz. pombe*, see Mitchison, 1970). The main advantage of working with fixed cells is that the range of available techniques can be extended by the use of autoradiography, staining, and microspectrophotometry.

Autoradiographs are usually made after a pulse of tritium-labeled precursor has been applied to a growing culture. The site and the extent of macromolecular synthesis is revealed after the unincorporated precursor has been removed by fixation, although it is possible to make autoradiographs of yeasts with an intact pool of precursor (Cummins and Mitchison, 1964). For some purposes it is sufficient to determine qualitatively whether or not cells are labeled, but grain counting is needed for quantitative work. The most widespread use of autoradiography in cell cycle work on higher eukaryotes has been in determining the timing of DNA synthesis after pulse treatment with tritiated thymidine. This is impossible with yeasts, since they lack thymidine kinase and do not incorporate thymidine into DNA. But this difficulty can be circumvented, at any rate in *S. cerevisiae*, by labeling all the nucleic acids with adenine and then selectively extracting the RNA (Williamson, 1965). Another promising possibility is to use the recently isolated mutants that incorporate thymidine monophosphate into DNA Brendel and Haynes, 1972; Jannsen *et al.* 1973; Wickner, 1974). Autoradiographs have also been used with yeasts to determine the rates and sites of synthesis of the cell wall (Johnson, 1965; Johnson and Gibson, 1966a,b) of RNA (Mitchison and Lark, 1962; Mitchison, 1963; Mitchison *et al.*, 1969) and of protein (Mitchison and Wilbur, 1962). The technique used for much of the work on RNA and protein synthesis is now obsolete, since synchronous cultures give results that are probably more accurate and certainly avoid laborious grain counting. It is worth pointing out that a precursor pulse, whether used for autoradiography or for bulk counting, requires care in interpretation. It is a sensitive way of measuring the rate of synthesis, but it usually involves assumptions about uptake mechanisms and pool size (Mitchison, 1971).

Staining techniques have mostly been used to examine cell wall growth. Streiblová and Wolf (1972) stained different regions of the wall in *Schiz. pombe* with the fluorescent stain primulin, and May (1962) showed the sites of cell wall extension in the same yeast with fluorescent antibody. Microspectrophotometry has seldom been used on yeast cells, mainly because of their small size, but there is one early article on RNA synthesis in *Schiz. pombe* employing this method (Mitchison and Walker, 1959).

Measurements on single dead cells have to be related to the cell cycle. The way that this has usually been done in yeasts is to position each cell in the cycle according to its morphology—length in the case of *Schiz. pombe*. The results from such an analysis are similar to those from age fractionation on a zonal rotor (see Section III,A,3) and have many of the same merits and disadvantages. There are other methods which rely on frequency distributions within cell populations (Mitchison, 1971), but they have seldom been used with yeasts (see, however, Mitchison and Walker, 1959; Kubitschek, 1971).

III.   Synchronous Cultures

There are two main ways of making synchronous cultures. In selection synchrony cells at one stage of the cycle are separated physically from the rest of an asynchronous culture and grown as a synchronous culture. In induction synchrony a whole culture is induced to divide synchronously by changes in the medium, changes in temperature, or treatment with chemical agents. In principle, induction methods give a high yield of synchronous cells but may distort the cycle, whereas selection methods give a low yield but are less likely to distort the cycle. In practice, however, there is an induction technique for budding yeasts which appears to cause little distortion of the cycle, and a selection technique (age fractionation) which gives a high yield although it does not produce a synchronous culture.

A. Selection Methods

1. VELOCITY SEPARATION IN TUBES

One of the best methods for making small synchronous cultures is gradient sedimentation in centrifuge tubes (Mitchison and Vincent, 1965). The principle is velocity separation by centrifuging a concentrated suspension of growing cells through a sucrose gradient. Cells are separated by size, and the smallest ones (which are normally at the start of the cycle) are removed and grown as a synchronous culture. The method is quick, easy, and requires only a minimum of apparatus (a standard centrifuge and a gradient former). It has been widely used both with yeasts (*Schiz. pombe, S. cerevisiae, Saccharomyces lactis*, and *Kluyveromyces fragilis*) and with other systems from mammalian cells to bacteria. Its main limitations are relatively low yield and, in some cases, transient perturbation of the cells by the selection procedure.

It may be useful to describe in greater detail the technique we use with *Schiz. pombe* (Mitchison, 1970). About 10^{10} cells are harvested from a 4-liter exponential-phase culture by collection on a 32-cm Whatman No. 50 filter paper in a holder similar to those used for membrane filters. This takes 10–15 minutes. Membrane filters or continuous-flow centrifuges are more expensive alternatives. Long periods of centrifuge collection should be avoided, since the cells then show a considerable lag in growth after resuspension in fresh medium.

The cells are scraped off the filter paper with a spatula, resuspended in medium to a final volume of 4–5 ml, and then layered on top of 80 ml of a linear gradient of 10–40% sucrose (in medium) made with a simple gradient former (Britten and Roberts, 1960) in a 100-ml centrifuge tube. The tube is centrifuged in a swing-out head until the top layer of cells has moved about halfway down the tube. This takes approximately 5 minutes at 500 g at 25 °C. The top layer of cells (5–10 ml) is then removed with a syringe with a long needle and added to warm medium to make the synchronous culture. The final yield is about 5 × 10^8 cells, 5% of those loaded on the gradient. The progress of the synchronous culture is monitored be measuring either the

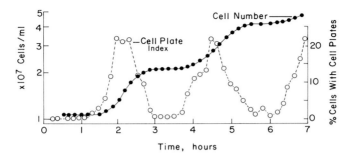

FIG. 2. Synchronous culture of *Schiz. pombe* prepared by velocity separation on a zonal rotor (MSE Type A). 3.5×10^9 small cells separated on the rotor and grown in minimal medium at 32°C. Two synchronous divisions are shown. Cell numbers were measured on a Coulter Counter. The cell plate index (percent cells with cell plates) was measured under a microscope. C. H. Sissons, unpublished data.

cell plate index or cell numbers or both. A culture prepared this way is very similar to that shown in Fig. 2 in which a zonal rotor has been used rather than a centrifuge tube, although the cell plate index can rise to 30% at the first division in a culture with good synchrony.

It is sometimes desirable to carry out a control in order to see whether the effects seen in a synchronous culture are the normal events of the cell cycle or whether they are artifacts caused by harvesting, centrifugation through sucrose, and resuspension. Such a control can be carried out by going through the normal procedure up to the moment when the top layer of cells is sucked out. At this point the whole tube is shaken up and a sample removed for resuspension. This gives an asynchronous culture which serves as a control for most of the artefacts that may be produced. But it does not cover all of them, since many of the cells are in higher concentrations of sucrose than the top cell layer used for synchronous cultures.

Cell clumping reduces the yield of single cells, and recently we have had to modify our medium in order to minimize clumping with strains of *Schiz. pombe*. The other difficulty that has sometimes arisen with other yeasts is that the top layer in the centrifuge may contain small cells that are non-viable. In this case a layer should be removed from just below the top.

The function of the sucrose gradient is to provide a dense medium on which the initial cell suspension can be layered, and also to stabilize the contents of the tube against stirring when the centrifuge is speeded up and slowed down. Its composition is relatively unimportant. We normally use sucrose, because it is inexpensive and convenient, but we have also used lactose (a sugar which is not metabolized) and glucose. Stetten and Lederberg (1973) made glycerol gradients, and if a gradient of low osmotic pressure is desirable, it should be possible to use Ficoll, dextrin, dextran, protein (e.g., serum albumin), Renografin, Urograffin, or Ludox (colloidal silica).

This technique can easily be scaled down by using a bench centrifuge, smaller tubes, and shallower sucrose gradients (5–15%). It can be scaled up to a limited extent by using more than one tube but, if large synchronous cultures are necessary, it is better to use a zonal rotor, as described in Section III,A,2.

2. Velocity Separation in Zonal Rotors

Synchronous cultures can be prepared on a large scale by separating cells according to size on a zonal rotor. The major advantage of this technique over the tube method is that much larger synchronous cultures can be obtained because of the greater capacity of zonal rotors. However, a zonal rotor is more expensive and cumbersome than a centrifuge tube and requires careful assembly and dismantling. The tube method therefore is more convenient, unless large cultures are required.

Zonal centrifugation has been used to obtain synchronous cultures of *S. cerevisiae* (Halvorson *et al.*, 1971b), *S. lactis* (Carter and Halvorson, 1973), and *Schiz. pombe* (Poole and Lloyd, 1973; B. L. A. Carter, unpublished). We describe the method of Halvorson *et al.* (1971b), since it has been used for both fission yeasts and budding yeasts and employs a high-capacity, 1500-ml, A XII zonal rotor (MSE Type A) rather than the 590-ml-capacity MSE HS zonal rotor used by Poole and Lloyd.

A sucrose gradient is introduced with a peristaltic pump into an MSE Type A zonal rotor running at 600 rpm. The choice of gradient is dependent on the type of yeast being fractionated. In practice, good results for diploid *S. cerevisiae* and for *Schiz. pombe* can be obtained with a linear 15–40% gradient, and for *S. lactis* and some haploid *S. cerevisiae* a linear 10–25% gradient may be used. The use of a gradient linear with volume (as opposed to rotor radius) means that an expensive gradient-making apparatus is unnecessary, and results in superior fractionation of the later stages of the cell cycle.

Exponential cultures for centrifugation are harvested as described earlier for tube separations. The harvested cells are scraped off the filter paper, resuspended in medium, introduced into the center of the rotor with a 50-ml syringe, and overlaid with 30 ml of distilled water to displace the cells from the central core. Once loaded, the rotor is accelerated to 2000 rpm, and migration of the cells is noted visually. When the fastest-migrating cells have migrated about four-fifths of the distance to the edge of the rotor (usually about 10 minutes), the rotor is decelerated to 600 rpm. To remove cells from the rotor, sucrose of the heaviest density is pumped into the outer edge of the rotor to displace cells from the center. Since cells are separated on the basis of size, the smallest cells early in the cycle are collected first. Later fractions can then be collected containing progressively larger cells. The first fractions are usually used to make synchronous cultures, although it is possible to use cells from the middle of the rotor or even the edge which contains large cells close to cell division. When large synchronous cultures are required, it is better to use the first fractions to inoculate the culture because, when high concentrations of cells are centrifuged, small cells may be pushed into the gradient by the larger cells as they begin to migrate and the small cells contaminate later fractions.

Using the small cells from the rotor, we have prepared synchronous cultures of budding yeasts with an inoculum of 4.5×10^{10} cells, which is about 100 times more than that produced by the tube method. Figure 2 shows a synchronous culture of *Schiz. pombe* made in this way with a lower initial inoculum of 3.5×10^9 cells (C. H. Sissons, unpublished). The exact limits of yield are uncertain, but it is likely that the yield from a 1500-ml rotor can be between 10 and 100 times greater than that from a single tube.

3. FRACTIONATION BY CELL AGE IN ZONAL ROTORS

In velocity sedimentation cells separate on the gradient mainly according to their size and therefore according to their age in the cell cycle. Successive fractions from a gradient on a zonal rotor (or, in principle, from a tube) give cells that are at successive stages in the cell cycle. These samples can be analyzed in order to follow the events of the cycle without generating a synchronous culture. This technique has been used to fractionate cultures of *S. cerevisiae* (Carter *et al.*, 1971; Sebastian *et al.*, 1971), *S. lactis* (Sebastian *et al.*, 1971; Carter and Halvorson, 1973), and *Schiz. pombe* (Wells and James, 1972; Poole and Lloyd, 1973; B. L. A. Carter, unpublished). The procedure described by Sebastian *et al.* (1971) was developed from the method of Halvorson *et al.* (1971b) described earlier. The methods of Poole and Lloyd (1973) and Wells and James (1972) are essentially similar to that of Sebastian *et al.* (1971), although different rotors were used: the MSE HS zonal rotor by Poole and Lloyd and the Sorvall SZ-14 by Wells and James.

When an MSE Type A rotor is used to fractionate exponential cultures according to cell age, it is chilled to 4°C and loaded with a chilled (4°C) sucrose gradient. In addition, the exponential culture is chilled when harvesting to arrest metabolism. An alternative is to add metabolic inhibitors. The cells (ca. 5×10^{10}) are suspended in 30 ml of chilled (4°C) water, loaded, and then centrifuged. When the fastest-migrating cells have traveled two-thirds of the distance to the edge of the rotor, it is decelerated and the cells are unloaded by pumping out. Cell fractions are collected in chilled tubes and can later be analyzed biochemically after removal of the sucrose by centrifugation. Cell counts are needed for each fraction, since the yield of cells varies along the gradient.

Gradient fractions can be positioned with respect to cell age by correlating cell size during the cycle with cell size in the fractions (measured, for example, with a Coulter Counter). Although the relation between the factors (cell size, shape, and density) that determine the exact separation of cells on the gradient is complex, Fig. 3 shows that for *S. cerevisiae* growing with a 2½-hour generation time cell age during the cycle is linearly related to fraction number from the zonal rotor. The situation with *Schiz. pombe* is complicated, because this organism increases in length for the first three-quarters of the cycle only. However, dry mass increases throughout the cycle, and so the density of the cell increases during the constant-volume stage

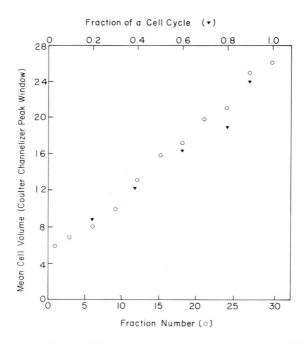

FIG. 3.   Mean cell volumes of *S. cerevisiae* from a synchronous culture at different stages of the cell cycle compared to mean cell volumes of cells in fractions removed sequentially from a zonal rotor. Samples were taken from a synchronous culture (prepared by velocity separation on a zonal rotor) at intervals during the cell cycle (generation time of 2.5 hours). Fractions from the zonal rotor were analyzed after separation of an exponential-phase culture. The mean cell volume of samples at different stages of the cycle (triangles) and of fractions from the zonal rotor (circles) was measured on a Coulter Counter. S. Sogin and B. L. A. Carter, unpublished data.

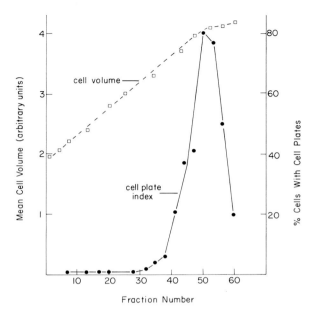

FIG. 4.   Size fractionation of an exponential culture of *Schiz. pombe* (strain 972*h*⁻) by zonal centrifugation. An exponential-phase culture grown in minimal medium was harvested by filtration and layered onto a chilled 15–40% sucrose gradient in an MSE Type A zonal rotor at 600 rpm. After centrifugation at 2000 rpm until the fastest-migrating cells had moved two-thirds of the distance to the edge of the rotor, the cells were pumped out from the center. Sixty fractions (15 ml each) were collected, and the mean cell volume of each fraction was measured with a Coulter Counter. The cell plate index (percent cells with cell plates) was measured under the microscope. B. L. A. Carter, unpublished data.

(Mitchison, 1957). While cells are separated mainly according to size on the zonal rotor, some separation is achieved because of density differences, and this must play a part in the separation of cells toward the end of the cell cycle. Figure 4 shows the separation of *Schiz. pombe* according to size, and also the cell plate index of fractions from the rotor. The maximum cell plate index (80%) of fractions from the rotor is higher than can be achieved in synchronous cultures. The cell plate index declines in the last fractions, because they contain large cells in which the center of the cell is beginning to pinch off, giving the appearance of doublets close to division.

An important criterion is separating cells according to cell age is that the fractions be homogeneous with respect to cell age. This has been shown to be true for *S. cerevisiae* by inoculating fractions from various positions on the rotor into medium and observing that synchronous growth ensues

(Sebastian *et al.*, 1971). The homogeneity of fractions is as good as that obtained from synchronous cultures.

This method of using velocity sedimentation has certain definite advantages over the alternative technique of growing a synchronous culture from a selected fraction. The yield is much higher, since all the original culture is used rather than 95% being discarded. It is also better and more convenient to apply many kinds of treatment, such as isotopic labeling (Sogin *et al.*, 1972) or mutagenesis (Dawes and Carter, 1974), to a single, unperturbed, asynchronous culture before age fractionation than to multiple samples from a synchronous culture. The main disadvantage is that only one cycle is analyzed in each experiment. This may make it harder to detect some cell cycle events. If there is a linear pattern of synthesis with one rate doubling per cycle, this pattern is easiest to resolve when there is a linear segment of maximum length (one cycle) on either side of the point of rate doubling. This requires at least two cycles. Age fractionation is also at a disadvantage in regard to events at the end of the cycle. DNA synthesis, for example, coincides with cell division in *Schiz. pombe*, so the increase in DNA per cell is split between the two ends of the gradient. Both techniques involve a possible disturbance of the cells by collection and centrifuging on a gradient. In age fractionation these are carried out in the cold after the action of an agent (if one has been used). The disturbance is probably minimal for most cell parameters, but it might occur, for example, for a few components of the pool. A control is to compare measurements before and after centrifugation. In synchronous cultures it is more likely that cells are perturbed, since they are grown after the gradient centrifugation. This can be controlled to some extent by comparing synchronous cultures with asynchronous ones made from shaken-up gradients, as described earlier, but the possibility of long-term perturbation should be kept in mind. The ideal situation is to use both age fractionation and selection-synchronized cultures, and to exploit the advantages of each method.

4. EQUILIBRIUM SEPARATION

The centrifugation techniques described in the previous three sections rely on velocity sedimentation in which the main separation is according to cell size and in which continued centrifugation pellets all the cells. Both fission yeasts and budding yeasts can vary in density through the cell cycle (Mitchison, 1957, 1958; Wiemken *et al.*, 1970; Hartwell, 1970; Nurse and Wiemken, 1974), so it is possible to use the alternative technique of gradient centrifugation in which the cells are centrifuged to equilibrium in an isopycnic gradient. Samples can then be extracted and grown as synchronous cultures in the same way as with velocity separation. This method has been applied by Hartwell (1970) to *S. cerevisiae* with gradients of 20–27%

Renografin containing 1% polyvinylpyrolidone. Dextrin gradients have been used on *S. cerevisiae* by Wiemken *et al.* (1970), and on *Candida utilis* by Nurse and Wiemken (1974).

Equilibrium centrifugation has the advantage that the gradients can carry heavier loads of cells than are possible with velocity centrifugation. Hartwell (1970) separated 5×10^{10} cells in a 14-ml gradient. However, the synchrony in the cultures does not appear to be as good as that from velocity separation, judging from the published figures. Much higher speeds of centrifugation are needed (up to 45,000 g), so it is desirable to use a swing-out head which acelerates and decelerates rapidly. The articles on these methods mention chilling during centrifugation, so there is the possibility of a temperature shock to the cells.

The cell cycle can be spread out in a zonal rotor on an equilibrium gradient in the same way as on a velocity gradient. This has been done by Poole and Lloyd (1973) with *Schiz pombe* with a linear gradient of 27–33% dextran. The distribution of cell volume (but not, surprisingly, cell length) across the rotor was about the same as that found with a velocity gradient of 10–40% sucrose.

B. Induction Methods

1. STARVATION AND REINOCULATION

Given the right conditions, starved cells in stationary phase can be induced to divide synchronously when inoculated into fresh medium. This is the basis of several methods that have been used on yeasts (references in Williamson, 1964; Mitchison, 1971), the best of which is probably that developed by Williamson and Scopes (1960, 1961) for *S. cerevisiae*. The technique is more complicated than a simple resuspension of stationary-phase cells in fresh medium, and involves both cycles of starvation and feeding and also separation of small from large cells. The outline that follows comes from a detailed description by Williamson (1964).

Shaken cultures of *S. cerevisiae* (N.C.Y.C. 239) are incubated for 10 days in a malt-extract medium. The cells are then washed and transferred at high density to growth medium at 4 °C. After 18 hours the supernatant containing small cells is decanted and discarded. The remaining cells are incubated with aeration at 25°C for 6 hours in a starvation medium with salts only. They are then stored overnight at 4 °C, given a further period of aeration at 25 °C the next day, and finally transferred to cold growth medium. The whole cycle of events is then repeated until there is a sufficiently homogeneous population of large cells to produce good synchrony when the cells are suspended in warm growth medium. Four to seven cycles are usually required.

Williamson (1964) describes alternative ways of removing the small cells by differential centrifugation or by foam flotation, which shorten the procedure. Another shorter version, which only takes 2–3 days, has been used by Tauro and Halvorson (1966). This version is also related to a technique used on *Schiz. pombe* by Sando (1963).

The merits of the method are that it gives a large yield of synchronized cells and that it involves little or no special apparatus. The degree of synchrony is about the same as that obtained by selection synchrony. There are two disadvantages. One is that it takes a long time and several handling operations, at any rate using the original procedure. The second is the doubt that exists for all induction methods about the degree to which the cell cycle in synchronized cultures is the same as the "normal" cycle of single cells in an asynchronous exponential culture. This normal cell cycle is of course a rather idealized concept, since cell characteristics can change as the medium changes during the exponential phase of batch cultures. It is also true that selection procedures can affect the initial stages of growth of synchronous cultures. But induction procedures by their very nature are more likely to distort the cell cycle. This method is less likely than most to raise this objection, since its rationale is to produce a homogeneous starting population. Tauro and Halvorson (1966) also showed that it gives synchronous cultures in which the timing of periodic enzyme synthesis is the same as that in synchronous cultures made by selection. The first cycle, however, is appreciably longer than subsequent cycles, and it is unwise to make an automatic assumption that events in this cycle, and perhaps in later cycles, are necessarily normal.

Synchrony can also be induced in chemostat cultures by periodic changes of the medium (Dawson, 1965; Müller and Dawson, 1968; von Meyenburg, 1969), but the relation between these cycles and normal cycles is even more problematic than it is with batch synchronous cultures.

2. INHIBITION OF DNA SYNTHESIS

A method of induction synchrony widely used with mammalian cells is to apply one (or two) periodic blocks to DNA synthesis inhibitors (Mitchison, 1971). The cells accumulate at the start of the S period during the period of the block and, when the block is released, they proceed synchronously through the S period and the following division.

This method was applied to *Schiz. pombe* by Mitchison and Creanor (1971a), using a 3- or 4-hour block with 2mM deoxyadenosine on a growing asynchronous culture (Fig. 5). DNA synthesis is inhibited after a lag of an hour and recommences slowly near the end of the block (Sissons *et al.*,

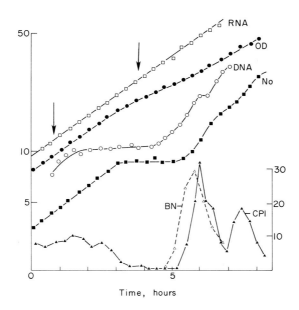

FIG. 5. Induction synchrony of *Schiz. pombe* produced by a DNA block. Abscissa, time in hours; left ordinate, arbitrary units. Effect of a 3-hour treatment with deoxyadenosine (2 mM) on various cell parameters of an exponential-phase culture. The treatment commences at the first arrow and finishes at the second arrow. RNA, one arbitrary unit (AU) = 1 μg/ml; OD, optical density at 595 nm, 1 AU = 0.0255 absorbance; DNA, 1 AU = 0.01 μg/ml; No, cell number, 1 AU = 0.715 × 10^6 cells/ml; CPI, cell plate index as percent on right ordinate; BN, proportion of binuclear cells without cell plates as 1.5 × percent on right ordinate. Results from three similar experiments (Mitchison and Creanor, 1971a).

1973). The cell plate index drops during the pulse from about 10% (the normal value in an asynchronous culture) to zero. It then rises to a peak value of about 30% just before the first synchronous division. This value is about the same as that found in a well-synchronized culture after selection from tubes. The second cell plate peak, however, occurs after only 1½ hours, 60% of the normal cycle time. One result of this shortened cycle time is that there is practically no plateau in cell numbers between the two synchronous divisions. The curve for binucleate cells without cell plates shows that the nucleus divides a little before the appearance of the cell plate, as it does in the normal cycle.

There is a sequence of three important events in the cell cycle—DNA synthesis, nuclear division, and cell division—and the results from many cell systems show that, if one of these events is blocked, so also are the subsequent ones. This method of synchronization works because DNA synthesis has been blocked and no cell can proceed to divide unless its DNA has been replicated. The residual divisions in the later stages of the block are from cells that had replicated their DNA by the time the block became effective. When the block is removed, cells proceed synchronously through this sequence of three events. But the situation with cell growth is different. Figure 5 shows that growth in RNA is unaffected by the inhibitor pulse, and optical density is slightly affected only after the end of the pulse. As a result, cells at the first synchronous division are 70–80% larger than normal dividing cells and also more variable in size. This suggests that growth is unsynchronized, although this statement has meaning only if discrete events can be found to mark the progress of growth during the cycle. Such events can perhaps be found in the periodic synthesis during the cycle of some enzymes, and it is significant that three enzymes that show periodic synthesis in cultures synchronized by selection do not show this pattern of synthesis after induction synchrony (Sissons *et al.*, 1973).

It is clear that the cell cycles after this kind of induction synchrony are different from normal cell cycles. But this distortion can be turned to an advantage, because the combined use of induction and selection synchrony is a powerful method for analyzing causal connections between events in the cell cycle. The fact that periodic enzyme synthesis is synchronized after selection but not after induction shows that the control of this synthesis can be temporarily dissociated from DNA synthesis and division. In contrast, steps in enzyme potential (rate of synthesis after derepression) are synchronized after induction, which suggests a causal link with DNA replication through gene-dosage control (Sissons *et al.*, 1973). This argument is expanded elsewhere (Mitchison, 1975). One further point about this and other methods of induction synchrony is that they can be convenient methods for synchronizing certain events, e.g., nuclear division, even though normal cycle relations are distorted.

Deoxyadenosine is not an effective DNA inhibitor in *S. cerevisiae* (B. L. A. Carter, unpublished), but hydroxyurea is. The work of Slater (1973) suggests that hydroxyurea could be used for budding yeasts in much the same way deoxyadenosine is used for *Schiz. pombe*, but the possibilities have not yet been fully explored. We have found (unpublished) that a block with hydroxyurea (11 mM) gives a degree of induction synchrony with *Schiz. pombe* that is as good or better than that from deoxyadenosine, but it also has undesirable side effects on enzyme induction.

Two other ways of reversibly blocking DNA synthesis and inducing synchrony can be used in special circumstances with *S. cerevisiae*. Haploid cells of the *a* mating type are blocked prior to the initiation of DNA synthesis by a mating pheromone, the α factor, produced by α- mating-type cells (Bücking-Throm *et al.*, 1973; Hartwell, 1973). During this block, the cells continue to enlarge. Cells can be synchronized after release from treatment with the α factor for a period usually longer than the generation time. The inhibition is released by transferring the cells to fresh medium. About 90 minutes after transfer there is a synchronous burst of division in most but not all of the cells. A second synchronous division occurs earlier than would be expected from the normal cycle time, as in induction synchrony with *Schiz. pombe*. The temperature-sensitive mutant *cdc4* is blocked in DNA initiation at the restrictive temperature, and pulse treatment at this temperature induces partial synchrony in DNA synthesis (Hereford and Hartwell, 1973).

3. REPETITIVE HEAT SHOCKS

For the last 20 years, Zeuthen and his colleagues have studied the effect of heat shocks (transfer to a high sublethal temperature) on division control mechanisms in the ciliate *Tetrahymena* (Zeuthen and Rasmussen, 1971). Repeated heat shocks applied either at short intervals or once per cycle induce synchrony in normal asynchronous cultures, and a hypothesis involving heat-labile "division proteins" has been put forward to explain this effect. This technique has been applied to *Schiz. pombe* by Kramhøft and Zeuthen (1971), and they found that five to six heat shocks (30 minutes at 41 °C every 130 minutes) produce a degree of synchrony similar to that occurring after selection. This throws an interesting light on the similarity of the mechanisms of division control in yeasts and ciliates, and it could conceivably be used as a routine method for producing synchronous cultures provided there was some automation in the temperature control of water baths. But the relation of these synchronous cultures to the normal cell cycle remains to be established, particularly since there is an environmental shock at the start of each cycle.

IV. Conclusions

Observation or time-lapse photography of single growing cells is still a good technique for following volume growth and the timing of budding or division. For nearly all other studies on the cell cycle, particularly those involving biochemical estimations or isotopic labeling, it is necessary to use synchronous cultures or age fractionation. Induction synchrony has its uses, particularly in determining the causal connections of cell cycle events, but it has largely been superseded by selection methods when the object is to produce normal synchronous cultures. The simplest method is selection of a cell fraction after velocity or equilibrium centrifugation in tubes. Zonal rotors can be used to increase the yield of selected cells. They can also be used to fractionate a population according to cell age or stage in the cycle. The ideal situation is to use both age fractionation and synchronous cultures so as to exploit the advantages of each method.

REFERENCES

Bayne-Jones, S., and Adolph, E. F. (1932). *J. Cell. Comp. Physiol.* **1**, 387–407.
Brendel, M., and Haynes, R. H. (1972). *Mol. Gen. Genet.* **117**, 39–44.
Britten, R. J., and Roberts, R. B. (1960). *Science* **131**, 32–34.
Bücking-Throm, E., Duntze, W., Hartwell, L. H., and Manney, T. R. (1973). *Exp. Cell Res.* **76**, 99–110.
Burns, V. W. (1956). *J. Cell. Comp. Physiol.* **47**, 357–375.
Cameron, I. L., and Padilla, G. M., eds. (1966). "Cell Synchrony." Academic Press, New York.
Carter, B. L. A., and Halvorson, H. O. (1973). *Exp. Cell Res.* **76**, 152–158.
Carter, B. L. A., Sebastian, J., and Halvorson, H. O. (1971). *Advan. Enzymol.* **9**, 253–263.
Cummins, J. E., and Mitchison, J. M. (1964). *Exp. Cell Res.* **34**, 406–409.
Dawes, I. W., and Carter, B. L. A. (1974). *Nature (London)* **250**, 709–712.
Dawson, P. S. S. (1965). *Can. J. Microbiol.* **11**, 893–903.
Duffus, J. H. (1971). *J. Inst. Brew., London* **77**, 500–508.
Halvorson, H. O., Carter, B. L. A., and Tauro, P. (1971a). *Advan. Microbial Physiol.* **6**, 47–106.
Halvorson, H. O., Carter, B. L. A., and Tauro, P. (1971b). *In* "Methods in Enzymology" (K. Moldave and G. Grossman, eds.), Vol. 21, Part D, pp. 462–471. Academic Press, New York.
Hartwell, L. H. (1970). *J. Bacteriol.* **104**, 1280–1285.
Hartwell, L. H. (1973). *Exp. Cell Res.* **76**, 111–117.
Hartwell, L. H., Culotti, J., and Reid, B. (1970). *Proc. Nat. Acad. Sci. U.S.* **66**, 352–359.
Hartwell, L. H., Culotti, J., Pringle, J. R., and Reid, B. J. (1974). *Science* **183**, 46–51.
Hereford, L. M., and Hartwell, L. H. (1973). *Nature (London), New Biol.* **244**, 129–131.
Jannsen, S., Witte, I, and Megnet, R. (1973). *Biochim. Biophys. Acta* **299**, 681–685.
Johnson, B. F. (1965). *Exp. Cell Res.* **39**, 613–624.
Johnson, B. F., and Gibson, E. J. (1966a). *Exp. Cell Res.* **41**, 297–306.
Johnson, B. F., and Gibson, E. J. (1966b). *Exp. Cell Res.* **41**, 580–591.
Knaysi, G. (1940). *J. Bacteriol.* **40**, 247–253.
Kramhøft, B., and Zeuthen, E. (1971). *C. R. Trav. Lab. Carlsberg* **38**, 351–368.
Kubitschek, H. (1971). *Biophys. J.* **11**, 124–126.
Lindegren, C. C., and Haddad, S. A. (1954). *Genetica* **27**, 45–53.
McCully, E. K., and Robinow, C. F. (1971). *J. Cell Sci.* **9**, 475–507.
May, J. W. (1962). *Exp. Cell Res.* **27**, 170–172.
Mitchison, J. M. (1957). *Exp. Cell Res.* **13**, 244–262.
Mitchison, J. M. (1958). *Exp. Cell Res.* **15**, 214–221.
Mitchison, J. M. (1963). *J. Cell. Comp. Physiol.* **62**, Suppl. 1, 1–13.
Mitchison, J. M. (1970). *In* "Methods in Cell Physiology" (D. M. Prescott, ed.), Vol. 4, pp. 131–165. Academic Press, New York.
Mitchison, J. M. (1971). "The Biology of the Cell Cycle." Cambridge Univ. Press, London and New York.
Mitchison, J. M. (1975). *In* "Cell Cycle Controls" (G. M. Padilla, I. L. Cameron, and A. M. Zimmerman, eds.), pp. 125–142. Academic Press, New York.

Mitchison, J. M., and Creanor, J. (1971a). *Exp. Cell Res.* **67**, 368–374.
Mitchison, J. M., and Creanor, J. (1971b). *Exp. Cell Res.* **69**, 244–247.
Mitchison, J. M., and Cummins, J. E. (1964). *Exp. Cell Res.* **35**, 394–401.
Mitchison, J. M., and Lark, K. G. (1962). *Exp. Cell Res.* **28**, 452–455.
Mitchison, J. M., and Vincent, W. S. (1965). *Nature (London)* **205**, 987–989.
Mitchison, J. M., and Walker, P. M. B. (1950). *Exp. Cell Res.* **16**, 49–58.
Mitchison, J. M., and Wilbur, K. M. (1962). *Exp. Cell Res.* **26**, 144–157.
Mitchison, J. M., Cummins, J. E., Gross, P. R., and Creanor, J. (1969). *Exp. Cell Res.* **57**, 411–422.
Müller, J., and Dawson, P. S. S. (1968). *Can. J. Microbiol.* **14**, 1115–1126.
Nurse, P., and Wiemken, A. (1974). *J. Bacteriol.* **117**, 1108–1116.
Padilla, G. M., Cameron, I. L., and Whitson, G. L. eds. (1969). "The Cell Cycle: Gene-Enzyme Interactions." Academic Press, New York.
Padilla, G. M., Cameron, I. L., and Zimmerman, A. M., eds. (1975). "Cell Cycle Controls." Academic Press, New York.
Poole, R. K., and Lloyd, D. (1973). *Biochem. J.* **136**, 195–207.
Sando, N. (1963). *J. Gen. Appl. Microbiol.* **9**, 233–241.
Sebastian, J., Carter, B. L. A., and Halvorson, H. O. (1971). *J. Bacteriol.* **108**, 1045–1050.
Slater, M. L. (1973). *J. Bacteriol.* **113**, 263–270.
Sissons, C. H., Mitchison, J. M., and Creanor, J. (1973). *Exp. Cell Res.* **82**, 63–72.
Sogin, S., Carter, B. L. A., and Halvorson, H. O. (1972). *J. Cell Biol.* **55**, 244a.
Stetten, G., and Lederberg, S. (1973). *J. Cell Biol.* **56**, 259–262.
Streiblová, E., and Wolf, A. (1972). *Z. Allg. Mikrobiol.* **12**, 673–684.
Swann, M. M. (1962). *Nature (London)* **193**, 1222–1227.
Tauro, P., and Halvorson, H. O. (1966). *J. Bacteriol.* **92**, 652–661.
von Meyenburg, H. K. (1969). *Arch. Mikrobiol.* **66**, 289–303.
Wells, J. R., and James, T. W. (1972). *Exp. Cell Res.* **75**, 465–474.
Wickner, R. B. (1974). *J. Bacteriol.* **117**, 252–260.
Wiemken, A., Matile, P., and Moore, H. (1970). *Arch. Mikrobiol.* **70**, 89–103.
Williamson, D. H. (1964). *In* "Synchrony in Cell Division and Growth" (E. Zeuthen, ed.), pp. 351–379 and 589–591. Wiley (Interscience), New York.
Williamson, D. H. (1965). *J. Cell Biol.* **25**, 517–528.
Williamson, D. H. (1966). *In* "Cell Synchrony" (I. L. Cameron and G. M. Padilla, eds.), pp. 81–101. Academic Press, New York.
Williamson, D. H., and Scopes, A. W. (1960). *Exp. Cell Res.* **30**, 338–349.
Williamson, D. H., and Scopes, A. W. (1961). *Symp. Soc. Gen. Microbiol.* **11**, 217–242.
Zeuthen, E., ed. (1964). "Synchrony in Cell Division and Growth." Wiley (Interscience), New York.
Zeuthen, E., and Rasmussen, L. (1971). *In* "Research in Protozoology" (T. T. Chen, ed.), Vol. IV, pp. 11–145. Pergamon, Oxford.

Chapter 12

Genetic Mapping in Yeast

R. K. MORTIMER AND D. C. HAWTHORNE

Division of Medical Physics, Donner Laboratory, University of California, Berkeley, California, and Department of Genetics, University of Washington, Seattle, Washington

I. Introduction

This chapter is devoted to genetic mapping in *Saccharomyces cerevisiae*; however, the principles and techniques are generally applicable to investigations with other yeasts. The status of genetic mapping in *Schizosaccharomyces pombe* is summarized by Leupold (1970). A limited number of linkage studies have been undertaken with *Kluyveromyces lactis* (Tingle *et al.*, 1968).

The pioneering endeavors toward the construction of linkage maps for *Saccharomyces* were undertaken by the Lindegrens (Lindegren, 1949; Lindegren and Lindegren, 1951). From these early maps, four chromosomes, defined by the centromere-linked genes, *ade1* (on chromosome I), *gal1* (on II), *α* (on III), and *ura3* (on V), have stood the test of time. For the next decade the linkage studies undertaken by the Lindegrens and their colleagues were devoted primarily to the demonstration of affinity, i.e., the preferential assortment of nonhomologous chromosomes at meiosis (Shult and Lindegren, 1957, 1959); however, there was an extension of the linkage group defined by the centromere marker *ura3*, with the addition of the genes *ch* (choline), *hi* (*his1*), *is* (*ilv1*), and *an* (*trp2*) (Lindegren *et al.*, 1959; Shult and Desborough, 1960; Desborough *et al.*, 1960). Renewed impetus for mapping studies came with the definition of five more centromeres by the genes *trp1* (on chromosome IV), *his2* (on VI), *leu1* (on VII), *pet1* (on VIII), and *his6* (on IX) (Hawthorne and Mortimer, 1960). The definition of three more centromeres, *met3* (on X), *met14* (on XI), and *asp5* (on XII), followed in short order (Lindegren *et al.*, 1962; Hwang *et al.*, 1963, 1964). Four more centromeres have been defined by *lys7* (on XIII), *pet8* (on XIV), *pet17* (on XV), and *tyr7* (on XVI) (Mortimer and Hawthorne, 1963, 1966; Hawthorne and Mortimer, 1968). Finally, from studies utilizing disomic stocks, there is evidence for a seventeenth chromosome, although no centromere marker has been discovered for the linkage group (Mortimer and Hawthorne, 1973).

The above mapping studies also established the location of many genes along the arms of the 17 chromosomes. The current maps define the location of approximately 175 genes and have a total length of more than 3000 centimorgans. The 17 linkage groups vary considerably in genetic length, i.e., from about 20 centimorgans for chromosomes I and XIV to greater than 300 centimorgans for chromosomes IV and VII. This variation in genetic lengths corresponds well to the size distributions calculated from studies on the physical separation of chromosomes by centrifugation through sucrose gradients (Blamire *et al.*, 1972; Petes and Fangman, 1972). Examination of the various fractions of the gradient by electron microscopy confirms the separation of DNA molecules according to their contour length (Finkelstein *et al.*, 1972; Petes *et al.*, 1973). Studies employing these techniques seem to be the most promising approach to physical confirmation of the chromosome number obtained by genetic investigations. Various cytologists employing light microscopy have failed to obtain consistent chromosome counts for *Saccharomyces*; with one exception (Fischer *et al.*, 1973), their estimates are low when compared to genetic data (Matile *et al.*, 1969). This controversy surrounding the nuclear cytology of yeast is not too surprising when one considers the small size of yeast chromosomes.

The techniques for constructing hybrids, inducing sporulation, dissecting asci, and scoring genetic traits are described in other chapters of this book and in an earlier publication (Mortimer and Hawthorne, 1969). Only specialized techniques that have been developed specifically for mapping purposes are described here.

II. Mapping Techniques

Four different approaches have been used in locating genes in *Saccharomyces*: (1) tetrad analysis, (2) random spore analysis, (3) mitotic segregation analysis, and (4) trisomic analysis. A fifth approach, haploidization, has been used with *Schiz. pombe* (Gutz, 1966; da Cunha, 1969), but has not proved very useful in *Saccharomyces* (Emeis 1966; Strömnaes, 1968).

A. Tetrad Analysis

In *Saccharomyces* the four spores from an ascus are the four products of an individual meiosis. By isolating and culturing the four spores and assaying and analyzing the resulting clones as a tetrad, one can obtain more information than can be obtained from random spore analysis. In particular, one can determine centromere linkage of a gene, detect chromatid interference, and recognize lethality associated with chromosome aberrations and mutations.

1. MONOFACTORIAL CROSSES: $A \times a$

A diploid resulting from a cross of wild-type (A) and mutant (a) strains is expected to yield asci that contain $2\,A{:}2\,a$ spores. This is the outcome predicted with the phenotype under the control of a single nuclear gene and normal meiosis. Nevertheless, all other combinations of $A{:}a$ may be encountered. Exclusively $4\,A{:}0\,a$ segregations indicate that the phenotypic difference between A and a is cytoplasmically inherited. Occasionally, a distribution of $4\,A{:}0\,a$, $3\,A{:}1\,a$, and $2\,A{:}2\,a$ segregations is encountered and this can indicate polyploidy, polysomy, polymeric gene control, or segregation of a suppressor. Additional crosses are necessary to distinguish among these alternatives. Alternatively, a distribution of $2\,A{:}2\,a$, $1\,A{:}3\,a$, and $0\,A{:}4\,a$ segregations is indicative of two or more complementary genes controlling the phenotype. Even when the phenotype is under the control of a single gene and the $2\,A{:}2\,a$ pattern predominates, it is usually accompanied by a low frequency of $3\,A{:}1\,a$ and $1\,A{:}3\,a$ segregations as a consequence of gene conversion (Table I). The few 4:0 and 0:4 segregations in Table I are most likely an expression of mitotic crossing-over in one of the divisions before meiosis.

2. TWO-FACTOR CROSSES: $AB \times ab$

If more than one heterozygous site is segregating in a 2:2 fashion, the various segregations can be considered in pairwise combinations, $AB \times ab$. Assuming 2:2 segregations for both A/a and B/b, only three types

TABLE I

SEGREGATION PATTERNS FOR AN ASSORTMENT
OF HETEROZYGOUS SITES IN SACCHAROMYCES[a]

Gene	Number of asci with segregation ratios:				
	4:0	3:1	2:2	1:3	0:4
arg4-1	0	39	1107	32	0
arg4-17	0	143	3456	130	0
his4	1	40	696	24	0
leu1-1	1	19	3676	34	0
leu2-1	0	13	742	6	0
lys1-1	4	85	3577	59	2
a/α	0	11	4460	17	0
met1	3	35	2497	15	0
thr1	0	113	3457	135	0
trp1	0	11	3713	4	0

[a]Unpublished data of S. Fogel, D. Hurst, and
R. Mortimer.

of asci are possible:

Parental ditype (PD):	AB	AB	ab	ab
Nonparental (NPD):	Ab	Ab	aB	aB
Tetratype (T):	AB	Ab	aB	ab

The relative frequencies of these three ascal classes depend on the positions of A and B relative to each other, or to their respective centromeres if A and B are on different chromosomes. The criterion used to establish that two genes are linked is PD/NPD > 1. The statistical significance of deviations from 1:1 for various PD/NPD ratios is presented in Table II. This table is taken from Perkins (1952).

For two genes on the same chromosome, the three tetrad classes are generated by crossovers in the interval between the genes. Shult and Lindegren (1956) developed an algebraic method to determine the consequences of various numbers of exchanges in the interval defined by A/a and B/b. If X represents a crossover event between A and B in asci in which A and B would otherwise have segregated as PD, NPD, or T, the following equations can be written to describe the resultant configurations:

$$X(PD) = T$$
$$X(NPD) = T$$

and

$$X(T) = \tfrac{1}{4} PD + \tfrac{1}{4} NPD + \tfrac{1}{2} T$$

These relations are obtained with the assumption of no sister strand ex-

TABLE II

SMALLEST NUMERICAL RATIOS SHOWING SIGNIFICANT DEVIATION IN ONE DIRECTION FROM 1:1[a]

Total number	Ratio attaining significance level (one-sided)			Total Number	Ratio attaining significance level (one-sided)		
	5%	2½%	1%		5%	2½%	1%
5	5:0	—	—	28	19:9	20:8	21:7
6	6:0	6:0	—	29	20:9	21:8	22:7
7	7:0	7:0	7:0	30	20:10	21:9	22:8
8	7:1	8:0	8:0	31	21:10	22:9	23:8
9	8:1	8:1	9:0	32	22:10	22:10	23:9
10	9:1	9:1	10:0	33	22:11	23:10	24:9
11	9:2	10:1	10:1	34	23:11	24:10	25:9
12	10:2	10:2	11:1	35	23:12	24:11	25:10
13	10:3	11:2	12:1	36	24:12	25:11	26:10
14	11:3	12:2	12:2	37	24:13	25:12	26:11
15	12:3	12:3	13:2	38	25:13	26:12	27:11
16	12:4	13:3	14:2	39	26:13	27:12	28:11
17	13:4	13:4	14:3	40	26:14	27:13	28:12
18	13:5	14:4	15:3	41	27:14	28:13	29:12
19	14:5	15:4	15:4	42	27:15	28:14	29:13
20	15:5	15:5	16:4	43	28:15	29:14	30:13
21	15:6	16:5	17:4	44	28:16	29:15	31:13
22	16:6	17:5	17:5	45	29:16	30:15	31:14
23	16:7	17:6	18:5	46	30:16	31:15	32:14
24	17:7	18:6	19:5	47	30:17	31:16	32:15
25	18:7	18:7	19:6	48	31:17	32:16	33:15
26	18:8	19:7	20:6	49	31:18	32:17	34:15
27	19:8	20:7	20:7	50	32:18	33:17	34:16

[a] From Perkins (1952).

changes and no chromatid interference. The fractions of PD, NPD, and T tetrads as a function of the number of exchanges r in the interval $A–B$ are presented in Table III.

A given sample of asci is distributed relative to r such that $p(r)$ defines the probability of r crossover events in $A–B$. The map distance x between A and B is defined as the average number of crossover events per strand times 100. Since each crossover involves only two of the four chromatids,

$$x = 100 \tfrac{1}{2} \sum_{r=0}^{\infty} r\, p\,(r)$$

If x is relatively short, $p(r)$ for $r > 2$ can be considered zero as a first approximation. Using the above equation and Table III, a minimum esti-

TABLE III

FRACTION OF VARIOUS ASCAL CLASSES VERSUS NUMBER OF EXCHANGES

Ascal class	Number of crossovers in A–B, r:								
	0	1	2	3	4	5	6	...	∞
PD	1	0	$\frac{1}{4}$	$\frac{1}{8}$	$\frac{3}{16}$	$\frac{5}{32}$	$\frac{11}{64}$	...	$\frac{1}{6}$
NPD	0	0	$\frac{1}{4}$	$\frac{1}{8}$	$\frac{3}{16}$	$\frac{5}{32}$	$\frac{11}{64}$	...	$\frac{1}{6}$
T	0	1	$\frac{2}{4}$	$\frac{6}{8}$	$\frac{10}{16}$	$\frac{22}{32}$	$\frac{42}{64}$	...	$\frac{4}{6}$

mate of the map distance can be derived in terms of the values of PD, NPD, and T as follows:

$$x = 100\,\tfrac{1}{2}\,\frac{T + 6NPD}{PD + NPD + T}$$

This equation was originally derived by Perkins (1949).

If three or more genes on a given chromosome segregate, their sequence relative to each other is determined by an examination of exchange patterns in individual asci. The sequence requiring the minimal number of exchanges is chosen as the most probable one.

3. CENTROMERE LINKAGE

The centromeres of homologous chromosomes segregate with the first division of meiosis. With the second division of meiosis, the centromeres split and the two chromatids are distributed to "sister spores." In yeast with ordered spore arrays, i.e., having asci in which sister spores can be recognized by their position, it is sometimes possible to retain the order of the spores during the dissection of the asci and thus have a direct measure of first-division segregation of the genes under investigation. In *Schiz. pombe*, the zygotic asci contain the spores in a linear order analogous to the classic *Neurospora crassa* pattern in that the pair of spores at each end are sisters, and thus a gene showing no recombination with the centromere gives the pattern *A A a a* (Leupold, 1950). Certain strains of *S. cerevisiae* form up to 40% linear asci with particular sporulation regimens (Hawthorne, 1955). In linear ascogenous cells the spindles for the second meiotic division are apparently parallel and overlapping; thus the resulting distribution of the nuclei is such that nonsister spores alternate in the ascus. Although the prevalent first-division segregation pattern is *A a A a*, as many as 10% of the asci reflect slippage of the nuclei, resulting in the arrays *A a a A* or *a A A a* for a gene at the centromere (Hawthorne and Mortimer, 1960).

Most crosses within *Saccharomyces* breeding stocks yield only oval or tetrahedral asci, and no order for sister spores can be demonstrated (Magni,

1961). In any case the use of snail gut juices to facilitate dissection of the spores by removing the ascus wall precludes the retention of an ordered spore array. Therefore, in practice, one includes one or more genes known to be close to their centromeres in crosses designed for mapping purposes. For most investigations it suffices to include either *trp1* or *pet8*; each shows less than 1% second-division segregation (Mortimer and Hawthorne, 1966). Alternatively, one includes at least three unlinked genes with second-division segregation frequencies in the order of 5–10% (e.g., *ade1*, *gal7*, *leu2*, *ura3*, *leu1*, *pet1*, *met3*, or *met14*) and then inspects each tetrad to decide which segregation pattern is the most probable first-division spore array for that ascus. In effect, one orders each ascus with the centromere markers, and then uses that array to decide whether the other genes in the cross show first- or second-division segregation (Mortimer and Hawthorne, 1969).

From our linkage studies with *Saccharomyces*, we have seen that, for genes located at increasingly greater distances from the centromere, second-division segregation frequency rises from zero to a maximum of 75–80% for genes located 40–70 centimorgans from the centromere, and then falls to 67% for genes at greater distances. The maximum reflects chiasma interference (Barratt *et al.*, 1954). Any gene that shows significantly less than 67% second-division segregation is assumed to be centromere-linked.

If one were to commence the investigation of a new yeast, from a breeding system distinct from *Saccharomyces*, which gave four-spored asci lacking an ordered array, the first step in the mapping of centromeres would be a search for gene pairs with PD/NPD/T ratios of 1:1: <4. The frequency of tetratype asci is dependent on the exchange frequencies between A and B and their centromeres. The relationship is $f(T_{ab}) = x + y - \frac{3}{2}xy$, where x and y are fractions of asci having a second-division segregation for A/a and B/b, respectively (Perkins, 1949). Both genes must be centromere-linked, i.e., both x and y must be less than $\frac{2}{3}$, for $f(T)$ to be less than $\frac{2}{3}$. Unless $f(T) = 0$, the cross must include a third centromere-linked gene C/c, with second-division segregation frequency z, in order to solve for x, y, and z. From the solution of the three simultaneous equations for $f(T_{ab})$, $f(T_{ac})$, and $f(T_{bc})$, one obtains three equations in the form:

$$x = \tfrac{2}{3} \left[1 \pm \sqrt{\frac{4 - 6T_{ab} - 6T_{ac} + 9T_{ab}T_{ac}}{4 - 6T_{bc}}} \right]$$

(Whitehouse, 1950; Fincham and Day, 1971).

With tetraploid hybrids it is possible to determine the second-division segregation frequency of a single centromere-linked gene A in the duplex state ($A/A/a/a$). If the four spores are true diploids (the usual case), a gene at the centromere will result in tetrads with only 4:0 or 2:2 segregations for dominant versus recessive phenotypes. Tetrads with 3:1 segregations result from crossovers between A and the centromere, and their frequency can be used to determine the percent second-division segregation for A (Roman et al., 1955; Leupold, 1956).

A tetraploid with the centromere-linked gene A in the triplex state $(A/A/A/a)$ should give only 4:0 tetrads. However, exceptional 3:1 asci can arise from gene conversion $A \rightarrow a$, or from a misassortment of chromosomes resulting in a monosome $(2n -1)$ and exposing a (Bruenn and Mortimer, 1970). If the monosome can in turn be sporulated, then one should observe the segregation of a lethal. The two surviving spores will be sister spores, thus providing a means to detect second-division segregation for any genes that are heterozygous (Takahashi, 1972a).

4.   The Definition of New Centromeres

If a gene newly introduced into the *Saccharomyces* breeding stock shows centromere linkage, the simplest procedure for placing it on the linkage maps is to cross it into a hybrid heterozygous for 8 to 10 centromere makers. Usually, only one or two crosses are needed to demonstrate linkage with one of the 16 centromere markers. To define a new centromere, the gene must be shown to be not linked to any of the established centromere markers. If the new gene lies across the centromere from the established marker, there is a possibility that they may not show significant linkage to one another. This is because of two factors: (1) the separation represents the sum of the two gene-centromere intervals, and (2) multiple crossovers are likely, since chiasma interference does not extend across the centromere (Hawthorne and Mortimer, 1960). When this problem arises, one should restrict the analysis to tetrads with first-division segregations for both genes (Catcheside, 1951). Any NPD tetrad for linked genes would require a four-strand double crossover within one of the gene-centromere intervals, a rare event among centromere-linked genes.

B.   Random Spore Analysis

Random spore analysis for detecting and measuring linkage in *Saccharomyces* can be used in situations in which one wishes to avoid micromanipulation, e.g., laboratory exercises (Bevan and Woods, 1963). The preparation of random spore samples involves the separation of spores from unsporulated diploid cells, as well as from one another. Generally, sporulating yeast are treated with an enzyme preparation, such as Glusulase (Endo Laboratories, Garden City, N.Y.), to digest the ascus walls, and then sonicated to disrupt the tetrads (Magni, 1963; Siddiqi, 1971). The spores then can be separated from the vegetative cells by phase separation (Emeis, 1958), selective killing (Zakharov and Inge-Vechtomov, 1964), or differential electrophoretic mobility (Resnick et al., 1967).

Even with the above separation procedures, the final sample may contain a small fraction of diploid vegetative cells. The diploid colonies can usually be recognized by their larger size; however, one can construct diploids heterozygous for recessive genes affecting colony color (ade1 or ade2 for red colonies) or morphology (rgh1 for rough colonies), and then pick only colonies expressing the recessive phenotype (Zakharov and Inge-Vechtomov, 1964). Alternatively, one can use diploids heterozygous for a recessive gene that confers resistance to a toxic agent, e.g., can1 for canavanine resistance. The sensitive diploid cells and one-half of the spores will be unable to grow on canavanine-containing medium. A variation of this approach has been to include two selective markers in the heterozygous condition, i.e., an ochre suppressor and an ochre suppressible allele of can1 (Mortimer and Hawthorne, 1973). With the latter approach, separation of the spores is unnecessary.

C. Mitotic Segregation

Diploid yeast cells heterozygous for a genetic marker, A/a, may yield A/A and a/a segregants during mitotic division. The spontaneous frequency per mitotic division depends on the distance of the gene from its centromere and is on the order of 10^{-4} to 10^{-5}. This frequency can be increased to several percent if the cells are exposed to x rays, ultraviolet light, or alkylating agents (Mortimer and Manney, 1971).

Two mechanisms contribute to mitotic segregation of heterozygous markers: mitotic crossing-over and mitotic gene conversion. One-half of the mitotic crossovers between the locus A/a and the centromere are expected to result in sectored colonies, A/A in one sector and a/a in the other, if there is no preferential assortment of exchange chromatids. All heterozygous sites distal to the crossover event sector together. Thus coincident sectoring of two markers is evidence for their location on the same chromosome arm.

A frequent practice is to construct diploids heterozygous for a "signal" marker such as an adenine gene that affects colony color. Both halves of sectored colonies can be isolated to another plate which then can be replica-plated to a series of plates deficient for the various markers present in the heterozygous condition. Coincident sectoring due to independent segregation of genes on different chromosomes occurs infrequently (< 1%). Thus a frequency of concomitant sectoring significantly above this level is the criterion for linkage. If the gene is proximal to the signal marker, this frequency can vary from a few percent to nearly 100%. For genes distal to the signal marker, the frequency will be in the 50–100% range. One usually does not find 100% concomitant sectoring for distal genes, because of the occurrence of mitotic gene conversion, as well as additional mitotic crossovers (Nakai and Mortimer, 1969).

Mitotic gene conversion is evidenced by homozygosity in one sector and persistent heterozygosity in the opposite sector (Roman, 1971). This event frequently occurs without sectoring of neighboring genes, and hence detracts from the efficiency of mitotic segregation in detecting linkage.

D. Trisomic Analysis

A diploid that carries an extra copy of one of the chromosomes is known as a trisome. Such trisomic strains result from a cross of a disome ($n + 1$) and a normal haploid. If the haploid carries a mutation on the disomic chromosome such that the resultant cross is $A/A/a$, the segregations of $A:a$ will be a distribution of 4:0, 3:1, and 2:2. Alternatively, if the mutation is on any of the other chromosomes, the segregations will be 2:2. Thus a given mutation can be either located on or excluded from a particular chromosome by crossing a strain bearing that mutation to a strain disomic for the chromosome in question. A variety of ways of constructing disomic strains has been described (Parry and Cox, 1971; Shaffer et al., 1971).

Triploids yield a low frequency of viable meiotic products that may contain one to five or more chromosomes in the disomic state. If a haploid carrying markers on most of the chromosomes plus unmapped markers is crossed to such aneuploids, the segregations of the mapped genes will determine which chromosomes are disomic and which are trisomic in the cross, i.e., 2:2 versus 4:0, 3:1, and 2:2. The segregations of the unmapped genes can then be considered. If 2:2, the gene is excluded from all chromosomes that are trisomic; if 4:0, 3:1, and 2:2, the gene is excluded from disomic chromosomes. A given cross can serve to exclude a gene from several chromosomes, so that relatively few such crosses are needed to narrow the location of a gene down to a specific chromosome (Mortimer and Hawthorne, 1973).

III. Genetic Map

A revised genetic map of *S. cerevisiae* is presented in Fig. 1. This map includes 28 genes not on the map in our previous publication (Mortimer and Hawthorne, 1973). These new linkages have been reported by several laboratories as follows: *cys 1* on chromosome I (S. de la Cruz Halos and S.

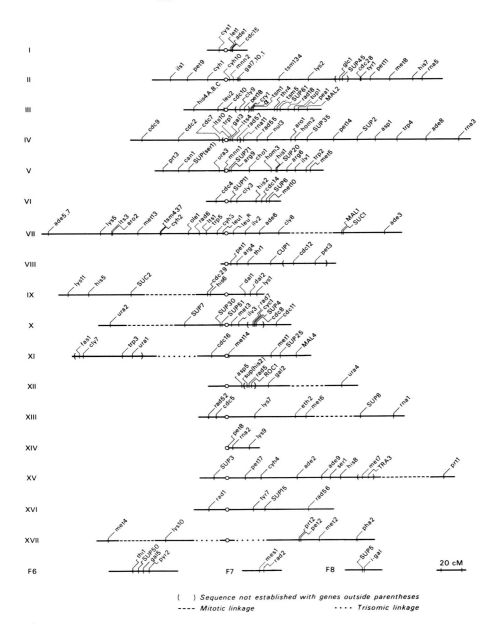

( ) Sequence not established with genes outside parentheses
---- Mitotic linkage   •••• Trisomic linkage

FIG. 1. Genetic map of *S. cerevisiae*.

Fogel, personal communication), *cyh10* on II, *lts4* and *lts10* on IV, *lts1* and *lts3* on VII (Singh and Manney, 1974), *mnn2* on II (L. Ballou and C. E. Ballou, personal communication), *cry1* on III (Skogerson *et al.*, 1973), *tsm1* and *tsm5* on III (G. R. Fink, personal communication), *tup1* on III (Wickner, 1973), *pea1* on III (Takahashi, 1972b), *rad57* on IV, *rad6* on VII, and *rad56* on XVI (Game and Mortimer, 1974), *cdc28* on II, *cdc29* on IX, and *cdc16* on XI (Hartwell *et al.*, 1973), *mnn1* on V (Antalis *et al.*, 1973), *cho1* on V (S. Fogel and B. Jensen, personal communication), *dal1* and *dal2* on IX (Lawther *et al.*, 1974), *rad7* on X (F. Sherman, personal communication), *fas1* on XI (Kühn *et al.*, 1972; Culbertson and Henry, 1973), *eth2* and *met6* on XIII (Masselot and de Robichon-Szulmajster, 1974), and *met7* and *TRA3* on XV (G. R. Fink, personal communication). Additionally, the location of *cdc10* has been changed to near the centromere on the left arm of III (G. R. Fink, personal communication), and *SUP31* has been deleted from VIII.

References

Antalis, C., Fogel, S., and Ballou, C. E. (1973). *J. Biol. Chem.* **248**, 4655–4659.

Barratt, R. W., Newmeyer, D., Perkins, D. D., and Garnjobst, L. (1954). *Advan. Genet.* **6**, 1–93.

Bevan, E. A., and Woods, R. A. (1963). In "Teaching Genetics" (C. D. Darlington and A. D. Bradshaw, eds.), p. 27–35. Oliver & Boyd, Edinburgh and London.

Blamire, J., Cryer, D. R., Finkelstein, D. B., and Marmur, J. (1972). *J. Mol. Biol.* **67**, 11–24.

Bruenn, J., and Mortimer, R. K. (1970). *J. Bacteriol.* **102**, 548–551.

Catcheside, D. G. (1951). "The Genetics of Micro-Organisms." Pitman, London.

Culbertson, M. R., and Henry, S. (1973). *Genetics* **75**, 441–458.

da Cunha, M. F. (1969). Ph.D. Thesis, Universidade do Rio de Janeiro.

Desborough, S., Shult, E. E., Yoshida, T., and Lindegren, C. C. (1960). *Genetics* **45**, 1467–1480.

Emeis, C. C. (1958). *Brauerei* **11**, 160–163.

Emeis, C. C. (1966). *Z. Naturforsch. B* **21**, 816–817.

Fincham, J. R. S., and Day, P. R. (1971). "Fungal Genetics." 3rd ed., Blackwell, Oxford.

Finkelstein, D. B., Blamire, J., and Marmur, J. (1972). *Nature (London), New Biol.* **240**, 279–281.

Fischer, P., Weingartner, B., and Wintersberger, U. (1973). *Exp. Cell Res.* **79**, 452–456.

Game, J. C., and Mortimer, R. K. (1974). *Mutat. Res.* **24**, 281–292.

Gutz, H. (1966). *J. Bacteriol.* **92**, 1567–1568.

Hartwell, L. H., Mortimer, R. K., Culotti, J., and Culotti, M. (1973). *Genetics* **74**, 267–286.

Hawthorne, D. C. (1955). *Genetics* **40**, 511–518.

Hawthorne, D. C., and Mortimer, R. K. (1960). *Genetics* **45**, 1085–1110.

Hawthorne, D. C., and Mortimer, R. K. (1963). *Genetics* **48**, 617–620.

Hawthorne, D. C., and Mortimer, R. K. (1968). *Genetics* **60**, 735–742.

Hwang, Y. L., Lindegren, G., and Lindegren, C. C. (1963). *Can. J. Genet. Cytol.* **5**, 290–298.

Hwang, Y. L., Lindegren, G., and Lindegren, C. C. (1964). *Can. J. Genet. Cytol.* **6**, 373–380.

Kühn, L., Castorph, H., and Schweizer, E. (1972). *Eur. J. Biochem.* **24**, 492–497.

Lawther, P., Reimer, E., Chojnacki, B., and Cooper, T. G. (1974). *J. Bacteriol.* **119**, 461–468.

Leupold, U. (1950). *C. R. Trav. Lab. Carlsberg* **24**, 381–480.

Leupold, U. (1956). *J. Genet.* **54**, 411–426.
Leupold, U. (1970). *In* "Methods in Cell Physiology" (D. Prescott, ed.), Vol. 4, p. 169–177. Academic Press, New York.
Lindegren, C. C. (1949). *Hereditas, Suppl.* 338–355.
Lindegren, C. C., and Lindegren, G. (1951). *Indian Phytopathol.* **4**, 11–20.
Lindegren, C. C., Lindegren, G., Shult, E. E., and Desborough, S. (1959). *Nature (London)* **183**, 800–802.
Lindegren, C. C., Lindegren, G., Shult, E. E., and Hwang, Y. L. (1962). *Nature (London)* **194**, 260–265.
Magni, G. E. (1961). *Atti. Ass. Genet. Ital.* **6**, 47–50.
Magni, G. E. (1963). *Proc. Nat. Acad. Sci. U.S.A.* **50**, 975–980.
Masselot, M., and de Robichon-Szulmalster, H. (1974). *Mol. Gen. Genet.* **129**, 339–348.
Matile, P. H., Moor, H., and Robinow, C. F. (1969). *In* "The Yeasts" (A. H. Rose and J. S. Harrison, eds.), Vol. 1, pp. 219–302. Academic Press, New York.
Mortimer, R. K., and Hawthorne, D. C. (1966). *Genetics* **53**, 165–173.
Mortimer, R. K., and Hawthorne, D. C. (1969). *In* "The Yeasts" (A. H. Rose and J. S. Harrison, eds.), Vol. 1, pp. 385–460. Academic Press, New York.
Mortimer, R. K., and Hawthorne, D. C. (1973). *Genetics* **74**, 33–54.
Mortimer, R. K., and Manney, T. R. (1971). *In* "Chemical Mutagens: Principles and Methods for their Detection" (A. Hollaender, ed.), Vol. 1, p. 289–310. Plenum, New York.
Nakai, S., and Mortimer, R. K. (1969). *Mol. Gen. Genet.* **103**, 329–338.
Parry, E. M., and Cox, B. S. (1971). *Genet. Res.* **16**, 333–340.
Perkins, D. D. (1949). *Genetics* **34**, 607–626.
Perkins, D. D. (1952). *Genetics* **38**, 187–197.
Petes, T. D., and Fangman, W. L. (1972). *Proc. Nat. Acad. Sci. U.S.A.* **69**, 1188–1191.
Petes, T. D., Byers, B., and Fangman, W. L. (1973). *Proc. Nat. Acad. Sci. U.S.A.* **70**, 3072–3076.
Resnick, M. A., Tippets, R. D., and Mortimer, R. K. (1967). *Science* **158**, 803–804.
Roman, H. (1971). *In* "Genetic Lectures" (R. Bogart, ed.), Vol. III, p. 43–59. Oregon State Univ. Press, Corvallis.
Roman, H., Phillips, M. M., and Sands, S. M. (1955). *Genetics* **40**, 546–561.
Shaffer, B., Brearley, I., Littlewood, R., and Fink, G. R. (1971). *Genetics* **67**, 483–495.
Shult, E. E., and Desborough, S. (1960). *Genetica* **31**, 147–187.
Shult, E. E., and Lindegren, C. C. (1956). *J. Genet.* **54**, 343–357.
Shult, E. E., and Lindegren, C. C. (1957). *Genetica* **29**, 58–82.
Shult, E. E., and Lindegren, C. C. (1959). *Can. J. Genet. Cytol.* **1**, 189–201.
Siddiqi, B. A. (1971). *Hereditas* **69**, 67–76.
Singh, A., and Manney, T. R. (1974). *Genetics* **77**, 651–659.
Skogerson, L., McLaughlin, C., and Wakatama, E. (1973). *J. Bacteriol.* **116**, 818–822.
Strömnaes, O. (1968). *Hereditas* **59**, 197–220.
Takahashi, T. (1972a). *Bull. Brew. Sci.* **18**, 37–48.
Takahashi, T. (1972b). *Bull. Brew. Sci.* **18**, 51–52.
Tingle, M., Herman, A., and Halvorson, H. O. (1968). *Genetics* **58**, 361–371.
Whitehouse, H. L. K. (1950). *Nature (London)* **165**, 892.
Wickner, R. B. (1973). *J. Bacteriol.* **117**, 252–260.
Zakharov, I. A., and Inge-Vechtomov, S. G. (1964). *Genetika* **8**, 112–118.

Chapter 13

The Use of Mutants in Metabolic Studies

F. LACROUTE

Institut de Biologie Moléculaire et Cellulaire, Strasbourg, France

I. Introduction

The great development of molecular biology during the last 30 years is a striking example of the linkage between progress in two areas of science, biochemistry and genetics. As in other areas of biochemistry, research on metabolic pathways has benefited from the possibilities offered by genetics to inactivate more or less specifically one cell function. This approach, first developed by Beadle and Tatum in *Neurospora*, has also been extensively and fruitfully used in yeasts. This chapter is designed to summarize the different ways one can use yeast mutants and to present selected examples of these uses. The examples are essentially chosen from our own studies on the pyrimidine pathway, not for a particular demonstrative value, but because the analysis of the difficulties encountered during the utilization of mutants strains, which is one of the purposes of this chapter, is generally not sufficiently detailed in published papers. We do not intend to present a summary of the results obtained through the use of mutants, but only a survey of the potential of the method. Some aspects of the utilization of mutants are described in other chapters of this book, especially the experimental uses of mutants (Chapters 14–15), and the use of *ts* mutants (Vol. 12 of this series); these aspects are therefore not developed here.

If one limits oneself to metabolic pathways, three principal problems can be approached experimentally by the production of mutants.

First, mutants can be used to confirm, to disprove, or even to discover the existence of a metabolic pathway. Second, they can be used to establish the occurrence within the cell of multiple parallel or cross-linked pathways. Third, they can be used to study the regulatory events involved in metabolic pathways and their mechanisms.

These three different aspects are presented after a short comment on the problems raised by the selection of mutants.

II.   Selection and Conservation of Mutants

The technique chosen depends on the type of mutant wanted and on the available methods of selection.

If one requires nonreverting mutations, it seems that the chances of success are better when one starts with spontaneous mutants. The probability can be very low; of more than 10,000 spontaneous *ura2* mutants studied, only 6 corresponded to large deletions of this gene (M. L. Bach, personnal communication). Nevertheless, at the *ade3* locus, Jones (1972) obtained two multisite deletions in 26 spontaneous mutants mapped. The difference in the rates reflects very likely a difference in the genetic background of the yeast strains used.

If one does not have a special requirement regarding the nature of the mutational event, any strong mutagen can be used, such as nitrous acid, N'-methyl-N′-nitrosoguanidine, or ultraviolet light. Nevertheless, it is generally better either to avoid the action of mutagens or to apply very low doses to decrease the probability of secondary mutations accompaning the mutation selected. This assumes that the selection of a few mutants in a huge cell population can be carried out, which happens only in a few cases, e.g., the selection of mutants resistant to a toxic compound, the selection of auxotrophic mutants by the nystatin technique provided they do not die during the starvation (Snow, 1966), the selection of red adenine mutants (Roman, 1956), and the selection of some pyrimidineless mutants (Bach and Lacroute, 1972). In the majority of other cases, one does not have such a strong selection, and it is necessary to increase the mutation rate with a mutagen to find enough mutants among the survivors. It is therefore very important to eliminate as much as possible, by backcrossing with a wild strain, all the secondary mutations accumulated. This is especially true in the case of metabolic studies in which one wants to correlate secondary regulatory changes with a primary enzymatic defect, and in which the occurrence of secondary mutations could completely confuse the issue.

The existence of secondary mutations is not a theoretical trap and occurs very frequently. For example, in a nitrous acid mutagenesis at 1% survival level, three different unlinked mutations involving the uracil-arginine pathway appeared in the same strain, one for carbamyl phosphate synthetase specific for the pyrimidine pathway, one for carbamyl phosphate synthetase specific for the arginine pathway, and one leaky mutation for another undetermined arginine gene.

Even when secondary mutations are avoided during the initial selection of a mutant strain, it happens frequently that they are selected during the cloning or the storage of the strain. A great proportion of mutations impair to different degrees the growth rate or survival after a long stationary phase. This gives a great selective advantage either to revertant (or suppressed) strains, which are easily discarded, or to secondary mutations, which alleviate the noxious effects of the primary mutation without suppressing the principal phenotype. The frequency of such mutations selected after the isolation of a mutant strain can be very high, e.g., when the accumulation of an intermediary metabolite is toxic for the cell. Such a case is found in the pyrimidine pathway for *ura4* mutants in which the accumulation of ureidosuccinic acid causes strong inhibition of growth (Fig. 1). Mutants belonging to this locus are more difficult to keep in stock for long storage since, if one is not especially cautious, all the cells may die. Moreover, the survivors are very often double mutants *ura2 ura4*, the *ura2* mutation preventing the accumulation of ureidosuccinic acid and therefore the toxic effect. It appears that none of the classic methods for keeping a stock is perfect. The majority of them give very good results for the wild strain, but fail for difficult mutants. Deep-freezing at $-90°C$ in complete liquid medium with 10% glycerol often seems to give good results. The best way to keep a mutant free of secondary mutations for long times is to conserve the mutation in an heterozygous diploid strain and to reisolate, after meiosis, spores containing the mutation each time one needs to work with it. While these secondary mutations correcting the poor growth or survival of a mutant are a burden for the biochemist when they occur without control, they may represent a very useful tool in determining regulatory interactions inside the cell.

In the detection of mutations useful for metabolic studies, all the classic methods may be useful at one time or another, depending on the kind of mutation wanted. The direct growth of colonies is observed when one looks for mutants resistant to a compound, or for survivors after contact with a

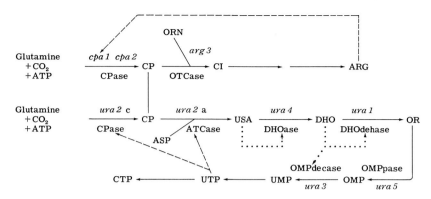

FIG. 1. Metabolic relationship between arginine and pyrimidine pathway. Solid lines represent reactions, dashed lines represent inhibitory effects (either direct or by repression), and dotted lines represent induction of enzymatic activity. The genes coding for each activity are indicated. CPase, Carbamyl phosphate synthetase; ATCase, aspartic transcarbamylase; DHOase, dihydroorotase; DHOdehase, dihydroorotic dehydrogenase; OMPpase, orotydilic acid pyrophosphorylase; OMPdecase, orotydilic acid decarboxylase; CP, carbamyl phosphate; USA, ureidosuccinic acid; DHO, dihydroorotic acid; OR, orotic acid; OMP, orotydilic acid; UMP, uridine 5'-phosphate; UTP, uridine triphosphate; CTP, cytidine triphosphate; OTCase, ornithine transcarbamylase; ORN ornithine; CI, citrulline; ARG, arginine.

killing agent. Velvet replica plating is used when one wants to obtain auxotrophic mutants or mutants sensitive to inhibition by a given compound. Growth or replica plating on a medium spread with an appropriate indicator strain is chosen when one wants to select mutants excreting a defined compound (one can see either the inhibition or the stimulation of growth of the indicator strain). Growth of colonies on a Millipore filter, followed by incubation on a radioactive substrate and autoradiography are used when one wants either mutants unable to incorporate starting from a strain that is able to incorporate, or conversely.

Since the great majority of mutations have a recessive effect, one starts generally from an haploid strain for mutant selection, but one must not restrict oneself to haploids when exploring all the selective possibilities of a system. This is especially true for mutants resistant to a compound for which one very often obtains interesting mutations by selecting them in a diploid strain. Since dominant mutations correspond generally to a modification of function, they are much less frequent than the recessive ones which correspond to a loss of function and would be missed if the selection were not made on a diploid.

When selecting mutations for resistance to a compound, one should use a very broad range of concentrations for the selection, including concentrations very close to the growth threshold of the wild type. Indeed, it often happens that the mutations most interesting to the biochemist are not those giving high resistance, which can be trivial (such as an enzymatic defect in the uptake or in the transformation of a toxic compound) but those giving a low resistance by changing affinity for the analog of an enzyme or increasing the rate of synthesis of the corresponding natural compound.

An interesting approach in determining the main level of action of a toxic compound inside a cell is to select at a low temperature (22 °C) mutants resistant to its action, and to detect (if any) the ones that are thermosensitive by replica plating to a high temperature (37 °C). If resistance is obtained by a change in the primary structure of an essential protein which interacts with the chemical, this change can often produce instability at the high temperature. The physiological behavior of the mutant at the high temperature gives an indication of the affected function. It is easy to discard thermosensitive mutants occurring independently of this resistance when the gene producing the resistance is recessive. One prepares heterozygous diploid by a cross of the thermosensitive mutant with the wild type and selects by resistance the homozygous diploid appearing by gene conversion. If thermosensitivity is independent of resistance, the converted clones will generally not be thermosensitive.

Another interesting method especially suited for the study of interpathway regulation and interaction is based on the selection of mutants inhibited by natural compounds of intermediate metabolism and is presented in Section IV.

III. Elucidation of a Metabolic Pathway

This is now required less and less frequently, since the great majority of the fundamental pathways are fairly well known from previous studies of biochemists and geneticists on bacteria and *Neurospora*. Although the basic schemes of biosynthesis or catabolism are very similar in all living organisms, some important details may differ, and genetics is still a useful tool for revealing them. In yeasts the pathways for biosynthesis of purines, pyrimidines, and amino acids have been analyzed with the help of mutants (see the review by de Robichon-Szulmajster and Surdin-Keryan, 1971). As stated before, we focus here on the pyrimidine pathway.

Independent mutageneses have consistently yielded four different loci characterized by an auxotrophy for pyrimidines (Mortimer and Hawthorne, 1966; Lacroute, 1968). Biochemical analysis of the enzymatic defects linked to either *ura1*, *ura2*, *ura3*, or *ura4* showed a correspondance of each locus with one enzymatic activity in the pathway (Fig. 1). In spite of a large number of independent mutants, two steps in pyrimidine biosynthesis were left without mutations for the corresponding activities: carbamyl phosphate synthetase and orotidine-5′-phosphate pyrophosphorylase. The way the mutants for these two last steps were obtained is interesting to detail, since it shows the reason to overlook mutants during classic selection as a result of the existence of isoenzymes in the cells.

Two enzymatic activities in yeast catalyze the biosynthesis of carbamyl phosphate; one is specific for the pyrimidine pathway, and the other is specific for the arginine pathway (Lacroute *et al.*, 1965). The first mutants for these two activities were obtained only by chance, but later Meuris *et al.* (1967) developed a technique allowing selection for these mutants. This selection is based on the fact that, if mutant strains for carbamyl phosphate synthetase specific for the arginine pathway can grow on minimal medium (since an activity corresponding to the pyrimidine pathway remains), their growth is completely blocked by uracil because of a strong feedback inhibition by UTP of the remaining activity. They grow normally if arginine is added with uracil. By growing colonies after mutagenesis on minimal medium and replica plating on uracil medium, one therefore selects the mutants for the enzyme of the arginine pathway in the uracil inhibited clones. To obtain mutants for the uracil pathway enzyme, one must, starting from a mutant for the arginine enzyme, grow survivors on uracil plus arginine, and replica plate on arginine. The strains unable to grow will belong either to the four uracil loci identified earlier, or to the locus for carbamyl phosphate synthetase specific for pyrimidines. The former are easy to discard after looking at the complementation pattern obtained in crosses with the different *ura* mutants. The situation corresponding to these two isoenzymatic activities is summarized in Fig. 1. Genetic analysis of mutants for carbamyl synthetase specific for the pyrimidine pathway has shown that they are in fact allelic to the locus coding for aspartic transcarbamylase (ATCase) activity. The fine structure of this locus reveals two adjacent regions, one responsible for ATCase activity, the other responsible for carbamyl phosphate synthetase activity (Denis–Duphil and Lacroute, 1971).

The last mutants to be obtained that completed identification of all the structural genes for the pyrimidine biosynthetic pathway were those for OMP pyrophosphorylase. The physiological basis of the technique used to select them is not yet understood. Schematically, the situation is the following. Ureidosuccinic acid inhibits cell growth when it accumulates in the cellular pool, either as the result of an enzymatic block occurring afterward, or because it is taken up from the growth medium. It appeared that, according to the location of a mutation in the pathway, strains were or were not resistant to externally added ureidosuccinic acid. The wild type, *ura2*, and *ura4* mutants were sensitive, but *ura1* was slightly resistant and *ura3* very resistant. This allowed direct selection of *ura1* and *ura3* mutants starting from a wild strain plated after mutagenesis on uracil and ureidosuccinic acid (Bach and Lacroute, 1972). Unexpectedly, a new type of mutant was obtained which had leaky growth on minimal medium but grew well on uracil. Complementation studies, as well as recombination measurements with the four known uracil loci, showed that this new type belonged to a new locus, ura5. Biochemical analysis revealed that it lacked orotydyllic acid (OMP) pyrophosphorylase activity. Clearly, this kind of mutant has not been obtained before, because the sensitivity of replica plating was not sufficient to reveal leaky mutants or, more likely, because they were discarded during the harvest in view of their poor phenotype. It remained to be explained why all the mutants obtained at this new locus presented a leaky phenotype. *In vitro* as well as *in vivo* experiments showed that in yeast uridine-5′-phosphate (UMP) pyrophosphorylase has weak OMP pyrophosphorylase activity, so that the situation was similar to the situation encountered with isoenzymes. It was impossible in particular to obtain a recombinant OMP pyrophosphorylase–less—UMP pyrophosphorylase–less strain, which showed that such a strain was lethal. This is consistent with the assumption that ura5 mutants are completely blocked for OMP pyrophosphorylase and that their leakiness is due to weak OMP pyrophosphorylase activity of UMP pyrophosphorylase (Jund and Lacroute, 1972).

IV. Study of Pathway Interactions

There are two principal possibilities for interactions between different biochemical pathways, interaction through common substrates of products, or interactions by cross-regulatory effects. These two aspects of interpathway links can very often be revealed by the systematic approach developed by Meuris *et al.* (1967). In Section III we saw an example of this method applied in revealing the existence of two isoenzymatic activities synthesizing carbamyl phosphate in yeast cells. The general observation on which the selection is based is that, when metabolic pathways share a biochemical substrate in common, they frequently develop during evolution either isoenzymatic activities, each one being regulated by the end product of the corresponding pathway, or a complex regulation of the key enzyme for the production of the common substrate involving as effectors the different end products (Woolfolk and Stadtman, 1964). The method consists of selecting mutants that are able to grow on minimal medium but are inhibited by addition to the growth medium of a normal product of the intermediary metabolism, e.g., an amino acid, a purine, or a pyrimidine. These mutants may result from the inactivation of one isoenzymatic activity by mutation, which makes the cell dependent on the other isoenzyme(s) to feed the pathway normally supplied by the mutated enzyme. When this is the case, the addition of end product(s) of the other pathway(s) will reveal the latent auxotrophy by inhibiting the biosynthesis or the activity of the remaining enzyme(s).

The power of analysis linked to the use of inhibited mutants is based on the possibility of localizing the pathway affected by the addition of the inhibiting metabolite by determining which metabolite relieves this inhibition. By this method it has been possible to obtain mutants for the two isoenzymes situated at the beginning of the aromatic amino acid pathway, one being inhibited and repressed by phenylalanine and the other by tyrosine (Meuris, 1973). Many other types of inhibited mutants have been obtained in yeast, but their physiological analysis has not yet been carried out. They may represent isoenzymatic systems, or they may be modified for an enzyme sensitive to cumulative feedback inhibition or regulated by concerted repression.

The accumulation in a mutant of an intermediary metabolite due to an enzymatic block is very often inhibitory for the cells. This inhibition can be trivial, but sometimes it reveals unsuspected regulatory links between different pathways. Such a case was found for the inhibition of growth due to the accumulation of ureidosuccinic acid in *ura4* mutants (Fig. 1). This inhibition was partly relieved by adenine, and it was seen later that inhibition of the wild type by externally added ureidosuccinic acid was also partly relieved by adenine. More detailed studies have proved that ureidosuccinic acid strongly inhibits one of the first five steps of the purine pathway, showing a new linkage between purine and pyrimidine biosynthesis (C. Korch, personal communication).

Another way in which mutants can be very helpful in the study of pathway cross-linking is analysis of the distribution inside the cell of metabolites common to different pathways. This channeling has not yet been thoroughly studied in yeasts but the work accomplished in *Neurospora* by Davis and co-workers can be cited as an example of the possibilities in this field (Davis, 1972).

V. Study of Regulatory Phenomena

It is in this area of metabolism that mutants are presently the most useful. The general picture of the different biochemical interconversions occurring in the cell is well established, but the much more complex problem of their regulation still remains unsolved. This regulation adjusts the rates of biosynthesis of the different metabolites to the levels needed by cell physiology. The first important aspect of the regulation of a metabolic pathway is the way in which the rates of biosynthesis of different enzymes behave under extreme conditions, such as strong repression or derepression, induction, or lack of induction.

The pyrimidine pathway revealed, through the utilization of different types of mutants, a complex scheme of regulation. The fact that during pyrimidine starvation intermediary enzymes had low specific activity in *ura2* mutants but high specific activity in *ura3* mutants implied that the biosynthesis of these enzymes was not repressed by the end product (UTP) but induced by the intermediate metabolites. Different complementary experiments gave the final scheme summarized in Fig. 1. The enzymatic complex aspartic transcarbamylase–carbamyl phosphate synthetase is regulated through feedback inhibition by UTP and repression by one end product (UTP or CTP). Dihydroorotase is induced by ureidosuccinic acid, dihydroorotic dehydrogenase and orotic-5′-phosphate decarboxylase are induced by dihydroorotic and orotic acids, and orotidine-5′-phosphate pyrophosphorylase is constitutive under all conditions studied (Jund and Lacroute, 1972). Even for the direct demonstration of this scheme, new mutants were useful. It was necessary to show that the same strain was induced for dihydroorotase when grown on ureidosuccinic acid, but not when grown on minimal medium. Unfortunately, ureidosuccinic acid was not taken up by the wild strain, so that mutants able to use it had to be selected, starting from a *ura2* strain, for the demonstration (Lacroute, 1968).

A second aspect of regulation is the inhibition or the activation of existing enzymes by cell metabolites. The most frequent case is the feedback inhibition of the activity of the first enzyme of a pathway by the corresponding end product. Different cases of mutations decreasing the efficiency of feedback inhibition have been described in yeast for amino acid pathways, the pyrimidine pathway, and the purine pathway. It is not necessary here to describe in detail the way to select such mutants and why they are of interest in the study of the regulation, since this is evident. Selection is generally done either by searching for mutants resistant to an analog of the end product of the pathway, or by looking for mutants excreting this end product, as revealed through the feeding of an appropriate indicator strain. Since this kind of mutation is generally dominant, it is a case in which selection starting from a diploid allows other types of mutations to be discarded when they are not wanted. Mutants not sensitive to feedback inhibition can be revealed if the other enzymes of the pathway are regulated by repression or semisequential induction. They can also reveal links with other pathways, since the excessive utilization of a common metabolic intermediate in the uninhibited pathway may starve other pathways and lead to their derepression.

The last, and maybe the most interesting, aspect of regulation that can be studied with mutants is mechanisms involved in the increase or decrease in the rate of biosynthesis of an enzyme. Such an approach, in spite of many attempts, has been completely unsuccessful in the pyrimidine pathway, where no regulatory mutant has yet been obtained. Fortunately, studies of other pathways have been more rewarding. Very often mutants have been obtained that behave as if repressorless, i.e., they synthesize the group of enzymes concerned in a derepressed manner and are recessive. Needless to say, one must be very cautious about the interpretation of such mutations. For example, a *ura5* mutant in the pyrimidine pathway which is able to grow on minimal medium is "derepressed" for the orotidine-5′-phosphate decarboxylase, even in excess of uracil, and could be mistaken, if studied superficially, for a regulatory mutant (Jund and Lacroute, 1972). In fact, to be sure of the correctness of the hypothesis, it would be necessary to achieve in eukaryotic organisms a direct *in vitro* binding assay of the product of the gene supposed to code for a repressor with the DNA bearing the regulated genes. This goal awaits the isolation of specific DNA regions in yeasts.

At present the two most interesting systems in yeasts for the analysis of regulatory mechanisms are the galactose pathway studied by Douglas and Hawthorne (1966) and the arginine pathway studied by Thuriaux *et al.* (1972). In the galactose utilization pathway, it has been shown that two types of gene control act successively: a negative control by a gene *i* coding for a protein which represses the synthesis of the product of a gene *gal4*; this gene *gal4* codes for an activator giving positive control of the structural genes *gal1*, *gal7*, and *gal10*; furthermore, mutants named *C*, equivalent to the operator constitutive mutants described in bacteria have been found for the gene *gal4*. The arginine pathway is subject to at least two different regulatory systems. One is specifically involved in the regulation of carbamyl phosphate synthetase of the arginine pathway. This system has shown an operator constitutive type of mutation strongly linked to the structural gene *cpa1* and a repressorlike mutation *cpR* (see Fig. 1). In this system the regulatory mutants have been selected by taking advantage of the fact that a *cpu* strain is normally inhibited by high concentrations of arginine, since it depends on carbamyl phosphate synthetase activity specific for the arginine pathway, which is repressed under these conditions. Regulatory mutants, with operator- as well as repressor-type mutations, have been found among the mutants resistant to this inhibition. The permeability mutants for arginine, which are also resistant to the inhibition, were avoided by the high concentration used (1 mg/ml).

Chapter 14

Isolation of Regulatory Mutants in Saccharomyces cerevisiae

HELEN GREER AND G. R. FINK

Department of Genetics, Development and Physiology, Cornell University, Ithaca, New York

I. Introduction

Regulatory mutations have been invaluable in the study of genetic control in *Escherichia coli*. Operator and repressor mutations have defined these regulatory genes and given insights into the physiology of regulation. By studying double mutants and dominance in heterozygotes, it is possible to obtain a picture of the regulatory circuit and regulatory interactions. These genetic studies allow construction of models which, because of the sophistication of the *E. coli* system, can be tested *in vitro*. Regulatory mutants have provided important reagents for the *in vitro* system. When purified *lac* repressor binds to *lac* O^+ DNA but not to O^c DNA, the conclusions seem clear and unambiguous, bolstering our confidence in the operon model for gene control. In addition, strains that are turned on for structural genes have facilitated the isolation of the mRNA and proteins coded for by these genes.

The control of gene expression is now under extensive investigation in a simple eukaryote, yeast. Many metabolic systems are known to be regulated, and attempts are being made to isolate mutations in the control elements themselves. A comparison between bacterial and yeast regulation is important for our understanding of eukaryote regulation. A difficult technical problem exists in the isolation of mutations that affect controlling elements or regulatory genes in most systems. Regulatory mutations do not usually alter the growth characteristics of the wild type, and therefore have no distinguishing phenotype. In order to identify regulatory mutants, one has to devise conditions under which regulatory mutations have a clear-cut phenotype. Several imaginative techniques were developed for bacteria. Certain idiosyncrasies of the yeast system preclude wholesale importation of these bacterial techniques. Genes of related function are not in operons, but are scattered. This makes it difficult to turn the whole pathway on or off with a single operator mutation. Selection for amino acid analog resistance often yields permease mutations. Many inducible fermentation systems require mitochondrial function. All these biological differences require special techniques suitable for yeast. It is the purpose of this article to review studies on gene regulation in yeast in order to identify those techniques that have been most successful in selecting regulatory mutants.

II.   Arginine

Arginine metabolism in yeast has been shown to be under the control of several different regulatory mechanisms. The existence of a negative type of control for enzyme production is supported by the isolation of two classes of regulatory mutations. These mutations resemble the classic operator and repressor mutations found in bacteria, although the actual molecular mechanisms involved in their mode of action have not yet been elucidated. Moreover, detailed mapping studies have not been made to show the position of the operator mutations relative to the structural genes they control. The presumptive operator mutations have been shown to be linked to a structural gene, but it is not clear whether the mutation defines a new gene or whether it lies within the structural gene itself. Clear understanding of the physiology of derepression of the arginine pathway was necessary before arginine regulatory mutants could be selected. Once the physiology was understood, it was possible for Wiame and his collaborators to use elegant selective techniques to obtain the desired mutants.

A.   Arginine Biosynthesis

The arginine biosynthetic pathway is summarized in Fig. 1. The control of the synthesis of the intermediate, carbamyl phosphate, has been studied by Wiame and co-workers. Yeast contain two distinct carbamyl phosphate synthetases—one involved in arginine biosynthesis (CPAse) and the other in pyrimidine biosynthesis (CPUse). CPUse activity depends on a gene defined by *cpu* mutations, and CPAse activity depends on two genes defined by *cpaI* and *cpaII* mutations.

In yeast, unlike *Neurospora*, the carbamyl phosphate produced from either enzyme can be used in either the arginine or pyrimidine pathway (Lacroute *et al.*, 1965; Wiame, 1968). This interchangeability between the two pathways is critical in the isolation of arginine regulatory mutants. *cpu* mutants do not have a pyrimidine requirement because they can use carbamyl phosphate made by CPAse. The addition of arginine to *cpu* mutants growing in minimal media results in repression (but not feedback inhibition) of CPAse (Thuriaux *et al.*, 1972). Carbamyl phosphate production is thus limited, and the growth rate is sharply reduced because the cells cannot make pyrimidines. Addition of uracil alleviates the growth inhibition. Regulatory mutants were isolated in *cpu* cells no longer sensitive to the addition of arginine (Thuriaux *et al.*, 1972). A high concentration of arginine (1000 mg/ml) was used to limit the number of permeability mutants obtained. By this procedure two different classes of regulatory mutations were isolated (*cpaIO* and *cpaR*). Both classes overcome repression and lead to constitutivity of CPAse activity (see Table I). The *cpaIO* mutations showed the following additional properties (Thuriaux *et al.*, 1972):

1. *cpaIO* is closely linked to one of the structural genes for carbamyl

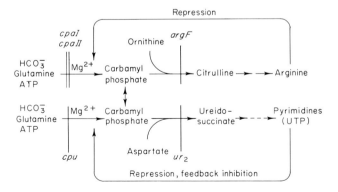

FIG. 1.   Synthesis and regulation of carbamyl phosphate in *S. cerevisiae*. (From Thuriaux *et al.*, 1972.)

TABLE I

MUTATIONS cpaIO AND cpaR[a,b]

Genotype	Growth rate	CPAse activity	
	M(am) + arg (150)	M(am) + ur	M(am) + ur + arg (1000)
cpu	Slow	0.60	0.06
cpu cpaIO	Normal	0.50	0.18–0.50
cpu cpaR	Normal	0.68–0.90	0.22–0.29
$\dfrac{cpu\ +\ +}{cpu\ cpaI\ cpaIO}$	Slow	0.40	0.05
$\dfrac{cpu\ +\ cpaIO}{cpu\ cpaI\ +}$	Normal	0.38	0.23
$\dfrac{cpu\ cpaR}{cpu\ +}$	Slow	0.52	0.06

[a] Adapted from Thuriaux et al. (1972).

[b] Growth on solid media. M(am) contains $0.02\ M\ (NH_4)_2SO_4$; ammonium is the nitrogen source. M(am) + arg (150) contains 150 μg arginine/ml. M(am) + ur contains 25 μg/ml uracil. M(am) + ur + arg (1000) contains 25 μg/ml uracil and 1000 μg/ml arginine.

phosphate synthetase (*cpaI*). Tetrad analysis from the cross (*cpu + cpaIO*) × (*cpu cpaI +*) gave 236/242 parental ditypes.

2. *cpaIO* is cis dominant and shows no trans effect.

The diploid *cpu + cpaIO/cpu cpaI +* overcomes sensitivity to arginine and makes CPAse constitutively, whereas the diploid *cpu + +/cpu cpaI cpaIO* is sensitive to arginine, CPAse being repressible (see Table I).

cpaIO mutations thus behave as constitutive operator mutations for the *cpaI* gene.

It is worth noting that the two genes necessary for arginine carbamyl phosphate synthetase, *cpaI* and *cpaII*, are not linked, and that therefore derepression of *cpaI* (by the *cpaIO* mutation) in some way leads to derepression of *cpaII*, or else *cpaII* is never limiting.

cpaR mutations are distinguishable from *cpaIO* mutations (Thuriaux et al., 1972).

1. *cpaR* mutations are not linked to *cpaI* (7 tetratypes/7 tetrads); *cpaII* (21 tetratypes/35 tetrads); or *argRI, -II, -III*—genes for repression of other arginine biosynthetic enzymes (9 tetratypes/12 tetrads).

2. *cpaR* is recessive. The diploid *cpu cpaR/cpu +* is sensitive to arginine addition (see Table I).

cpaR mutations do not affect permeability as determined by arginine uptake experiments. Measurements of arginine pool size are the same as for the wild type. Mutations in the *CPAR* gene seem to affect a repressor for the CPAse enzyme, perhaps a corepressor or an aporepressor.

Other arginine biosynthetic enzymes also appear to be regulated by some type of negative control. Ornithine transcarbamylase (OTC) converts ornithine to citrulline and is strongly repressed by the addition of arginine (Bechet *et al.*, 1962). Mutants have been isolated in which OTC is no longer subject to arginine repression (Bechet *et al.*, 1970).

These nonrepressible mutants were isolated by selecting cells that were resistant to the arginine analog canavanine. The cell can overcome canavanine inhibition either by (1) increasing its arginine pool, which effectively dilutes the canavanine, or by (2) preventing canavanine uptake. The arginine pool can be increased only by counteracting the control mechanisms that normally prevent an excess of arginine from being made. Only permease mutants are resistant to high concentrations of canavanine (Grenson *et al.*, 1966).

To avoid permease mutants and to enrich for repression mutants, it is important to start with cells whose arginine pool size is dependent on the level of repression. This was accomplished by growing an arginine auxotroph on an arginine intermediate, ornithine or citrulline. To grow on these intermediates cells must have reasonably high enzyme levels (Bechet *et al.*, 1970). Addition of the false corepressor, canavanine, inhibits growth on these intermediates by repressing the synthesis of arginine biosynthetic enzymes. In actual practice regulatory mutations were obtained on minimal ammonia media with ornithine (200 μg/ml) and low levels of canavanine (8 μg/ml). Low doses of canavanine allow the growth of regulatory mutants and inhibit the growth of normally regulated strains, whereas high doses inhibit all but permease mutants.

Canavanine-resistant mutants (*argR*) were categorized as regulatory mutants by the following criteria (Bechet *et al.*, 1970).

1. OTC synthesis of mutants, unlike that of wild-type cells, was not repressed by the addition of exogenous arginine. The absolute level of OTC was also higher (approximately 2-fold) than that in wild type after growth on just minimal media (see Table II).

TABLE II

ACTIVITY AND REPRESSION OF ORNITHINE
TRANSCARBAMYLASE[a]

Genotype	OTC activity[b]	
	Min(am)	Min(am) + arg (5 mM)
Wild type	37	1
argR	80–97	65–84

[a] From Bechet *et al.* (1970).
[b] Min(am) contains ammonia as the only nitrogen source. The activity is measured on permeabilized cells.

2. *argR* mutations affect the synthesis of five of the nine enzymes involved in arginine biosynthesis: OTC, acetylglutamate kinase, phosphoacetyl-glutamate reductase, *N*-acetylglutamic semialdehyde transaminase, and arginosuccinate synthetase. (Thuriaux *et al.*, 1972; Wiame, 1971; Bechet *et al.*, 1964, 1970).

3. Permeability, as measured by arginine uptake, and growth rate are the same for the wild type as for *argR* mutants.

4. The arginine pool size is greater than in the wild type on ammonia, and the same as the wild type on ammonia plus arginine (Ramos *et al.*, 1970).

5. *argR* mutations are recessive, as determined by crosses with the wild type.

6. Each mutation segregated as a single gene, as determined by tetrad analysis.

7. Complementation tests defined three genes—*argRI, -II,* and *-III.* There is no additivity of repression. None of the *argR* mutants are linked to the OTC structural gene (*argF*) or to each other.

The recessive constitutive nature of *argR* mutations implicates these genes in repressor modification. It is suggested that *argR* genes are involved in OTC regulation by affecting an aporepressor made up of several different polypeptides. It is also conceivable that *argR* genes help in production of the true corepressor by modifying arginine (Bechet *et al.*, 1970).

argR mutations define a regulatory circuit distinct from *cpaR* mutations. *argR* mutants are not constitutive for CPAse. However, the systems are not completely independent because under repressing conditions, *cpaR* mutants have three times higher an OTCase level, than the wild type (Thuriaux *et al.*, 1972). (This effect is not the result of increased levels of CPAse activity in *cpaR* mutants, because higher levels of OTCase are not observed in cells containing *cpaIO*, the operator constitutive mutation for CPAse.)

B.   Arginine Degradation

Wild-type yeast cells can use arginine or ornithine as a sole nitrogen source. Growth on ornithine requires constitutive synthesis of ornithine transaminase (the third enzyme in the pathway, pyrroline carboxylate dehydrogenase, is normally constitutive), whereas growth on arginine requires constitutive synthesis of ornithine transaminase and arginase. These catabolic enzymes appear to be under positive control, since they require the presence of the *argR* product (Wiame, 1971).

Regulatory mutations, resulting in constitutive synthesis of ornithine transaminase and arginase were isolated as mutations that overcome the absence of induction by permitting growth on arginine or ornithine as sole nitrogen source in *argR⁻* cells (Wiame, 1971). Cells with two *argR⁻* mutations were used to prevent reversion to *ARGR⁺*.

When ornithine was used as sole nitrogen source, operator mutations, $(cargB^+O^-)$, for ornithine transaminase were isolated as determined by:

1. A 100-fold increase in the level of ornithine transaminase compared to that in the wild type when grown in minimal media.

2. No effect on arginase levels.

3. Close linkage to the structural gene for ornithine transaminase $(cargB)$.

4. Dominance of the mutation.

5. Heterozygous diploid containing half the level of ornithine transaminase compared to the haploid (owing to a dilution factor of 2) .

When arginine was used as sole nitrogen source, the selection procedure demands mutations to constitutivity of two enzymes. Therefore it was necessary to start with a strain already constitutive for ornithine transaminase in order to obtain mutations regulating arginase synthesis. By plating cells constitutive for ornithine transaminase on arginine, operator mutations for arginase were obtained. Cells containing the O^c mutation had the following properties:

1. They were constitutive for arginase and not for ornithine transaminase.

2. They contained half the level of arginase in the diploid heterozygous for the operator mutation, as compared to the haploid strain.

3. The operator constitutive mutation was closely linked to the structural gene for arginase $(cargA)$.

Another class of mutations, $cargR$, was obtained from the ornithine $argR^-$ selection. Members of this class behave like repressor mutations in that they are constitutive for both arginase and ornithine transaminase (unlinked genes), and are recessive. The existence of $cargR$ mutations supports the hypothesis that the apparent positive control exerted by $ARGR^+$ on these catabolic enzymes is the result of two negative controls. The $ARGR^+$ product in some way may be repressing a repressor, either by inhibiting activity of a gene coding for the repressor of the catabolic enzymes, or by deactivating the repressor (Wiame, 1971).

C.   Controls Linking Arginine Catabolic and Anabolic Enzymes

A cell must be able to control its degradative and synthetic pathways in such a way as to avoid continual recycling of a synthesized and degraded product.

The $ARGR$ product appears to be active in both the arginine anabolic pathway (which represses enzymes necessary for arginine synthesis) and in the arginine catabolic pathway (which results in induction of enzymes that break down arginine). The presumptive heteropolypeptidic aporepressor $ARGR$ (coded for by $argRI$, $-II$, and $-III$) thus seems to be a common negative controlling element involved in repression of arginine biosynthetic enzymes and the induction of arginine degradative enzymes (Wiame, 1971).

The pathway enzymes themselves have been shown to interact with each other and to cause inhibition of certain enzymatic activity. OTC (an anabolic enzyme) is strongly inhibited by arginase (a catabolic enzyme) (Messenguy *et al.*, 1971). Arginase binds to OTC with the result that OTC is inhibited but arginase is not affected. This binding is controlled by high levels of arginine and ornithine.

The levels of both arginine and ornithine seem to be important in the control of pathway enzymes. Arginine represses OTC (Bechet *et al.*, 1962) and some biosynthetic enzymes (Bechet *et al.*, 1964); feedback inhibits *N*-acetylglutamate kinase (DeDeken, 1963) and, together with ornithine, stimulates binding of arginase to OTC (Bechet *et al.*, 1970). In addition, exogenous ornithine itself causes the repression of OTC and CPAse activity by acting either directly or indirectly as a corepressor. If *argR* cells are grown on ammonia plus ornithine, there is a small arginine pool because CPAse is repressed and, if *cpaO* cells are used, the arginine pool size is small because of repression of OTC or some later enzyme (Ramos *et al.*, 1970). Ornithine is also involved in catabolic control through its action as an inducer for arginase and ornithine transaminase (Ramos *et al.*, 1970).

III. Galactose Degradation

In yeast the three structural genes conferring the ability to ferment galactose are closely linked. The *GAL1* locus specifies the galactokinase, *GAL7* specifies the transferase, and *GAL10* specifies the epimerase (Douglas and Hawthorne, 1964). These genes seem to be regulated by an unlinked gene cluster which consists of locus *i* which codes for a repressor, locus *c* which is the site for binding the repressor, and locus *GAL4* which is a structural gene controlled by *c* and whose product in turn positively controls the *GAL1, -7, -10* gene cluster (Douglas and Hawthorne, 1966). Evidence for this regulatory model comes from the isolation of different types of control mutations.

Recessive regulatory mutations that permit constitutive synthesis of the galactose enzymes were isolated by starting with a petite strain which was blocked in induction (Douglas and Pelroy, 1963). In the wild type the synthesis of galactokinase, transferase, and epimerase is inducible. Attempts were made to isolate strains that synthesized these enzymes in the absence of the inducer, galactose. Initial efforts to obtain regulatory mutants utilized a strain containing a *gal3* mutation. *gal3* strains have a long lag period before induction of the galactose enzymes begins. The idea was to obtain mutations that allow rapid growth of *gal3* on galactose, because they confer constitutive synthesis of kinase, transferase, and epimerase. The overwhelming number of gal+ cells obtained in this way are *gal+* phenocopies which revert to the *gal3* phenotype when grown in the absence of galactose and then retested. This problem was overcome by starting with a strain that had, in addition to *gal3*, a petite (respiratory-deficient) mutation. The ability of *gal3* cells to grow on galactose is a function of the respiratory state of the cells (Douglas and Pelroy, 1963); *gal3 rho⁻* cells give no colonies on galactose. The stable gal⁺ revertants from this strain were called i⁻. Genetic and biochemical analysis indicates that these are recessive constitutive mutations. They are not linked to *gal3*, or to the *GAL1, -7, -10* complex.

The properties of these *i⁻* mutations indicate the existence of a repressor molecule formally analogous to that outlined in the operon model for gene control (Monod and Jacob, 1961). Under normal conditions the cells are *i⁺* and therefore make repressor, and the galactose enzymes are inducible. If there is a mutation *i⁻* which prevents synthesis of repressor, the galactose enzymes are made constitutively.

The finding that an *i⁻* mutation is suppressed by a nonsense suppressor supports the idea that the *i* gene codes for a protein. A suppressor of an ochre leucine auxotroph also suppresses an *i⁻* mutation (Douglas and Hawthorne, 1972).

The existence of the structural gene complex and an unlinked repressor gene prompted a search for dominant, constitutive mutations that would define an operator site linked to the galactose complex. However, only recessive *i⁻* mutations were obtained by isolating mutants constitutive for the three enzymes from a *gal3* petite strain (Douglas and Pelroy, 1963). Operator mutants were sought that would prevent synthesis of the three-gene complex by starting with a galactose-sensitive (epimerase, *gal10*, deficient) strain and isolating galactose-resistant mutants. All such galactose-resistant revertants had acquired a *gal1* (kinase) or *gal4* (deficient for the three enzymes) mutation, and none had acquired an operator (O^0) or promoter mutation (Douglas and Hawthorne, 1964).

Mutations that simultaneously prevent synthesis of all three galactose enzymes have been obtained, although they do not map near the galactose structural genes. gal⁻ cells were isolated after ultraviolet treatment by screening an inducible *GAL⁺* petite strain for gal⁻ clones (Douglas and Hawthorne, 1964). The gal⁻ phenotype is easier to observe in repiratory-deficient cells. Phenotypically similar mutations have been isolated in the *E. coli gal* and *lac* operons, and these have been shown to be polar mutations in the first structural gene of the operon, which prevent synthesis of all the downstream proteins. However, phenotypically similar, pleiotropic, negative mutations in yeast, called *gal4*, are unlinked to the three galactose

genes. The *gal4* phenotype is the result of a single recessive mutation. *gal4 i*⁺
and *gal4 i*⁻ behave similarly. Therefore the *gal4* lesion does not result from
interference with permeation or inducer metabolism, or from production of
an unantagonizable repressor. *gal4* mutations show interallelic complementation, and some can be suppressed by external suppressors (Douglas and
Hawthorne, 1966). The *gal4* gene can be interpreted as producing a cytoplasmic product necessary for *GAL1, -7, -10* expression (Douglas and
Hawthorne, 1964).

   The repressor (*i* gene product), was shown to control negatively the synthesis of the positive regulator from the *GAL4* gene via an operator locus,
called *c*, adjacent to *GAL4*. The *c* region was identified by isolation of dominant constitutive mutants not linked to the *GAL* structural genes (Douglas
and Hawthorne, 1966). They were selected in a similar fashion as described
above (Douglas and Pelroy, 1963) for the *i*⁻ repressor isolation by using a
gal3 petite strain, except that the starting strain was a diploid, used in order
to avoid obtaining recessive *i*⁻ mutations. The diploid was homozygous
for the mating-type locus to ensure mating and sporulation. Mutations to
growth on galactose medium were obtained spontaneously and with nitrosoguanidine. The mutants fell into two classes: normally regulated (inducible)
and abnormally regulated (constitutive). The class of gal⁺ clones that were
inducible proved to be *gal3* → *GAL3* revertants. The other gal⁺ clones were
crossed with the gal⁺ diploid, *i*⁺/*i*⁺ *gal3*/*gal3*. The tetraploids that were
constitutive (i.e., their levels were equivalent to those of fully induced
strains) were assumed to contain a dominant constitutive mutation donated
by the mutant diploid. (The mutant diploid gal⁺ character was also dominant
in the triploid when crossed with a haploid *i*⁺ *gal3* strain). Tetrad analysis
from the sporulated tetraploids showed that the dominant constitutive
mutants still contained the *gal3* mutation. The new mutation, designated *C*,
segregates as a single marker (2:2), is not linked to *GAL1* (structural gene for
kinase) or to the *i* genes, but is closely linked to *GAL4* (81/82 tetrads gave
the parental ditype class in the cross *galC* × *gal4*). Dominant constitutive
synthesis was shown to require *C* and the + allele of *gal4* (*GAL4*) to be cis to
each other. This behavior suggests that the *c* region is analogous to a bacterial
operator for the *GAL4* structural gene. Bacterial systems have been shown
to contain operator regions adjacent to the structural genes of a given
pathway and subject to repressor action. In the yeast galactose complex, no
such operators have been found (see above). However, the *GAL4* product
seems to have evolved to act as an intermediary (via action on the *c* region)
between the repressor and the galactose structural genes (Douglas and
Hawthorne, 1966).

   Mutations phenotypically similar to *i*ˢ mutations in the *E. coli* lactose
operon (Willson *et al.*, 1964) have been obtained in yeast. These super-

repressible i^s mutants are believed to possess an altered repressor which is still able to bind to the operator but which is no longer susceptible to inducer action. The isolation of these uninducible gal⁻ mutants in yeast was obtained by two different procedures (Douglas and Hawthorne, 1972). One scheme involved starting with a petite (respiratory-deficient) diploid which was homozygous for the wild-type *gal* genes and for mating type. Mutants were obtained by screening for gal⁻ clones by replica plating after nitrosoguanidine mutagenesis. The second procedure gave gal⁻ mutants by isolating galactose-resistant cells from a galactose-sensitive (epimerase-deficient, *gal10*) haploid stock.

The mutants, called i^s, from both isolation procedures are uninducible. The enzyme levels of i^s strains are not inducible or are only partially inducible (5–30% enzyme levels of an i^- strain). None of the mutations are linked to the galactose structural genes (analysis gave 31 tetratypes/47 tetrads). However, they are closely linked to mutations at the i locus (tetrad analysis of the cross $i^s \times i^-$ gave only parental ditypes). The i^s mutations are dominant, giving galactose nonfermenters when heterozygous with i^+ or i^- (recessive constitutive) galactose fermenters. However, the mutation is not expressed when crossed to a strain containing a C mutation (dominant constitutive).

Revertants of the i^s mutations gave constitutive mutations which produced 50–100% transferase levels of i^- cells. Some of the revertants are recessive, closely linked to the original mutation, and therefore were classified as containing the double $i^s i^-$ mutations (preventing active repressor synthesis); whereas the one revertant that was dominant and closely linked to c (the repressor binding site) was concluded to be $i^S C$ (preventing repressor binding) (Douglas and Hawthorne, 1972).

IV.   Acid Phosphatase Biosynthesis

Yeast contain two species of acid phosphatases, as shown by differing K_m values and heat stability. One of these enzymes is coded by a gene defined by *phoC* mutations, and is produced constitutively, whereas the other acid phosphatase is repressible by inorganic phosphate (Toh-e *et al.*, 1973). Mutations in any of four different complementation groups *phoB*, *-E*, *-D*, or *-S* result in cells lacking repressible acid phosphatase. *phoD* or *phoS* mutations also cannot produce the repressible alkaline phosphatase enzyme (Toh-e *et al.*, 1973). *phoR* mutations specify a repressor for repressible acid phosphatase and alkaline phosphatase enzymes, and *phoS* mutations define a regulatory gene for *phoR* expression and produce the *phoR* product constitutively (Toh-e *et al.*, 1973). Operator-like mutations, *PHOO*, have been isolated for the *PHOD* gene, believed to be a positive regulator for repressible acid phosphatase production (A. Toh-e and Y. Oshima, personal communication).

Mutations that affect the acid phosphatase systems were obtained by Toh-e and co-workers using the diazo-coupling procedure of Dorn (1965). Cells were mutagenized with ultraviolet light, spread on high (repressed condition) or low (derepressed) P_i plates, incubated, and overlaid with soft agar with a staining solution containing the diazo-coupling reagent (naphthyl phosphate, Fast blue salt, acetate buffer). Cells containing acid phosphatase cleave the phosphate, allowing the formation of the colored diazo compound which stains the colonies dark red. Mutant colonies lacking the enzyme are white. The staining does not kill the cells, and therefore the desired mutations can be isolated directly.

Mutants lacking constitutive acid phosphatase were selected as unstained colonies on high P_i (Toh-e *et al.*, 1973). These *phoC* mutations are recessive, segregate 2:2, and retain repressible acid phosphatase as tested by specific enzyme activity of cells grown on low P_i.

In order to study the regulation of repressible acid phosphatase, a method had to be devised to eliminate expression of constitutive acid phosphatase. Therefore mutants lacking repressible acid phosphatase were selected on low P_i in strains containing a *phoC* mutation (Toh-e *et al.*, 1973). These mutations, *phoB*, *-E*, *-D*, and *-S*, are all recessive and segregate 2:2. *phoE* is linked to *phoC*, and all the other genes are not linked to these or to each other. Acid phosphatase activity is lowered severalfold in *phoB*, *-E*, *-D*, or *-S* as compared to the wild type on both low and high P_i. In addition, on low P_i, alkaline phosphatase levels are decreased in *phoD* and *phoS* cells. This suggests the possibility of common regulation or the sharing of some common product for acid and alkaline phosphatases.

phoS mutations are believed to be constitutive for repressor (*phoR* product) synthesis. They are distinguishable from mutations in a phosphatase structural gene, which would give rise to the same phenotype, by the fact that there is acid phosphatase activity in the double mutant *phoS phoR*. *phoC* strains carrying *phoS phoR* are constitutive for acid phosphatase on low or high P_i, whereas *phoC* strains carrying *phoB phoR*, *phoD phoR*, or *phoE phoR* have no phosphatase activity. In addition, *phoR PHOS/PHOR phoS* makes acid phosphatase on low P_i but not on high P_i, whereas *phoR phoS/PHOR phoS* does not produce enzyme on either low or high P_i. These results support a regulatory role for the *PHOS* gene. A model for its mode of action explains the *phoS* phenotype as constitutive synthesis of the repressor (*PHOR* product), such that the phosphatases cannot be derepressed, and the *phoR phoS* phenotype as constitutive synthesis of an inactive repressor such that repressible phosphatases are made (Toh-e *et al.*, 1973).

The *PHOD* gene also appears to be a regulatory gene. This is supported by the isolation of temperature-sensitive (*ts*) *phoD* mutants unable to make acid phosphatase at the nonpermissive temperature (A. Toh-e and Y. Oshima, personal communication). Acid phosphatases derived from cell-free extracts at the permissive temperature from the *ts phoD* mutant and wild-type *PHOD* cells were shown to have the same thermolability and K_m values. Enzymes from revertants of regular *phoD* mutants also gave similar values when compared to wild-type phosphatase (A. Toh-e and Y. Oshima, personal communication). These results suggest that *PHOD* does not code for a structural gene. *phoD* seems to exert some type of positive control in the phosphatase system. The fact that *phoR phoD* does not make repressible acid phosphatase may mean that *PHOR* controls phosphatase enzyme production via the *PHOD* gene product.

Mutants constitutive for repressible acid phosphatase were selected in strains containing a *phoC* mutation (Toh-e *et al.*, 1973). Constitutivity due to either true reversion or intragenic suppression of *phoC*, extragenic suppression of *phoC*, mutation in a regulatory gene, or mutation in an operator (O^c), were distinguished by dominance and segregation patterns in crosses with *PHOC* and *phoC* strains. Mutations which were recessive for constitutivity when crossed to *phoC*, and gave $4 + :0 -$, $3 + :1 -$, and $2 + :2 -$ segregation for constitutivity on high P_i when crossed to *PHOC*, were classified as regulatory mutations, *phoR*. *phoR* cells make both acid and alkaline phosphatase on high P_i, and make higher levels on low P_i than a *PHOR* strain (with *phoC*). *PHOR* seems to code for a repressor which exerts negative control over repressible phosphatases. Mutations that were dominant for constitutivity, and gave $4 + :0 -$, $3 + :1 -$, and $2 + :2 -$ segregation patterns on high P_i when crossed to *PHOC*, were either operator-like or extragenic suppressor mutations. Suppressor mutations when crossed to a *phoC phoB* strain should give a $2 + :2 -$ segregation pattern on high P_i. The existence of operator mutations, *PHOO*, was suggested by the behavior of one mutation which showed a dissection pattern of 12 asci ($1 + :3 -$) and 1 ascus ($2 + :2 -$) on high P_i, and all $2 + :2 -$ on low P_i.

PHOO resembles in certain respects an O^c-type operator constitutive mutation for the *phoD* gene (A. Toh-e and Y. Oshima, personal communication). *PHOO* is closely linked to *phoD* (8 tetratypes/714 tetrads). *PHOO* acts in cis. On high P_i acid phosphatase is made by *PHOO PHOD/phoO PHOD* but is not made by *PHOO phoD/phoO PHOD*, whereas both diploids make the enzyme on low P_i. *PHOO* is only semidominant to *phoO*. A *phoO/PHOO* diploid on high P_i gives an intermediate level of phosphatase activity (0.015) as compared to a *PHOO* haploid (0.076) and a *phoO* haploid (0.001). It is this partial dominance that distinguishes it from the classic O^c mutations which are completely dominant.

The model proposed by Toh-e and Oshima (pers. commun.) maintains that the *phoO* locus is the recognition site adjacent to *PHOD* for the *PHOR* product. Thus the genes *phoR*, *phoO*, and *phoD* seem to be analogous, respectively, to genes *i*, *c*, and *GAL4* in the galactose regulatory system. Therefore both the phosphatase and galactose genes are examples of regulator genes mediating the effect of repressors on structural genes. In fact, a single model explains the regulation of these two systems. A positive regulatory element is required for the expression of structural genes. This element is under the control of a repressor. Mutations in the repressor gene lead to constitutive synthesis of the structural gene product. Mutations in the positive regulatory element lead to the absence of structural gene function. This model may have some general significance. An analogous type of control exists in prokaryotic systems, e.g., phage lambda, in which structural genes are under the control of the positive regulator, *N* protein, which in turn is under the negative control of the repressor, *cI* protein (Thomas, 1971).

V.   Pyrimidine Biosynthesis

The enzymes that catalyze the pathway for pyrimidine biosynthesis and the genes that control their synthesis are shown in Fig. 2. Five genes have been identified as the structural genes for the enzymes involved (Lacroute, 1968; Jund and Lacroute, 1972): *ura1* for dihydroorotic dehydrogenase, *ura2-C* for carbamyl phosphate synthetase, *ura2-B* for aspartic transcarbamylase, *ura3* for orotydilic acid decarboxylase, *ura4* for dihyroorotase, *ura5* for orotydilic acid (OMP) pyrophosphorylase.

The enzymes for the first two steps, carbamyl phosphate synthetase and aspartic transcarbamylase, are controlled by both feedback inhibition and repression. Almost all of the subsequent enzymes are regulated in an unusual fashion, in that they are not subject to pyrimidine repression but rather each is induced in a sequential fashion by pathway intermediates (Lacroute, 1968). OMP pyrophosphorylase, however, does not seem to be regulated by the same general scheme as the other enzymes. Physiological conditions shown to repress, derepress, or induce the other pyrimidine enzymes do not significantly affect the activity of OMP pyrophosphorylase (Jund and Lacroute, 1972).

The sequential induction of enzymes by their substrates in the pyrimidine pathway has complicated the search for regulatory mutants. Because of the unusual mode of induction, mutations in structural genes can masquerade as regulatory mutants. For example, mutations that block the production of ureidosuccinic acid (*ura2-B*) have the phenotype of strains carrying mutations in a positive regulatory element (see Fig. 2), because subsequent pathway enzymes are at low levels. Mutations in feedback resistance (in the same *ura2-B* gene) which overproduce ureidosuccinate have constitutive levels of subsequent pathway enzymes. Perhaps the technology worked out in *Neurospora* by Gross (1965) will help to overcome these difficulties.

Attempts to understand the regulation of the pyrimidine pathway have involved obtaining mutants resistant to 5-fluoropyrimidines (Jund and Lacroute, 1970). Haploid cells were mutagenized with ultraviolet light and plated on minimal media containing different concentrations of the analogs.

Table III and Fig. 3 summarize the properties and pathway blocks of the different noncomplementing groups of mutations obtained (Jund and Lacroute, 1970, 1972). None of the cells mutated such that the enzymes showed activity for the physiological substrate but not for the analog.

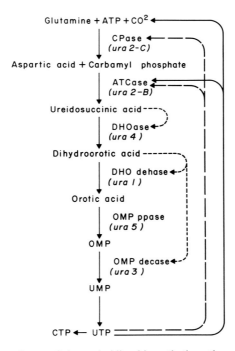

FIG. 2.   Regulation scheme of the pyrimidine biosynthetic pathway. The different enzymatic steps and the names of the corresponding mutants are presented. Solid lines, repression; dotted lines, induction; dashed lines, feedback inhibition. (Adapted from Lacroute, 1968; and Jund and Lacroute, 1972.)

TABLE III: PROPERTIES OF 5-FLUOROPYRIMIDINE-RESISTANT MUTANTS

Genotype	5-Fluoropyrimidine resistance	Dominance[a]	Defective in incorporation of ^{14}C precursors	Pyrimidine excretion	Enzyme defect
FUR1 Three phenotypes:					
1-1	5-Fluorouracil, 5-Fluorouridine, and, 5-Fluorocytosine[b]	R	Uracil and cytosine	Uracil and uridine	UMP pyrophosphorylase (uracil → UMP)
1-5	5-Fluorouracil, 5-Fluorouridine, and, 5-Fluorocytosine[b]	D	Uracil and cytosine	—	Not known
1-6	5-Fluorouracil, and 5-Fluorocytosine[b]	R	Uracil and cytosine	Uracil	UMP pyrophosphorylase
FUR2 (allelic to *URA2*)	5-Fluorouracil, 5-Fluorouridine, and 5-Fluorocytosine	D	Cytosine, uridine, and uracil	Uracil	Desensitization of aspartic transcarbamylase and carbamyl synthetase to feedback inhibition by UTP
FUR3	5-Fluorouracil, 5-Fluorouridine, and 5-Fluorocytosine	D	—	Small amounts of pyrimidines	Not known
FUR4	5-Fluorouracil and 5-Fluorouridine	R	Uracil	—	Uracil-specific permease
FUI1	5-Fluorouridine	R	—	—	Uridine-specific permease
FCY1	5-Fluorocytosine	R	Cytosine	None	Cytosine deaminase
FCY2	5-Fluorocytosine	R	Cytosine	None	Cytosine-specific permease
FOR1	5-Fluoroorotic acid and 5-Fluorouracil	R	Orotic acid, cytosine, and adenine	Uracil	Increases levels of OMP pyrophosphorylase and OMP decarboxylase

[a] R, recessive; D, dominant.
[b] *FUR1-1* and *-1-6* resistance to 5-fluorocytosine results from the fact that cytosine must be converted to uracil to be used by yeast cells.

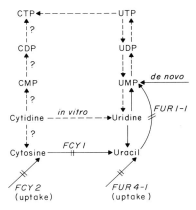

FIG. 3. Schematic representation of pyrimidine utilization in *S. cerevisiae*. Broken lines indicate plausible pathways. Reactions blocked in the different mutants are shown. (From Jund and Lacroute, 1970.)

The resistance of the mutants was deduced to be due to a defective enzyme in the pathway, loss of feedback inhibition, or loss of a specific permease. None of the mutants appear to be regulatory mutants that result in constitutive (derepressed) synthesis of pyrimidine pathway enzymes. The increase in levels of OMP pyrophosphorylase and OMP decarboxylase in *for1* mutants (see Table IV) is believed not to be great enough to account for the high degree of 5-fluorouracil resistance. It is postulated that the *for1* mutation, rather than directly affecting OMP pyrophosphorylase regulation, causes an increase in some intermediate regulatory product (Jund and Lacroute, 1972).

TABLE IV

INFLUENCE OF THE for1 MUTATION ON
OMP PYROPHOSPHORYLASE (OMPPASE) AND
OMP DECARBOXYLASE (OMPDECASE)[a]

	Specific activity[b]	
	OMPPase	OMPdecase
Wild type	29	6.5
for1	40	10.5

[a]From Jund and Lacroute (1972).
[b]Expressed as nanomoles of substrate transformed per minute per milligram of protein.

VI.   Methionine Biosynthesis

The biosynthesis of methionine involves synthesis of a four-carbon skeleton, sulfur incorporation, and synthesis and incorporation of the methyl groups (Cherest *et al.*, 1969). The methionine pathway (see Fig. 4) is regulated by end product inhibition and repression. Studies of regulation mechanisms are complicated by the existence of controls overlapping threonine, cysteine, and methionine. In addition, there is a large number

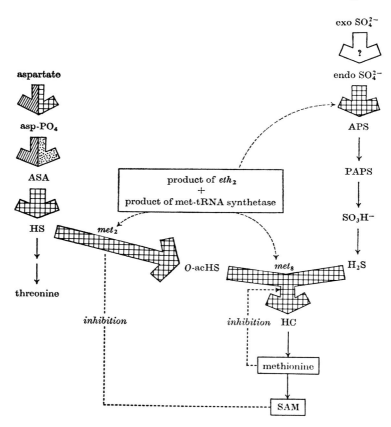

FIG. 4.   The effects of methionine or *S*-adenosylmethionine on synthesis and activity of enzymes directly involved in threonine and methionine biosynthesis. Cross-hatching, derepression in methionine limitation; dotted area, derepression in excess methionine; diagonal hatching, derepression in threonine limitation. asp-PO_4, β-aspartyl phosphate; ASA, aspartate semialdehyde; HS, homoserine; *O*-acHS, *O*-acetylhomoserine; APS, adenine monophosphate SO_4^{-2}; PAPS, phosphadenosine phosphosulfate; HC, homocysteine; and SAM, *S*-adenosylmethionine. (From de Robichon-Szulmajster and Surdin-Kerjan, 1971.)

of methionine genes which have not been associated with steps in methionine biosynthesis.

Homoserine-O-transacetylase (homoserine $\rightarrow$ O-acetylhomoserine), homocysteine synthetase (O-acetylhomoserine + H_2S $\rightarrow$ homocysteine), adenosine triphosphate sulfurylase (ATP + SO_4^{-2} $\rightarrow$ adenine monophosphate $-$ SO_4^{-2}), sulfite reductase (SO_3H^- $\rightarrow$ H_2S), asparatokinase (aspartate $\rightarrow$ aspartyl phosphate), and homoserine dehydrogenase (aspartate semialdehyde $\rightarrow$ homoserine) are all repressible by methionine (Cherest *et al.*, 1969, 1971). These enzymes have been divided into two groups defined by the type of regulation (Cherest *et al.*, 1971). The group I enzymes are homoserine-O-transacetylase (coded for by *met2*), homocysteine synthetase (coded for by *met8*), adenosine triphosphate sulfurylase, and sulfite reductase. The *met2* and *met8* genes are unlinked. The group II enzymes, which are common to methionine and threonine synthesis, are aspartokinase and homoserine dehydrogenase. The genes coding for group I enzymes are under the control of *eth2* and *mes1* genes, whereas the genes coding for group II enzymes are not.

Regulation of methionine biosynthesis has been studied with the aid of DL-ethionine, a methionine analog. Ethionine-resistant mutants were obtained by three successive transfers in liquid media containing 2×10^{-3} M ethionine, and then by replica-plating onto solid media containing 2×10^{-2} M ethionine (Cherest and de Robichon-Szulmajster, 1966). Three different types of mutations, mapping in different genes, were obtained by this procedure. Amino acid permease mutants (*aap*) were isolated, which were resistant to 2×10^{-2} M ethionine as well as being resistant to other amino acid analogs such as p-fluorophenylalanine and canavanine. There are two classes of mutations that confer resistance specifically to ethionine (2×10^{-3} M). The *ETH1*^{r} mutation is dominant to the wild-type *eth1*^{s}, while the *eth2*^{r} mutation is recessive to the wild-type *ETH1*^{s}. The *ETH1*^{r} phenotype, observable only in conjunction with *eth2*^{r}, increases the level of ethionine resistance of the *eth2*^{r} allele (Cherest and de Robichon-Szulmajster, 1966).

The *eth2* mutation has been shown to lower the repressibility by methionine of only group I enzymes (Cherest *et al.*, 1969, 1971). The state of *eth1* gene does not affect the repressibility of these enzymes (see Table V).

Repression and dominance results suggest that the *eth2* gene product is involved in pleiotropic repression of group I methionine enzymes, at least two of whose structural genes are unlinked. The simplest model is that the *eth2* gene product is an aporepressor (Cherest *et al.*, 1971). The fact that the *eth2-2* allele is suppressible by nonsense suppressors suggests that the *ETH2* gene codes for a protein (Masselot and de Robichon-Szulmajster, 1972).

TABLE V

HOMOSERINE-O-TRANSACETYLASE, HOMOCYSTEINE SYNTHETASE, ATP SULFURYLASE, AND HOMOSERINE DEHYDROGENASE REPRESSION IN DIFFERENT STRAINS OF *S. cerevisiae*[a]

Genotype	Average repression (%)[b]			
	Homoserine-O-transacetylase	Homocysteine synthetase	ATP sulfurylase	Homoserine dehydro-genase
eth1ˢ eth2ˢ	66	87	93	41–45
eth1ʳ eth2ˢ	66	87	93	n.t.
eth1ˢ eth2ʳ	3	31	81	n.t.
eth1ʳ eth2ʳ	3	31	81	35–52

[a]Compiled from data in Cherest *et al.* (1969).
[b]Repression was measured by growing strains in minimal media with and without 2×10^{-3} M DL-methionine. n.t., Not tested.

Methionyl-tRNA has also been implicated in regulation of methionine biosynthesis (de Robichon-Szulmajster and Surdin-Kerjan, 1971). A *ts* mutant, *ts296*, isolated by McLaughlin and Hartwell (1969), was shown to have a defective tRNA synthetase, as well as being derepressed for only the group I enzymes at both permissive and nonpermissive temperatures (Cherest *et al.*, 1971). Conditions that lead to the absence of repression (0.1 mM methionine) in the *ts* mutant have been correlated with a decrease in bulk tRNA^{met} charging, whereas conditions that cause repression (4 mM

TABLE VI

ANALOG-RESISTANT AND -SENSITIVE STRAINS

Gene	Dominance[a]	Defect
TRA1	D	Phosphoribosyl transferase (*his1*)
tra2	R	Permeation
tra3	R	Regulation
tra4	R	?
tra5	R	Regulation
tra6	R	?
tra7	R	?
tra8	R	Permeation
tra9	R	?
aas1	R	Regulation
aas2	R	Regulation
ass3	R	Regulation

[a]D, dominant; R, recessive.

methionine) concomitantly restore high levels of methionyl-tRNA (Cherest *et al.*, 1971). However, in wild-type cells no such correlation between enzyme and methionyl-tRNA levels exists, suggesting that perhaps a minor iso-accepting tRNA^{met} species is involved in methionine regulation (Cherest *et al.*, 1971). Taken together, the evidence suggests that a minor methionyl-tRNA, or some derivative, behaves as the corepressor for group I enzymes.

The results showing that group II enzymes are repressed by methionine, but are not synthesized coordinately with group I enzymes, are not affected by limitation of the tRNA^{met} charging, and are not regulated by the *eth2* gene imply the existence of an as yet undiscovered additional repressor(s), aporepressor(s), or corepressor(s) operative in methionine biosynthesis (Cherest *et al.*, 1969, 1971).

VII. Histidine Biosynthesis

There are seven independently segregating genes for histidine biosynthesis (Fink, 1965). Three of the ten steps in the biosynthetic pathway are controlled by the *HIS4* gene cluster (Fink, 1966). The pathway is under dual control: feedback inhibition and repression. The first enzyme in the pathway is feedback-inhibited by histidine (Korch and Snow, 1973; Messenguy and Fink, 1973). All the genes are under unusual repression control. Starvation of the cells for arginine, histidine, lysine, or tryptophan leads to derepression of all histidine enzymes. The study of regulatory mutants has shown that this bizarre behavior is the manifestation of a general control over amino acid biosynthesis.

The factors involved in the isolation of regulatory mutants were studied by D. Yep and G. R. Fink (manuscript in preparation). Mutants resistant to the histidine analog 1,2,4-triazole-3-alanine (TRA) were obtained on minimal medium plus 0.25 mM TRA (see Table VI). Over 100 mutants were obtained in this way. The mutations leading to TRA resistance in these strains defined five genes, *TRA2, -6, -7, -8,* and *-9*. Mutations in these genes do not lead to regulatory defects.

The absence of regulatory mutants among TRA-resistant strains selected on standard yeast nitrogen base medium is a consequence of some unusual aspects of yeast physiology. Amino acids enter yeast through two genetically distinct permease systems (Crabell and Grenson, 1970). Histidine and TRA can enter the cell by the general amino acid permease which can take up any amino acid, or by the histidine-specific permease (which can take up only histidine and related compounds). Each of these systems is controlled by separate genes. A third gene *AAP* seems to govern both the specific and the general permeases. Mutations in this gene drastically reduce amino acid uptake through both permease systems. Over 90% of the histidine entering the cell appears to come in through the general permease system.

The general amino acid permease is inhibited by ammonia (Grenson *et al.*, 1970), whereas the specific is not. In strains derived from S288C, the initial rates of amino acid uptake through the general amino acid permease are hardly affected by ammonia (Rytka, 1975; D. Yep and G. R. Fink, unpublished data). Any mutation that lowers the uptake of TRA on ammonia can appear as a TRA-resistant mutant. Such mutations could be in the gene for the general amino acid permease (GAP) or any gene affecting the ammonia sensitivity of the general amino acid permease.

Three methods were used to study the defects of TRA-resistant mutants obtained on ammonia—direct assay of the permeases, analog sensitivity, and tests of allelism with known *gap* mutants. By direct permease assay *tra2* and *tra8* mutants had reduced levels of histidine uptake through general amino acid permease when tested on ammonia but not when tested on proline. *tra2, -6, -7, -8,* and *-9* show increased resistance (as compared with the wild type) to *p*-fluorophenylalanine and ethionine. In addition, *tra6, -7, -8,* and *-9* are resistant to high concentrations of histidine. The wild type is sensitive to concentrations of histidine in excess of 6 mM, whereas *tra* mutants are resistant to 30 mM histidine. None of these mutants is allelic to the *gap* mutant of Grenson. These data indicate that *tra2* and *tra8* are mutations that lower the levels of the general amino acid permease on medium containing ammonia. The resistance of these strains to TRA is explained by their permease defects. The resistance of the other strains is unexplained, since both histidine uptake and histidine regulation appear normal. They could have altered uptake of the analog but not of histidine. This possibility cannot be tested, because the radioactive analog is not available. Alternatively, these strains could have an altered intracellular distribution of the analog.

TRA-resistant mutants were isolated in the wild type on minimal medium with proline as the nitrogen source. The *tra2* and *tra8* types (and perhaps other permease types) should not appear on this medium. Of 40 TRA-resistant mutants isolated in this way, 18 map in the *HIS1* gene. These mutants were designated *tra1*. This gene codes for phosphoribosyl transferase, the enzyme catalyzing the first step in the pathway of histidine biosynthesis. All the *TRA1* mutations are dominant to *tra1* $^+$, and some *TRA1* mutants excrete histidine. Direct biochemical analysis showed that these mutations did not significantly affect the catalytic site of phosphoribosyl transferase but did alter the sensitivity of the enzyme to feedback inhibition by histidine (Messenguy and Fink, 1973). These results provide strong evidence that the molecular consequence of the *TRA1* mutation is the loss of feedback sensitivity of the first enzyme. The loss of feedback control destroys a strain's ability to regulate the size of its histidine pool. Direct measurements of the histidine pool showed that some *TRA1* mutants have an internal pool more than 10 times that found in the wild type. TRA resistance of *TRA1* mutants results from dilution of the TRA by the high internal histidine pool.

Feedback-resistant mutants can be eliminated by selecting TRA-resistant mutations in a *his1* mutant. This strain has a phosphoribosyl transferase partially defective in the catalytic reaction. In the *his1* strain, mutations affecting the feedback site should not lead to an increased pool of histidine. Selection using the *his1* mutant yielded several mutants with regulatory defects, and no feedback-resistant types. The mutations causing these defects involve two genes—*tra3* and *tra5*. All alleles at these loci lead to constitutive synthesis not only of histidine biosynthetic enzymes, but of arginine, lysine, and tryptophan enzymes as well.

Schuerch *et al.* (1974) report 5-methyltryptophan-sensitive strains are unable to derepress the enzymes of histidine, arginine, and tryptophan biosynthesis. Wolfner *et al.* (1975) extended these observations and showed that these amino acid analog–sensitive (*aas*) strains are also unable to derepress for lysine biosynthesis. Moreover, these studies showed that *TRA* and *AAS* genes act in concert to regulate arginine, histidine, lysine, and tryptophan biosynthesis. In the absence of the *TRA3* product, all these pathways are constitutively derepressed; in the absence of the *AAS1* product, the same pathways cannot be derepressed.

The studies of Wolfner *et al.* (1975) indicate that the *TRA* and *AAS* genes are, respectively, negative and positive control elements in a system of general control of amino acid biosynthesis. A model based on their genetic and biochemical studies pictures the *TRA* gene product(s) as preventing the synthesis of amino acid biosynthetic enzymes (Fig. 5). The *AAS* gene product inactivates the *TRA3* gene product if the intracellular pool of any of the amino acids (arginine, histidine, lysine, tryptophan) is low. Under repressed conditions the *AAS* gene product is itself inactivated by the amino acids under *AAS* and *TRA* control. This model explains the fact that the *aas tra* double mutant is constitutive, like the *tra3* mutant itself. If the *AAS* product were a sigma-like element necessary for expression of the structural genes, then the double mutant would be unable to derepress and would have the phenotype of the *aas* mutant.

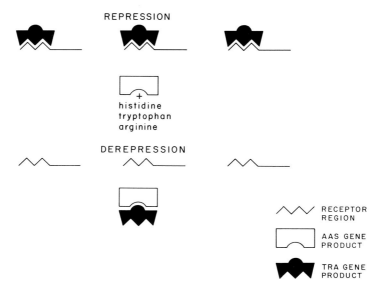

FIG. 5   A model for the role of the *AAS1* and *TRA3* genes in the regulation of amino acid biosynthesis. Under REPRESSION: The *TRA3* gene product prevents expression of a number of structural genes. Repression could be at the level of transcription, translation, or processing of the messages. The *AAS1* gene product is inactivated in the presence of arginine, histidine, lysine, and tryptophan. The net effect is to have only a low level of gene expression in the presence of all four amino acids. Under DEREPRESSION: If cells are starved for one of the amino acids the *AAS1* product is now functional and inactivates the *TRA3* product. The net effect is to allow full derepression of the structural genes. The phenotypes of the *tra3* and *aas1* mutants are compatible with this model.

Genetic studies of regulatory mutants have been greatly facilitated by a petri plate assay which indicates the state of derepression. A leaky histidine auxotroph can grow slightly on minimal medium, because it can derepress its biosynthetic pathway. When this histidine mutation is coupled with a regulatory mutation like *aas1*, which prevents derepression, the resulting double mutant is a complete histidine auxotroph; when it is coupled with a mutation like *tra3*, which is constitutive, the double mutant grows better than the leaky strain without regulatory mutations. Figure 6 is a graphic illustration of the differential growth responses caused by regulatory mutations.

tra3 and *tra5* mutations lead to defects in the cell cycle. Strains carrying these mutations are constitutive at 23 °C and temperature-sensitive for growth at 37 °C. Studies have shown that *tra3* and *tra5* mutants cannot synthesize DNA and cannot initiate a new cell cycle at the nonpermissive temperature. These defects indicate that the *tra* mutants are blocked early

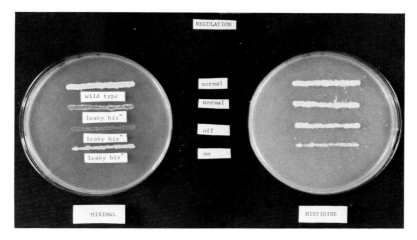

FIG. 6. Visualization of regulatory phenotypes on a petri plate. Wild-type and a leaky histidine auxotroph in three different regulatory states were replicated to minimal and minimal plus histidine medium. The strains, from top to bottom, are wild type (S288C), *his1-29*, *his1-29 aas1-1*, *his1-29 tra3-1*. All strains grow well on histidine medium. On minimal medium the growth of the leaky *his1* is dependent on the state of the regulatory genes.

during G_1 in the cell cycle. On the basis of regulatory and cell cycle defects, Wolfner *et al.* (1975) have hypothesized that the *TRA* genes are sensor genes which integrate amino acid biosynthesis into the cell cycle.

The derepression of arginine, lysine, and tryptophan pathways by histidine starvation is difficult to understand on physiological grounds, but is explicable on the basis of a supercontrol over a large number of pathways. Rather than sensing each pathway, yeast cells may recognize a single signal which relays information on the state of a large segment of metabolism. The control of catabolic pathways by cyclic AMP in bacteria is a good example of this type of general control. In yeast the signals from general control systems may feed into the cell cycle at G_1 to prevent the cell from initiating a new cycle if the proper nutrients are unavailable. The co-regulation of arginine, histidine, lysine, and tryptophan may be a manifestation of this cell cycle control.

REFERENCES

Bechet, J., Wiame, J. M., and Grenson, M. (1962). *Arch. Int. Physiol. Biochim.* **70**, 564–565.
Bechet, J., Wiame, J. M., and Grenson M. (1964). *Arch. Int. Physiol. Biochim.* **73**, 137–139.
Bechet, J., Grenson, M., and Wiame, J. M. (1970). *Eur. J. Biochem.* **12**, 31–39.
Cherest, H., and de Robichon-Szulmajster, H. (1966). *Genetics* **54**, 981–991.
Cherest, H., Eichler, F.- and de Robichon-Szulmajster, H. (1969). *J. Bacteriol.* **97**, 328–336.

Cherest, H., Surdin-Kerjan, Y., and de Robichon-Szulmajster, H. (1971). *J. Bacteriol.* **106**, 758–772.
Crabell, M., and Grenson, M. (1970). *Eur. J. Biochem.* **14**, 197–204.
DeDeken, R. H. (1963). *Biochim. Biophys. Acta* **78**, 606–616.
de Robichon-Szulmajster, H., and Surdin-Kerjan, Y. (1971). In "'The Yeasts" (A. H. Rose and J. S. Harrison, eds.), Vol. 2, pp. 335–418, Academic Press, New York.
Dorn, G. (1965). *Genet. Res.* **6**, 13–26.
Douglas, H. C., and Hawthorne, D. C. (1964). *Genetics* **49**, 837–844.
Douglas, H. C., and Hawthorne, D. C. (1966). *Genetics* **54**, 911–916.
Douglas, H. C., and Hawthorne, D. C. (1972). *J. Bacteriol.* **109**, 1139–1143.
Douglas, H. C., and Pelroy, G. (1963). *Biochim. Biophys. Acta* **68**, 155–156.
Fink, G. R. (1965). *Science* **146**, 525–527.
Fink. G. R. (1966). *Genetics* **53**, 445–459.
Grenson, M., Mousset, M., Wiame, J. M., and Bechet, J. (1966). *Biochim. Biophys. Acta.* **127**, 325–338.
Grenson, M., Hou, C., and Crebell, M. (1970). *J. Bacteriol.* **103**, 770–777.
Gross, S. R. (1965). *Proc. Nat. Acad. Sci. U. S.* **54**, 1538–1546.
Jund, R., and Lacroute, F. (1970). *J. Bacteriol.* **102**, 607–615.
Jund, R., and Lacroute, F. (1972). *J. Bacteriol.* **109**, 196–202.
Korch, C., and Snow, R. (1973). *Genetics* **74**, 287–305.
Lacroute, F. (1968). *J. Bacteriol.* **95**, 824–832.
Lacroute, F., Piérard, A., Grenson, M., and Wiame, J. M. (1965). *J. Gen. Microbiol.* **40**, 127–142.
McLaughlin, C. S., and Hartwell, L. (1969). *Genetics* **61**, 557–566.
Masselot, M., and de Robichon-Szulmajster, H. (1972). *Genetics* **71**, 535–550.
Messenguy, F., and Fink, G. R. (1973). "Genes, Proteins and Evolution", pp. 85–94. Plenum, New York.
Messenguy, F., Pennickx, M., and Wiame, J. M. (1971). *Eur. J. Biochem.* **22**, 277–286.
Monod, J., and Jacob, F. (1961). *Cold Spring Harbor Symp. Quant. Biol.* **26**, 389–401.
Ramos, F., Thuriaux, P., Wiame, J. M., and Bechet, J. (1970). *Eur. J. Biochem.* **12**, 40–47.
Rytka, J. (1975). *J. Bacteriol.* **121**, 562–570.
Schuerch, A., Miozzari, J., and Hutter, R. (1974). *J. Bacteriol.* **117**, 1131–1140.
Thomas, R. (1971). *In* "The Bacteriophage Lambda" (A. D. Hershey, ed.), pp. 211–220. Cold Spring Harbor Lab., Cold Spring Harbor, New York.
Thuriaux, P., Ramos, F., Piérard, A., Grenson, M., and Wiame, J. M. (1972). *J. Mol. Biol.* **67**, 277–287.
Toh-e, A., Ueda, Y., Kakimoto, S., and Oshima, Y. (1973). *J. Bacteriol.* **113**, 727–738.
Wiame, J. M. (1968). *In* "Biochemical Evolution" (N. Van Thoai and J. Roche, eds.), pp. 371–395. Gordon & Breach, New York.
Wiame, J. M. (1971). *Curr. Top. Cell. Regul.* **4**, 1–38.
Willson, C., Perrin, D., Cohn, M., Jacob, F., and Monod, J. (1964). *J. Mol. Biol.* **8**, 582–592.
Wolfner, M., Yep, D., Messenguy, F., and Fink, G. R. (1975). Submitted for publication.

Chapter 15

Methods for Selecting Auxotrophic and Temperature-Sensitive Mutants in Yeasts

BARBARA SHAFFER LITTLEWOOD

Department of Physiological Chemistry, School of Medicine,
University of Wisconsin, Madison, Wisconsin

I. Introduction 273
II. Nystatin Selection 275
    A. General Procedure for Selecting Auxotrophic Mutants 275
    B. Protocol for Selecting Auxotrophs in *S. cerevisiae* 276
    C. Selection of *ts* Mutants 277
    D. Possible Complications 278
    E. Efficiency of Nystatin Selection 278
III. Tritium Suicide Selection 279
    A. General Procedure for Selecting *ts* Mutants 279
    B. Protocol for Selecting *ts* mutants in *S. cerevisiae* 281
    C. Selection of Auxotrophs 282
    D. Possible Complications 282
    E. Efficiency of Tritium Suicide Selection 283
IV. Fatty Acid-less Death in *S. cerevisiae* 283
    References 284

I. Introduction

    Mutant selection is strictly defined as those processes that enrich a culture for desired mutants by decreasing the number of viable nonmutant cells. In a more general sense, considerable pressure toward increasing the percentage of specific mutants can be exerted at each of the four basic steps in mutant isolation: choice of parental strain, mutagenesis, enrichment of the culture for auxotrophic or temperature-sensitive (*ts*) mutants, and identification of the desired mutants in the remaining cell population.

    This chapter describes general enrichment procedures applicable to the selection of a wide variety of auxotrophic and *ts* mutants in yeasts. Although used primarily for *Saccharomyces cerevisiae*, these techniques appear to be usable with most species of yeast. In an attempt to detail the potentials and pitfalls of these selection techniques and to assist those applying the techniques to less commonly used species of yeast, generalized procedures for selecting auxotrophs with nystatin and for selecting *ts* mutants by tritium suicide are given. These are followed by specific protocols used for *S. cerevisiae*.

These techniques have been applied primarily to select strains carrying nuclear mutations. Methods for obtaining mutants with lesions in the cytoplasmic genome are given by Linnane *et al.* (1968), Coen *et al.* (1969), and Deutsch *et al.* (1974).

The yeast used most extensively when isolating auxotrophic and *ts* mutants is *S. cerevisiae* (e.g., Hartwell, 1967, Hartwell *et al.*, 1974; Shaffer *et al.*, 1969; Korch and Snow, 1973). General techniques for working with this organism are described by Fink (1970) and in the laboratory manual for a course in methods in yeast genetics (Cold Spring Harbor Laboratory of Quantitative Biology, Cold Spring Harbor, N.Y.). A table of genetic markers and associated gene products in *S. cerevisiae* is presented in the *CRC Handbook of Biochemistry* (Sober, 1970); a current genetic map is in Mortimer and Hawthorne (1973). A variety of auxotrophic and *ts* mutants is also available in *Schizosaccharomyces pombe* (e.g., Leupold, 1970; Bonatti *et al.*, 1972). Leupold (1970) and Mitchison (1970) describe procedures for working with this yeast.

Auxotrophic and *ts* mutations are usually recessive and therefore should be induced in a haploid strain. It is advisable to use a standard strain carrying at least one easily identified marker mutation. For *S. cerevisiae*, standard wild-type strains as well as a wide variety of mutants suitable for use in genetic analyses are available from the Yeast Genetics Stock Center (Dr. John Bassel, Donner Laboratory, University of California, Berkeley, Calif). Leupold (1970) describes the standard strain of *Schiz. pombe*.

In certain cases the use of a particular parental strain may facilitate the recovery and/or characterization of the resulting mutants. In *S. cerevisiae*, this is the case for the isolation of mutants carrying conditional-lethal defects in sporulation (Esposito and Esposito, 1969) or DNA metabolism (Jannsen *et al.*, 1973; Wickner, 1974), and mutants with nonsense mutations in essential functions (Rasse-Messenguy and Fink, 1973).

Protocols for using a variety of mutagens and the specificity of their mutagenic action are described by Prakash and Sherman (1973).

In addition to the methods described here, inositolless death (Leupold, 1970) and treatment with 2-deoxyglucose (Megnet, 1965) have been used to select auxotrophs in *Schiz. pombe*.

II. Nystatin Selection

Nystatin, a polyene macrolide antibiotic produced by *Streptomyces*, is inhibitory to yeasts and other molds, but is ineffective against bacteria and viruses. In sensitive organisms, nystatin binds to sterols (primarily ergosterol) in the cell membrane (Lampen *et al.*, 1962), causing distortions of the membrane and subsequent leakage of essential metabolites. Metabolic activities in cell-free systems are virtually unaffected by polyenes (Hamilton-Miller, 1973). Hamilton-Miller (1973) recently published an extensive review of the chemistry and biology of polyene antibiotics.

A. General Procedure for Selecting Auxotrophic Mutants

1. *Outgrowth of mutagenized culture.* After mutagenesis the cells are grown in minimal medium containing the substances required by the desired auxotrophs, or in YEPD (Section III,B,1). Minimal medium is usually more appropriate, as undesired auxotrophs will not grow. This growth period must be long enough for the cells to recover from the mutagenic treatment and for new mutations, i.e., auxotrophy, to be expressed (Snow, 1966).

2. *Nitrogen starvation.* Growth of the "outgrown" culture is stopped by starving for nitrogen (by the elimination of ammonium sulfate from unsupplemented minimal medium). Cell growth must cease completely at this stage, and each strain should be tested for the length of time this requires. The strain studied by Thouvenot and Bourgeois (1971) continued to grow slowly for 10 hours in nitrogen-free medium. Glucose starvation is ineffective (Snow, 1966).

3. *Growth in minimal medium with a nitrogen source.* The cells are transferred to minimal medium with ammonium sulfate but without the growth requirements of the desired auxotrophs. This is a critical step in the procedure. After this step the wild-type and undesired mutant cells must be growing vigorously, while the desired mutants remain inactive and therefore refractory to nystatin action. To ensure that the undesired cells are in log phase, the culture should be monitored by absorbance or hemocytometer count and incubation continued until a 10–15% increase in cell number is observed.

Careful supplementation of this medium can increase the selectivity of the method. For example, if histidine-requiring mutants are desired, the medium should contain all amino acids except histidine (Snow, 1966;

Thouvenot and Bourgeois, 1971). In particular cases additional specificity may be possible by adding intermediates in the metabolic pathway being studied, which are produced prior to the desired metabolic block; for example, if one desires only those histidine-requiring mutants that cannot convert histidinol to histidine, the addition of histidinol (plus the other amino acids) during this incubation should eliminate unwanted types of histidine mutants. (Such directed selection is limited to those available intermediates to which yeast is freely permeable.)

4. *Nystatin treatment*.   Nystatin is routinely used at 10 μg/ml. At this drug concentration a cell concentration of 10^6 to 10^7 cells/ml is optimal for *S. cerevisiae* (Snow, 1966). Exposure times of 1–1.5 hours are sufficient; at 30 °C drug binding to the cells plateaus at 80 minutes (Lampen *et al.*, 1959). Nystatin binding is severalfold higher at pH 4 than at pH 6 and above; the pH of the minimal medium used in this step should be adjusted accordingly.

There is a rather fine balance between the total amount of cell death and the extent of mutant enrichment. If exposure to the polyene is prolonged (Snow, 1966), or the drug concentration too high (Thouvenot and Bourgeois, 1971), even nongrowing mutant cells will be killed. If the selection proves unsuccessful using the protocol described in Section II,B, a reconstruction experiment may be informative. A mixture of the parental strain and a characterized isogenic mutant can be treated with nystatin under varying conditions to determine the optimal time and drug concentration for enrichment for the mutant strain (Moat *et al.*, 1959; Thouvenot and Bourgeois, 1971).

After nystatin treatment the cells are washed and plated on solid minimal medium containing the growth requirements of the desired mutants. Auxotrophs are easily detected by replica plating to minimal medium lacking the growth substance or by the phloxine-B method (Horn and Wilkie, 1966).

5. *Second nystatin treatment*.   Repetition of the above procedure increases the efficiency of selection for auxotrophs. Instead of plating the cells from step 4, one returns them to minimal medium and steps 1 through 4 are repeated.

B.   Protocol for Selecting Auxotrophs in *S. cerevisiae*

Fink (1970) suggests the following specific conditions for selecting auxotrophs in *S. cerevisiae* (strain S288C).

1. *Outgrowth of mutagenized culture*.   One milliliter of a mutagenized culture (washed several times with sterile water if chemical mutagens are used) is diluted in 9 ml of liquid minimal medium (Table I) supplemented with the growth requirement of the desired auxotrophs. The culture is shaken for 2 days at 30 °C.

TABLE I

MEDIA FOR NYSTATIN SELECTION

A. Minimal medium without $(NH_4)_2SO_4$[a]

The following three $100 \times$ solutions are prepared. To make 1 liter of $1 \times$ medium, 10 ml of each solution plus 20 gm glucose are diluted to 1 liter with water.

1. Vitamins, trace elements, and salts $(100\times)$: $MgSO_4 \cdot 7H_2O$, 50 gm; NaCl, 10 gm; $FeCl_3 \cdot 6H_2O$, 5 mg; $ZnSO_4 \cdot 7H_2O$, 7mg; H_3BO_3, 1 mg; $CuSO_4 \cdot 5H_2O$, 1 mg; KI, 1 mg; thiamine, 40 mg; pyridoxine, 40 mg; pantothenate, 40 mg; inositol, 200 mg; biotin, 0.2 mg. Add water to 1 liter.

2. Phosphate buffer $(100\times)$: KH_2PO_4, 87.5 gm; K_2HPO_4, 12.5 g. Add water to 1 liter.

3. Calcium chloride $(100\times)$: $CaCl_2 \cdot 2H_2O$, 10 gm. Add water to 1 liter.

B. Minimal medium with $(NH_4)_2SO_4$[a]

One liter of $1 \times$ medium is prepared by adding 1 gm $(NH_4)_2SO_4$ to 1 liter of the medium described above.

[a] Media $(1 \times)$ are sterilized by autoclaving.

2. *Nitrogen starvation.* One milliliter of the "outgrown" culture is washed twice with sterile water and the cells suspended in 1 ml of minimal medium without ammonium sulfate (Table I) and without the desired growth factor. The culture is shaken overnight at 30 °C.

3. *Growth in minimal medium with a nitrogen source.* The cells are collected by centrifugation, suspended in 0.9 ml of minimal medium containing 1 gm per liter ammonium sulfate (Table I) and shaken for 6 hours at 30 °C.

4. *Nystatin treatment.* The stock solution of nystatin (100 μg/ml; prepared fresh daily) is prepared by dissolving 1 mg of nystatin in 1 ml of 95% ethanol and diluting the ethanol solution with 9 ml of sterile water. The solution will be opalescent.

One-tenth milliliter of the 100 μg/ml nystatin solution is added to the 0.9-ml culture, and incubation is continued for 1 hour. The cells are then washed twice with sterile water and suspended iň 1 ml of minimal medium containing the growth substances required for the repetition of step 1.

5. *Second nystatin treatment.* Steps 1 through 4 are repeated, except that the incubation time in the presence of nystatin (step 4) is increased to 1.5 hours. After nystatin treatment the cells are washed twice and suspended in sterile water. This suspension may be diluted and plated; unused portions can be stored at 4 °C.

C. Selection of *ts* Mutants

Nystatin has not been used extensively to select *ts* mutants in yeasts, although such selection should be possible. Nystatin binding to yeast cells is temperature-dependent; binding is high at both 45° and 30 °C, and approaches zero at 4 °C (Lampen *et al.*, 1959). To select for *ts* mutants, steps 3 and 4 in the procedure in Section II,B should be performed at the nonpermissive temperature. Butow *et al.* (1973) used nystatin to select for cold-sensitive respiratory-deficient mutants in *S. cerevisiae.*

D.   Possible Complications

1. *Respiratory-deficient mutants.*   Many yeast cultures contain a significant proportion of respiratory-deficient (petite) cells, and mutagenesis may increase this percentage (Fink, 1970). These cells grow somewhat slowly and are often refractory to selection techniques. Consequently, nystatin-treated populations often contain a high percentage of petite cells (Moat *et al.*, 1959; Bard, 1972) which do not carry the desired auxotrophic or *ts* mutation. These cells can be eliminated by plating the nystatin-selected culture directly onto medium containing a nonfermentable compound such as glycerol as the carbon source (Fink, 1970), in which petites cannot form colonies. Alternatively, the respiratory competence of clones can be tested by replica plating to glycerol medium or by the tetrazolium overlay procedure (Ogur *et al.*, 1957).

2. *Identical mutants.*   When several mutants are obtained from a nystatin-treated culture, there is no guarantee that the isolates represent independent mutational events. This is especially true when two cycles of growth and nystatin treatment are used. This problem can be eliminated by splitting the mutagenized culture into several aliquots, each of which is taken through the selection procedure; only one mutant is saved from each aliquot.

3. *Nystatin-resistant mutants.*   Subjecting a culture to nystatin treatment is of course the method used to isolate nystatin-resistant mutants (Woods, 1971). While the appearance of drug-resistant mutants does not seem to have created problems, the sensitivity of the parental strain being used should be verified before using this procedure.

E.   Efficiency of Nystatin Selection

Very efficient nystatin selection of auxotrophs from mutagenized prototrophic cultures of *S. cerevisiae* has been reported. After treating a mutagenized culture with 10 µg/ml nystatin for 1 hour, Snow (1966) found that 10% of the survivors were auxotrophs. Thouvenot and Bourgeois (1971) report similar success. In both instances a wide variety of auxotrophs requiring different amino acids were obtained. When selecting specifically for histidine auxotrophs, Snow (1966) found that 5–10% of the survivors had the desired phenotype.

Stromnäes and Mortimer (1968) adapted nystatin selection to the elimination of spontaneous prototrophic revertants in a predominantly auxotrophic culture. Gaillardin *et al.* (1973) reported successful nystatin selection of auxotrophs in *Candida lipolytica*.

III. Tritium Suicide Selection

Tritium suicide is the death of cells caused by the decay of tritium incorporated into macromolecular components. The mechanism of cell death in eukaryotic organisms is not completely understood; it appears to result from a combination of factors including β-irradiation of the nucleus and damage to labeled molecules by transmutation (Burki and Okada, 1968). Tritium decay in the cell nucleus (in either DNA or RNA) is more efficient in killing cells than decay in the cytoplasm; in both mammalian (Burki and Okada, 1968) and bacterial systems (Bockrath *et al.*, 1968), labeled thymidine produces more killing per disintegration than does uridine which in turn is more efficient than any individual amino acid.

A.   General Procedure for Selecting *ts* Mutants

1. *Outgrowth of mutagenized cells.*   After mutagenesis the cells are grown in rich medium at the permissive temperature. This growth period must be long enough for the cells to recover from the mutagenic treatment and to allow the mutant gene product to become predominant in the cells (this may take several generations for macromolecules with slow turnover).

2. *Preparation of the cells for tritium labeling.*   For maximal incorporation of ³H precursors, nonmutant cells must be growing exponentially in minimal medium. The "outgrown" culture is diluted in minimal medium ($\sim 10^5$ to 10^6 cells/ml) and grown to log phase at the permissive temperature. Growth should be monitored by absorbance or hemocytometer count until a 10–20% increase in cell number is observed.

The log-phase culture is then shifted to the nonpermissive temperature and incubated until the desired block in macromolecular synthesis is expected to have been expressed, and growth of the *ts* mutants inhibited. The time required for *ts* mutations to be expressed is not always obvious. In *S. cerevisiae* times ranging from 2 minutes to several hours are required to halt RNA or protein synthesis in mutants with well-characterized *ts* defects in the production of these macromolecules (e.g., Hartwell, 1967; Hartwell and McLaughlin, 1969) and to inhibit some *ts* cell cycle mutants

(Hartwell *et al.*, 1974). However, before choosing a long "shut-off" period, consideration should be given to the difficulties encountered in characterizing mutants that are inhibited only after long periods of incubation at the nonpermissive temperature, or mutants in which metabolism is slowed but never completely stopped (also, see Section III,D,3).

3. *Tritium labeling.* By careful choice of the ^{3}H precursor, tritium suicide selection can be made to select for specific classes of *ts* mutants. Mutants defective in the synthesis of particular macromolecules are selected by labeling with a precursor of that molecule at the nonpermissive temperature. In bacteria (Tocchini-Valentini and Mattocia, 1968) and yeasts (Section III,E), labeling with an amino acid at high temperature selects strongly for heat-sensitive mutants with defects in protein synthesis. Similarly, incorporation of ^{3}H-uridine enriches for mutants with *ts* defects in RNA metabolism (Reid, 1971). It appears that any ^{3}H compound that can be incorporated at high levels and which does not leak out of the cells during storage (step 4) will produce efficient killing. Wild-type cells of *S. cerevisiae* are impermeable to both thymine and thymidine; mutants that can incorporate dTMP (Jannsen *et al.*, 1973; Wickner, 1974) may prove useful for tritium suicide following the incorporation of base-labeled precursor.

A small quantity of high specific activity ^{3}H precursor is added to the culture at the nonpermissive temperature. With either ^{3}H-uridine or a mixture of ^{3}H-amino acids, the final specific activity of the culture medium should be approximately 50 μCi/ml. Changes in the specific activity of the cells are estimated from changes in the ratio of trichloroacetic acid (TCA)-precipitable counts per minute per milliliter of labeled culture to absorbance of culture. Incubation with ^{3}H precursor is continued until the specific activity of the cells no longer increases. This plateau is reached between 2.5 and 3.5 hours for both ^{3}H-uridine and a mixture of ^{3}H-amino acids when *S. cerevisiae* is labeled at 38 °C (Littlewood and Davies, 1973). Examples of specific activities (disintegrations per minute per viable cell) obtained are given in Table II. Growth and labeling are terminated by quickly cooling the culture to 4 °C.

4. *Tritium suicide.* The highly radioactive cells are placed in a non-growth-supporting medium and stored at 4 °C. Medium lacking a carbon source is recommended, as some yeast strains grow slowly at 4 °C and growth dilutes the label. The viability of the stored culture is assayed every 2–3 days until 90–95% killing has occurred, or the viability no longer decreases. There was no loss in viability in an unlabeled culture of *S. cerevisiae* stored under these conditions (Littlewood and Davies, 1973).

Representative suicide data are given in Table II. While killing efficiencies vary with labeling temperatures and with the parental strain, 1 week of storage should allow sufficient killing of cells labeled to 20–50 dpm per viable cell.

TABLE II

REPRESENTATIVE SPECIFIC ACTIVITIES AND SUICIDE TIMES FOR *S. cerevisiae*

Strain	Labeling temperature (°C)	Precursor	Disintegrations per minute per viable cell[a]	Storage time for 99% killing (hours)
A224A	38	5-³H-uridine	54	111
	38	³H-amino acids[b]	52	110
	38	³H-amino acids	27	281
Y166	18	5-³H-uridine	13	160
	18	³H-amino acids	20	111

[a]At the end of the labeling period, culture aliquots were placed in cold 10% TCA and the acid-precipitable material collected on Whatman GF/C glass fiber filters which were washed with TCA, dried, and counted in toluene-based liquid scintillation fluid. Counting efficiency was estimated at 10% (Person and Lewis, 1962; Littlewood and Davies, 1973).

[b]Tritiated reconstituted protein hydrolysate (yeast profile, Schwarz/Mann), a mixture of 15 L-amino acids.

Taking the suicide to its absolute limit is probably inadvisable. The desired mutants may have incorporated a small amount of label and eventually may be killed. Littlewood and Davies (1973) found that with prolonged storage the percentage of *ts* mutants no longer increased, although a small amount of cell death occurred.

After suicide the survivors are plated to rich medium at the permissive temperature. Temperature sensitivity can be scored by replica plating or streaking (Fink, 1970) to rich medium at the nonpermissive temperature.

B. Protocol for Selecting *ts* Mutants in *S. cerevisiae*

Littlewood and Davies (1973) suggest the following procedure for selecting *ts* mutants in *S. cerevisiae*. In this procedure 28°C is the permissive temperature; 38°C is the nonpermissive temperature. A mixture of ³H-amino acids can be substituted for the ³H-uridine with no other alterations in the method.

1. *Outgrowth of the mutagenized culture.* One milliliter of a mutagenized culture (washed several times with sterile water if chemical mutagens were used) is diluted in 4 ml YEPD (per liter: 10 gm yeast extract, 20 gm Bacto peptone, 20 gm glucose) and grown overnight at 28°C with shaking.

2. *Preparation of the cells for tritium labeling.* The "outgrown" culture is diluted to A_{550} of 0.1 (1 cm path length) in 15 ml of minimal medium (per liter: 7 gm Difco yeast nitrogen base without amino acids, 20 gm glucose) in a flask, and incubation continued for 4–5 hours. A 2.0-ml portion of the culture is transferred to a screw-cap tube. (The tube culture will be radio-

actively labeled; the flask culture will be used to follow culture growth during the labeling period.) Both the tube and the flask are transferred to 38 °C and incubated with shaking for 30 minutes.

3. *Tritium labeling.*   5-^{3}H-uridine (20 Ci/mmole) is added to the tube culture to a final level of 50 μCi/ml. Incubation is continued at 38 °C with shaking.

Every 30 minutes a 0.05-ml sample of the labeled culture is placed in 1 ml of 10% TCA at 4 °C, and the acid-precipitable radioactivity assayed (Table II). At similar intervals the absorbance of the flask culture is read. Labeling is continued for 2.5 hours, or until the specific activity of the cells no longer increases. Labeling is stopped by placing the tube culture in ice; the flask culture is discarded.

The chilled, labeled cells are washed twice with sterile water and suspended in 2 ml of 0.7% Difco yeast nitrogen base without amino acids (no carbon source added). To determine the initial specific activity, a 0.05-ml sample is removed for counting, and a second 0.05-ml sample is diluted with sterile water and plated on YEPD at the permissive temperature. The remaining labeled culture is stored at 4 °C.

4. *Tritium suicide.*   Every 2 days a 0.05-ml sample is removed from the refrigerated culture, diluted with sterile water, and plated on YEPD at the permissive temperature. Several dilutions are plated at each time point (i.e., $10^{-2}, 10^{-3}, 10^{-4}$). When the viability decreases below 10%, the selection is completed. The remaining culture is then plated on YEPD at appropriate dilutions, and the survivors tested for temperature sensitivity.

C.   Selection of Auxotrophs

Because the cells are labeled in minimal medium, the procedure outlined above selects strongly for auxotrophs. More efficient enrichment for specific auxotrophs can be obtained by (1) supplementing the minimal media (steps 2 and 3) with nonradioactive amino acids not required by the desired mutants, (2) performing all procedures at the optimal growth temperature for the parental strain, and (3) using as the ^{3}H precursor either ^{3}H-uridine or a single amino acid otherwise missing from the medium and not required by the desired mutants.

D.   Possible Complications

1. *Respiratory-deficient mutants.*   See Section II,D,1.
2. *Identical mutants.*   See Section II,D,2.
3. *Temperature sensitivity and cell death.*   For the majority of *ts* mutants, short exposure to the nonpermissive temperature inhibits cell growth but

does not kill the cells. However, death at the nonpermissive temperature is expected for certain classes of *ts* mutants, including mutants in lipid metabolism (Henry, 1973), osmotic mutants (Medoff *et al.*, 1972), and some cell cycle mutants (Hartwell, 1971a,b). Such strains cannot be selected by tritium suicide or any enrichment procedure involving exposure to the nonpermissive temperature.

E.   Efficiency of Tritium Suicide Selection

The tritium suicide procedure described here enriches the culture for both *ts* and auxotrophic mutants. Labeling with either ³H-uridine or a mixture of ³H-amino acids to approximately 50 dpm per viable cell. Littlewood and Davies (1973) found at least a 10-fold increase in both auxotrophic and heat-sensitive mutants when the suicide reached 99% killing. At this point 36% of the survivors were identifiable as either auxotrophic or *ts* mutants; a significant proportion of the additional survivors were slow growers (some probably petites) which appeared to be mutants, but which were not further characterized. Labeling with the amino acid mixture at high temperature should enrich specifically for *ts* mutants defective in protein synthesis. At the nonpermissive temperature protein synthesis stopped completely within 2 hours in 27 of 137 *ts* mutants isolated by amino acid tritium suicide, as compared to two such mutants among 68 *ts* mutants isolated without using an enrichment procedure.

Selecting with ³H-uridine, Gorman, Young, and Bock (unpublished results) obtained mutants of *S. cerevisiae* with both dominant and recessive *ts* alterations in the activity of the UAA suppressor *SUP7*. The parental strain, which carries *SUP7* plus two UAA mutations, is generally resistant to mutant isolation, as many induced mutations will be suppressed.

S. Henry (personal communication) attempted to select mutants with defects in the synthesis of complex lipids by incorporating ³H fatty acids into a fatty acid auxotroph of *S. cerevisiae*. Although suicide occurred, enrichment for the desired mutants was not significant, possibly because of the substantial pool of free fatty acids in the parental cells.

IV.   Fatty Acid-less Death in *S. cerevisiae*

Certain fatty acid auxotrophs of *S. cerevisiae* die when starved for fatty acids under otherwise growth-supporting conditions. Death is prevented by inhibition of protein synthesis. Henry (1973, and personal communication) recently developed a technique for selecting auxotrophs and certain *ts* mutants by fatty acid–less death. Mutagenized *ole1 fas1* cells are placed under conditions in which the parent strain grows but the desired mutants do not grow (as in Sections II,A,3 and III,A,2). Once growth of the desired mutants is stopped, the culture is starved for fatty acids for several hours. The growing cells in the population undergo fatty acid–less death, thus enriching for the desired mutants. Selection is most effective at low cell densities ($\sim 2 \times 10^4$ cells/ml), where cross-feeding is not significant.

After 12–18 hours of fatty acid–less death in minimal medium, S. Henry (personal communication) found enrichment for several different auxotrophs; enrichment was greatest for cells requiring methionine (28.6-fold), leucine (23.4-fold), and lysine (14.4-fold). A 1.8-fold enrichment for *ts* mutants was found after a 30 minutes "shut-off" period (Section III,A,2) followed by 8 hours of fatty acid–less death at the nonpermissive temperature; this low enrichment may reflect death of certain *ts* mutants (Section III,D,3) after long exposure to the nonpermissive temperature.

References

Bard, M. (1972). *J. Bacteriol.* **111**, 649–657.
Bockrath, R., Person, S., and Funk, F. (1968). *Biophys. J.* **8**, 1027–1036.
Bonatti, S., Simili, M., and Abbondandolo, A. (1972). *J. Bacteriol.* **109**, 484–491.
Burki, H. J., and Okada, S. (1968). *Biophys. J.* **8**, 445–456.
Butow, R., Ferguson, M., and Cedarbaum, A. (1973). *Biochemistry* **12**, 158–164.
Coen, D., Deutsch, J., Netter, P., Petrochilo, E., and Slonimski, P. P. (1969). *Symp. Soc. Exp. Biol.* **24**, 449–496.
Deutsch, J., Dujon, B., Netter, P., Petrochilo, E., Slonimski, P. P., Bolotin-Fukuhara, M., and Coen, D. (1974). *Genetics* **76**, 195–219.
Esposito, M. S., and Esposito, R. E. (1969). *Genetics* **61**, 79–89.
Fink, G. R. (1970). *In* "Methods in Enzymology" (H. Tabor and C. W. Tabor, eds.), Vol. 17A, pp. 59–78. Academic Press, New York.
Gaillardin, C. M., Charby, V., and Heslot, H. (1973). *Arch. Mikrobiol.* **92**, 69–83.
Hamilton-Miller, J. M. T. (1973). *Bacteriol. Rev.* **37**, 166–196.
Hartwell, L. H. (1967). *J. Bacteriol.* **93**, 1662–1670.
Hartwell, L. H. (1971a). *J. Mol. Biol.* **59**, 183–194.
Hartwell, L. H. (1971b). *Exp. Cell Res.* **69**, 265–276.
Hartwell, L. H., and McLaughlin, C. S. (1969). *Proc. Nat. Acad. Sci. U.S.* **62**, 468–474.
Hartwell, L. H., Culotti, J., Pringle, J. R., and Reid, B. J. (1974). *Science* **183**, 46–51.
Henry, S. (1973). *J. Bacteriol.* **116**, 1293–1303.
Horn, P., and Wilkie, D. (1966). *J. Bacteriol.* **91**, 1388.
Jannsen, S., Witte, I., and Megnet, R. (1973). *Biochim. Biophys. Acta* **299**, 681–685.
Korch, C. T., and Snow, R. (1973). *Genetics* **74**, 287–305.
Lampen, J. O., Morgan, E. R., Slocum, A., and Arnow, P. (1959). *J. Bacteriol.* **78**, 282–289.
Lampen, J. O., Arnow, P. M., Borowska, Z., and Laskin, A. I. (1962). *J. Bacteriol.* **84**, 1152–1160.
Leupold, U. (1970). *In* "Methods in Cell Physiology" (D. M. Prescott, ed.), Vol. 4, pp. 169–177. Academic Press, New York.

Linnane, A. W., Saunders, G. W., Gingold, E. B., and Lukins, H. B. (1968). *Proc. Nat. Acad. Sci. U.S.* **59**, 903–910.

Littlewood, B. S., and Davies, J. E. (1973). *Mutat. Res.* **17**, 315–322.

Medoff, G., Kobayashi, G. S., Kwan, C. N., Schlessinger, D., and Venkov, P. (1972). *Proc. Nat. Acad. Sci. U.S.* **69**, 196–199.

Megnet, R. (1965). *Mutat. Res.* **2**, 328–331.

Mitchison, J. M. (1970). *In* "Methods in Cell Physiology" (D. M. Prescott, ed.), Vol. 4, pp. 131–165. Academic Press, New York.

Moat, A. G., Peters, N., and Srb, A. M. (1959). *J. Bacteriol.* **77**, 673–677.

Mortimer, R. K., and Hawthorne, D. C. (1973). *Genetics* **74**, 33–54.

Ogur, M., St. John, R., and Nagai, S. (1957). *Science* **125**, 928–929.

Person, S., and Lewis, H. L. (1962). *Biophys. J.* **2**, 451–463.

Prakash, L., and Sherman, F. (1973). *J. Mol. Biol.* **79**, 65–82.

Rasse-Messenguy, F., and Fink, G. R. (1973). *Genetics* **75**, 459–464.

Reid, P. (1971). *Biochem. Biophys. Res. Commun.* **44**, 737–740.

Shaffer, B., Rytka, J., and Fink, G. R. (1969). *Proc. Nat. Acad. Sci. U.S.* **63**, 1198–1205.

Snow, R. (1966). *Nature (London)* **211**, 206–207.

Sober, H. A., ed. (1970). "Handbook of Biochemistry." Chem. Rubber Publ. Co., Cleveland, Ohio.

Strömnaes, O., and Mortimer, R. K. (1968). *J. Bacteriol.* **95**, 197–200.

Thouvenot, D. R., and Bourgeois, C. M. (1971). *Ann. Inst. Pasteur, Paris* **120**, 617–625.

Tocchini-Valentini, G. P., and Mattoccia, E. (1968). *Proc. Nat. Acad. Sci. U.S.* **61**, 146–151.

Wickner, R. B. (1974). *J. Bacteriol.* **117**, 252–260.

Woods, R. A. (1971). *J. Bacteriol.* **108**, 69–73.

Chapter 16

Isolation and Characterization of Mutants of *Saccharomyces cerevisiae* Able to Grow after Inhibition of *d*TMP Synthesis

M. BRENDEL,[1] W. W. FÄTH,[1] AND W. LASKOWSKI[2]

I. Introduction

When bacteria are inhibited or genetically blocked in thymidylate biosynthesis, this condition can be overcome by exogenously supplied thymine (Thy) or deoxythymidine (dThd) since these organisms can take up Thy and dThd and utilize them for thymidylate synthesis via the enzymes thymidine phosphorylase (tpp) and thymidine kinase (tk), or via tk alone (O'Donovan and Neuhard, 1970).

[1] Fachbereich Biologie der J. W. Goethe-Universität, Frankfurt am Main, Germany.
[2] Zentralinstitut 5 der Freien Universität, Berlin, Germany.

The yeast *Saccharomyces cerevisiae* obviously lacks tk (Grivell and Jackson, 1968) and also seems to take up poorly both Thy and dThd (Grenson, 1969; Jannsen *et al.*, 1968; Lochmann, 1965). It is therefore not possible in this organism to overcome a blockage in thymidylate biosynthesis by offering either molecule. Hence, if blockage of dTMP synthesis is to be overcome in yeast, one must offer deoxythymidine phosphoric acids, at least in the form of deoxythymidine 5'-monophosphate (dTMP). In 1968, Jannsen *et al.* reported DNA-specific labeling by exogenous dTMP in strain 211 of *S. cerevisiae*. Further experiments verified the DNA-specific nature of this labeling (Jannsen *et al.*, 1970; Brendel and Haynes, 1972), but showed rather poor efficiency of utilization of the offered dTMP molecules (Fäth and Brendel, 1974). DNA-specific labeling is achieved with ^{3}H- or ^{14}C-labeled dTMP and also with dTMP-^{32}P when derepression of acidic phosphatase is inhibited (Brendel and Haynes, 1973). In order to obtain more economical utilization of dTMP in *S. cerevisiae*, it was necessary to develop a screening procedure for the isolation of mutants able to grow with exogenous dTMP after inhibition of dTMP biosynthesis. Such mutants would allow further screening for low requirers of exogenous dTMP.

II. Mutants Able to Grow after Inhibition of dTMP Synthesis

A. Reversible Inhibition

In bacteria thymidylate synthesis may be inhibited by folic acid antagonists such as aminopterin (APT), which strongly interferes with dihydrofolic acid (DHFA) reductase, thus preventing DHFA from being reduced to tetrahydrofolic acid (THFA) (O'Donovan and Neuhard, 1970; Brown, 1970; Kit, 1970). Thymidylate biosynthesis by thymidylate synthetase (ts) is known to be the only C_1 transfer reaction in which THFA is oxidized (Brown, 1970). All the other C_1 transfer reactions leave THFA in its reduced state, i.e., they are THFA-conservative. Thus the continued action of ts in the presence of APT rapidly depletes the supply of THFA within the cell, thereby stopping growth.

In yeasts total inhibition of growth could not be achieved by APT alone (Laskowski and Lehmann-Brauns, 1973; Fäth and Brendel, 1974). However, this is possible by the simultaneous use of APT and a sulfonamide (e.g., sulfanilamide, SAA), which interfere synergistically with THFA production. In the presence of APT + SAA, the cell obviously is not able to perform any step of THFA-dependent metabolism once ts has oxidized the remaining THFA to DHFA. Therefore one would expect restoration of growth in yeasts inhibited by APT + SAA after addition of dTMP plus those products requiring THFA-conservative metabolism for synthesis (Brown, 1970).

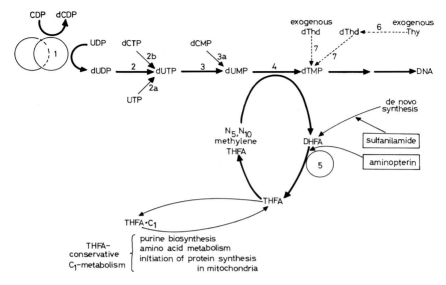

FIG. 1. Putative folic acid metabolism and its relation to thymidylate biosynthesis in yeast. The scheme was designed according to the data given by O'Donovan and Neuhard (1970), Kit (1970), Hartman (1970), Brown (1970), and Baglioni and Colombo (1970). The numbers given represent the following enzymes: 1, ribonucleoside diphosphate reductase (B1 subunit); 2, nucleoside diphosphate kinase; 2a, ribonucleoside triphosphate reductase; 2b, dCTP deaminase; 3, dUTP pyrophosphatase; 3a, dCMP deaminase; 4, thymidylate synthetase; 5, DHFA reductase; 6, thymidine phosphorylase; 7, thymidine kinase. Reactions 1 to 4 constitute the putative main pathway, and reactions 2a, 2b, and 3a possible shunts for thymidylate biosynthesis. Reactions 6 and 7 (broken lines) have not been found in yeasts but were shown to exist in prokaryotes. CDP, dCMP, dCDP, dCTP, Ribo- and deoxyribonucleoside phosphoric acids of cytosine; UTP, UDP, dUMP, dUTP, ribo- and deoxyribonucleoside phosphoric acids of uracil; dTMP, deoxythymidine-5' monophosphate; Thy, thymine; dThd, deoxythymidine; DHFA/THFA, dihydro-/tetrahydrofolic acid.

Among these are adenine (to compensate for the inhibition of purine biosynthesis) and the amino acids glutamic acid, glycine, and methionine (Fig. 1).

In the yeast *S. cerevisiae*, however, an offer of the above-mentioned substances does not lead to restoration of growth in an APT + SAA-inhibited culture (Fig. 2a). But mutants can be found that are able to grow under these conditions (Fig. 2b).

B. Isolation Procedure

Between 10^5 and 10^8 stationary cells, depending on strain and ploidy, are plated on medium C, the composition of which is given in Table I. A few large colonies will arise after a 3–6 days' incubation at 30°C. Among these

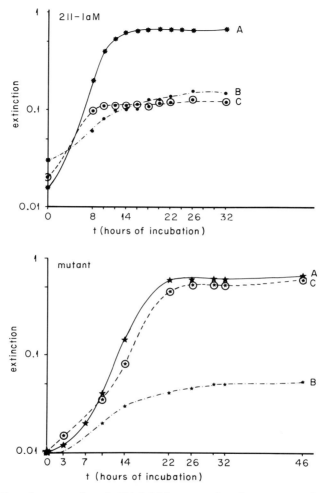

FIG. 2. Growth curves of strain 211-1aM (upper part) and a *typ* mutant (lower part) in media A, B, and C (from Laskowski and Lehmann-Brauns, 1973).

one will find mutants able to synthesize DNA from external dTMP only (Fäth *et al.*, 1974). Such mutants are called *TMP-per* by Jannsen *et al.* (1973), *typ* by Laskowski and Lehmann-Brauns (1973), and *tup* by Wickner (1974). *TMP-per* mutants have not yet been genetically characterized. Growth curves of a *typ* mutant and the corresponding parental strain in liquid media A, B, and C are shown in Fig. 2.

TABLE I

MEDIUM FOR THE ISOLATION OF *typ* MUTANTS[a]

Component	Compound	Medium
I	Yeast nitrogen base without amino acids (Difco), 6.7 gm	
	Casamino acids without vitamins (Difco), 2.0 gm	
	D(+)-Glucose, 20 gm	A \ B \ C
	Adenine, 50 mg	
	Bacto agar (Difco), 20 gm	
II	APT, 20 mg	
	SAA, 4–6 gm	
III	dTMP, 10 mg	

[a] Dissolve the compounds of component I in 400 ml of distilled water and autoclave. Allow mixture to cool to 60°C and add: for medium B, a filter-sterilized solution of SAA in 600 ml of distilled water and a solution of 20 mg APT in 1 ml of dimethyl sulfoxide; for medium C, add to medium B an adequate amount of filter-sterilized dTMP. Prepare plates immediately. Liquid media A, B, and C are prepared in a similar fashion, the agar being omitted.

III. Characterization of *typ* or *tup* Mutants

A. Genetic Characterization

So far, all *typ* or *tup* mutants have been found to be recessive, and nearly all are respiration-deficient (*petite*). The latter may be a result of the procedure of isolation, since both folic acid analogs and dimethyl sulfoxide are known to be inducers of cytoplasmic petite mutations (Wintersberger and Hirsch, 1973a,b; Yee *et al.*, 1972). The mutants isolated by Wickner (1974) fall into four, and those isolated by Laskowski and Lehmann-Brauns into three, complementation groups. Whether these complementation groups are identical or similar has not yet been established. Their localization in the yeast genome seems to vary widely, *typ* mutants apparently being linked (Laskowski and Lehmann-Brauns, 1973), while *tup* mutants seem to be located on several chromosomes (Wickner, 1974).

B. Biochemical Characterization

typ or *tup* mutants take up and incorporate labeled dTMP specifically into DNA in the presence as well as in the absence of APT and SAA, provided the cells are supplied with a medium sufficiently rich in inorganic phosphate, e.g., medium A or C.

Strain 211-1aMT2, a well-characterized *typl* mutant, requires approximately 30 µg dTMP/ml to grow optimally in the presence of APT + SAA (Fäth *et al.*, 1974). This suffices to obtain cellular DNA whose Thy content is totally derived from exogenous dTMP. When labeling is performed in the same medium without APT + SAA and adenine, a three times the amount of exogenous dTMP is needed to obtain the same result.

IV. Isolation of dTMP Low Requirers (*typ tlr* mutants)

dTMP as a crucial parameter for growth versus no growth in the presence of APT + SAA makes it possible to select for mutants with a low requirement for dTMP. This can simply be achieved by the use of gradient plates containing the inhibitory medium described above and a linear gradient of dTMP. Approximately 10^6 *typ* mutant cells are plated onto freshly prepared gradient plates, and the plates are then incubated for 4–6 days at 30 °C. Single colonies growing beyond the borderline of dense growth are putative *typ tlr* mutant colonies (Fäth *et al.*, 1974).

The *typ tlr* mutants so far obtained require approximately 10 µg dTMP/ml for optimal growth in the presence of APT + SAA, i.e., they utilize exogenous dTMP three times as well as the originally isolated haploid *typ* mutants. This means that, to obtain cellular DNA maximally labeled with exogenous dTMP, one has to offer only five times the amount of macromolecular cellular Thy content under APT + SAA conditions (Fäth *et al.*, 1974).

V. Mutants Auxotrophic for dTMP (*typ tmp* Mutants)

A. Theory

The rationale for the isolation of such mutants may be derived from Fig. 1. If adenine is omitted from the inhibitory medium described above and only dTMP and the essential amino acids are offered, cells in which thymidylate biosynthesis via ts is possible would not be expected to grow. Cells in which thymidylate biosynthesis is partially or totally blocked by a malfunction of ts itself or by a poor supply of its substrate (Fig. 1), however, would be expected to grow. In such mutant cells the amount of THFA oxidized by ts is more or less drastically reduced. Thus they are more able to perform the THFA-conservative biosynthesis of purine nucleotides and have a growth advantage over parents prototrophic for dTMP.

B. Isolation Procedure

A stationary culture of any *typ* or *tup* mutant is prepared in medium A (without adenine). The cells are collected, washed three times in phosphate buffer (0.067 *M*, pH 7), and then treated as follows. They are suspended in a 10^{-7} dilution of ethyl methanesulfonate (EMS) in phosphate buffer to give a titer of 10^7 cells/ml and incubated for 2 hours at 30 °C with continuous shaking. Then an equal volume of 10% sodium thiosulfate is added, and neutralization of EMS takes place for 15 minutes. The mutagenized cells are pelleted, washed three times in phosphate buffer, and resuspended in liquid medium A (without adenine) supplemented with 20–30 μg dTMP/ml to a final titer of approximately 10^7 cells/ml. Incubation takes place for 4 hours at 30 °C. Then 10^5 to 10^6 cells are plated onto solid medium C described in Table I (but without adenine) and incubated for 4–7 days at 30 °C.

One will find many small colonies and a few large ones on the plates, the large ones being potential dTMP auxotrophs. These are removed and streaked onto solid medium A which additionally contains 20–30 μg dTMP/ml. Incubation takes place for 1–2 days at 30°C. This master plate is replica-plated onto solid medium A. Clones not growing on this medium are dTMP auxotrophs.

VI. dTMP Auxotrophs with a Low Requirement for dTMP

The isolation procedure of *typ tmp tlr* mutants is analogous to that described in Section IV, with the exception that medium-A plates with a linear gradient of dTMP are used.

RERERENCES

Baglioni, C., and Colombo, B. (1970). *In* "Metabolic Pathways" (D. M. Greenberg, ed.), 3rd ed., Vol. 4, pp. 278–349. Academic Press, New York.

Brendel, M., and Haynes, R. H. (1972). *Mol. Gen. Genet.* **117**, 39–44.

Brendel, M., and Haynes, R. H. (1973). *Mol. Gen. Genet.* **126**, 337–348.

Brown, G. M. (1970). *In* "Metabolic Pathways" (D. M. Greenberg, ed.), 3rd ed., Vol. 4, pp. 369–410. Academic Press, New York.

Fäth, W. W., and Brendel, M. (1974). *Mol. Gen. Genet.* **131**, 57–67.

Fäth, W. W., Brendel, M., Laskowski, W., and Lehmann-Brauns, E. (1974). *Mol. Gen. Genet.* **132**, 335–345.

Grenson, M. (1969). *Eur. J. Biochem.* **11**, 249–260.

Grivell, A. R., and Jackson, J. F. (1968). *J. Gen. Microbiol.* **54**, 307–317.

Hartman, S. C. (1970). *In* "Metabolic Pathways" (D. M. Greenberg, ed.), 3rd ed., Vol. 4, pp. 1–58. Academic Press, New York.

Jannsen, S., Lochmann, E.-R., and Laskowski, W. (1968). *Z. Naturforsch. B* **23**, 1500–1507.

Jannsen, S., Lochmann, E.-R., and Megnet, R. (1970). *FEBS (Fed. Eur. Biochem. Soc.) Lett.* **8**, 113–115.

Jannsen, S., Witte, I., and Megnet, R. (1973). *Biochim. Biophys. Acta* **299**, 681–685.
Kit, S. (1970). *In* "Metabolic Pathways" (D. M. Greenberg, ed.), 3rd ed., Vol. 4, pp. 70–252. Academic Press, New York.
Laskowski, W., and Lehmann-Brauns, E. (1973). *Mol, Gen. Genet.* **125**, 275–277.
Lochmann, E.-R. (1965). *Naturwissenschaften* **52**, 498.
O'Donovan, G. A., and Neuhard, J. (1970). *Bacteriol. Rev.* **34**, 278–343.
Wickner, R. B. (1974). *J. Bacteriol.* **117**, 252–260.
Wintersberger, U., and Hirsch, J. (1973a). *Mol. Gen. Genet.* **126**, 61–70.
Wintersberger, U., and Hirsch, J. (1973b). *Mol. Gen. Genet.* **126**, 71–74.
Yee, B., Tsuyumu, S., and Adams, B. G. (1972). *Biochem. Biophys. Res. Commun.* **49**, 1336–1342.

Chapter 17

Mutants of Saccharomyces cerevisiae That Incorporate Deoxythymidine 5'-Monophosphate into DNA in Vivo

REED B. WICKNER

Laboratory of Biochemical Pharmacology,
National Institute of Arthritis, Metabolism, and Digestive Diseases,
National Institutes of Health,
Bethesda, Maryland

I. Introduction

The DNA of *Escherichia coli* is conveniently labeled *in vivo* with thymine or thymidine, and that of animal cells with thymidine. However, *Saccharomyces cerevisiae* apparently lacks thymidine kinase (Grivell and Jackson, 1968) and, as a consequence, does not utilize either thymine or thymidine as a precursor for DNA synthesis. Jannsen and co-workers (1968, 1970) showed that certain strains of *S. cerevisiae* incorporate small amounts of deoxythymidine 5'-monophosphate (dTMP) specifically into DNA. This strain was also studied by Brendel and Haynes (1972). However, only a few percent of the thymine residues of the cellular DNA of this strain are derived from exogeneous dTMP. Thus it was desirable to isolate strains that incorporate large amounts of exogenous dTMP into cellular DNA.

This article describes methods for isolating strains of *S. cerevisiae* that incorporate exogenous dTMP into DNA (called *tup* for d*T*MP *up*take), methods for labeling DNA with dTMP or deoxybromouridine 5'-monophosphate (dBrUMP) using these strains, and the results of genetic analysis of these strains (Jannsen *et al.*, 1973; Laskowski and Lehmann-Brauns, 1973; Wickner, 1974a). The method of mutant isolation described here is essentially the same as that utilized by Jannsen *et al.* (1973) and by Laskowski and Lehmann-Brauns (1973) and, so far as is known, the properties of the mutants isolated by these workers are similar to those described here (see also Wickner, 1974a). Jannsen *et al.* (1973) refer to their mutants of this type as *dTMP-per* and Laskowski and Lehmann-Brauns (1973) use the designation *typ*.

II. Materials

A. Media

Complete medium (YPD) contained: yeast extract (Difco), 1%; peptone (Difco), 2%; dextrose, 2%; and agar, 2%. Synthetic (minimal) medium (Wickerham, 1946) contained: yeast nitrogen base (Difco) without amino acids, 0.67%; dextrose, 2%; agar, 2%. It was supplemented where appropriate with: uracil, 24 mg per liter; adenine sulfate, 24 mg per liter; tyrosine, 36 mg per liter; histidine, 24 mg per liter; lysine, 36 mg per liter; tryptophan, 24 mg per liter; leucine, 36 mg per liter; or methionine, 24 mg per liter. The medium used for labeling cells (YPD-2) was essentially that of Brendel and Haynes (1972): yeast extract, 0.25%, peptone, 1%; and dextrose, 2%. Sulfanilamide (Sigma Chemical Company, St. Louis; 6 mg/ml, added prior to sterilization), aminopterin (K & K Laboratories, ICN, Cleveland; 50 μg/ml), and dTMP (100 μg/ml, unless otherwise noted) were added as indicated. Filter-sterilized aminopterin and dTMP were added to cooled plates or broth.

B. Yeast Strains

Saccharomyces cerevisiae A364A (*a adel ade2 his7 gall ural lys2 tyrl*) (Hartwell *et al.*, 1970) was obtained from Dennis Cryer.

C.   Radioactive Compounds

dTMP methyl-^{3}H and dTMP-^{32}P were obtained from New England Nuclear Corporation. dTMP-^{32}P was freed of ^{32}PO$_4$ by chromatography on Dowex-1-Cl (Lehman *et al.*, 1958). To a 5-ml column of Dowex-1-Cl washed extensively with water was applied 50 nmoles of dTMP-^{32}P (20 Ci/mole). Contaminating ^{32}PO$_4$ was eluted with 0.01 N HCl, and dTMP-^{32}P was eluted with 0.01 N HCl containing 0.01 N LiCl. The purified dTMP-^{32}P was 98% Norite-adsorbable.

dBrUMP was prepared by bromination of dUMP by a procedure parallel to that of Howard *et al.* (1969). dBrUMP-^{3}H was prepared similarly from dUMP- 6a^{3}H (Schwarz/Mann).

D.   Genetic Methods

Diploids from mass matings were isolated on minimal medium lacking the complementary requirements of their parents. Tetrad dissection and genetic analysis were performed by classic methods (Hawthorne and Mortimer, 1960; Johnston and Mortimer, 1959; Mortimer and Hawthorne, 1966). The usual replica plating method using velveteen was not satisfactory for scoring dTMP uptake, because it gave too heavy and nonuniform an inoculum. Instead, cells from spore colonies growing on a master plate were transferred with a 32-point inoculator (Melrose Machine Shop, 176 Fairview Road, Woodlyn, Pa.) to a polypropylene block containing 32 water-filled wells (home-made), and cell suspensions were produced in the wells by agitating the inoculator with the points in the wells. By using the 32-point inoculator, drops of the suspensions were then transferred to YPD-2 medium supplemented with sulfanilamide and aminopterin with or without 100 μg dTMP/ml. Plates could be read best after 3 or 4 days of growth at 30 °C.

III.   Isolation of Mutants Incorporating dTMP into DNA *in Vivo*

The strategy used to isolate thymine-utilizing strains of *E. coli* is to block endogenous synthesis with folate analogs and supply exogenous thymine (Okada *et al.*, 1961 Stacey and Simson, 1965). While yeast is not sensitive to folate analogs (Nickerson and Webb, 1956) such as aminopterin, its growth is blocked by a combination of aminopterin and the *p*-aminobenzoic acid analog sulfanilamide (Table I), which blocks endogenous folate synthesis. Mutants arise only if dTMP is supplied (Jannsen *et al.*, 1973; Laskowski and

TABLE I

SELECTION AND GROWTH OF *tup* MUTANTS

Medium	Line no.	Additions					Growth[a]	
		Sulfanilamide, 6 mg/ml, plus aminopterin, 50 µg/ml	dTMP, 100 µg/ml	Thymidine, 30 µg/ml	Thymine, 30 µg/ml	Methionine, 15 µg/ml	A364A	A364A-T108[b]
YPD-2	1	−	+ or −	+ or −	+ or −		+	+
	2	+	−	−	−		−	−
	3[c]	+	+	−	−		−	+
	4	+	−	+	−		−	−
	5	+	−	−	+		−	−
Minimal complete medium	6	−	+ or −	+ or −	+ or −	+ or −	+	+
	7	+	−	−	−	+ or −	−	−
	8	+	+	−	−	+	−	+
	9[c]	+	+	+	−		−	−
	10	+	−	+	−	+ or −	−	−
	11	+	−	−	+	+ or −	−	−

[a] Strains A364A and A364A-T108, a *tup* mutant derived from A364A, were grown overnight on YPD slants and suspended in water at about 10⁸ cells/ml; drops of the suspension were placed on plates containing the indicated additions. Growth was observed after 48–72 hours at 30°C.

[b] Results shown here for A364A-T108 were typical of all mutants tested, including at least one representative of each complementation group.

[c] Conditions used for mutant selection.

Lehmann-Brauns, 1973; Wickner, 1974a). Neither thymine nor thymidine will suffice. On minimal medium methionine and adenine, whose bio-synthesis also require folate, must also be supplied. Trimethoprim cannot be substituted for aminopterin, and the pH of the medium must be between 4 and 6.

Spontaneous mutants, selected from strain A364A as in Table I (line 3 or 9), required 100 μg dTMP/ml for growth in the presence of aminopterin and sulfanilamide on YPD-2 medium (Table I, lines 2 and 3), but did not require dTMP for growth in the absence of drugs. Attempts to select mutants from strain A364A in the presence of only 5–20 μg dTMP/ml did not produce mutants with significantly lower requirements for dTMP in the presence of drugs. (One mutant, A36A-Tll, selected on plates with 100 μg dTMP/ml, has a relatively low dTMP requirement, 10 μg/ml.)

IV. Incorporation of Nucleotides by *tup* Mutants

Strains isolated as described above incorporate base-labeled dTMP into acid-insoluble material which is completely DNase-sensitive but completely resistant to digestion with pronase, RNase, or alkali (Wickner, 1974a). Hydrolysis shows that all the base label is in thymine (Jannsen *et al.*, 1973). If one uses dTMP-α-^{32}P, extensive labeling of RNA also occurs because the dTMP-α-^{32}P is split to form thymidine and ^{32}PO$_4$ (Wickner, 1974a; Brendel and Haynes, 1973). This problem is at least partially relieved by using mutants derived from phosphatase-deficient strains (Toh-e *et al.*, 1973; Wickner, 1974a).

Cellular DNA may be labeled by growth in either YPD-2 or minimal complete medium with or without drugs. After a pulse of dTMP-^{3}H in the absence of drugs, a delay of 30 minutes was observed before a constant rate of labelling was reached. Long-term labeling experiments, in which pool effects can be neglected, show that *tup* mutants derive about 40% of cellular thymine residues from extracellular dTMP in the absence of drugs, and nearly all thymine residues from outside in the presence of drugs (Wickner, 1974a). Cryer *et al.* (1973) found that *tup1* and *tup2* strains, labeled in the absence of drugs, incorporated label into both nuclear and mitochondrial DNA, but mitochondrial DNA was preferentially labeled.

Tup mutants can also incorporate dBrUMP-^{3}H into DNA (R. Wickner, unpublished data) and become sensitive to 313-nm light as a result (Jannsen *et al.*, 1973). This property may be useful in the isolation of mutants temper-ature-sensitive for DNA synthesis. dTMP-less death has also been reported (Jannsen *et al.*, 1973).

In a typical experiment to label DNA, a *tup* – strain is innoculated at less than 10^5 cells/ml into YPD-2 medium containing 100μg/ml of dTMP-^{3}H (5 cpm/pmole). After 2 days' growth with shaking at 30 °C, cells are harvested. Typically, 20–80% of cellular thymine residues are derived from the extracellular dTMP. DNA may then be extracted.

V. Genetic Analysis

Please see Wickner (1974a) for a fuller discussion of material in this section.

While *tup* mutants are easily isolated, several problems are encountered in working with these strains.

Twenty-nine spontaneous mutants of strain A364A were isolated, which were able to grow on either YPD-2 or minimal medium plus methionine and adenine in the presence of sulfanilamide, aminopterin, and dTMP as described above, and grew well in the absence of drugs. Diploid strains, heterozygous for the ability to incorporate dTMP, failed to grow with the drugs and dTMP. Thus the dTMP uptake of these mutants is a recessive trait. Of these, 17 mutants showed consistent 2:2 segregation of dTMP uptake. Segregation analysis was complicated by the fact that sporulation of some diploids, obtained from two non-dTMP-incorporating strains, resulted in the segregation of strains that could incorporate dTMP. Furthermore, it became clear that certain background genes were also essential in both parents if one was to observe 2:2 segregation of dTMP uptake. Those strains showing segregation consistent with a single nuclear gene were examined by complementation analysis. Four complementation groups were found and confirmed, in some cases, by allele tests (Wickner, 1974a).

In the process of performing segregation, complementation, and allele tests, it was noted that α strains carrying a dTMP uptake (*tup*) mutation in complementation group 1 failed to mate normally. These strains showed increased efficiency in mating with α strains and decreased efficiency in mating with *a* strains, so that α *tup1* strains mated about equally well with *a* and α tester strains. The rare diploids formed from α *tup1* $\times$ *a tup* $+$ sporulated normally, whereas those from the cross α *tup1* $\times$ α *tup* $+$ failed to sporulate. This finding is consistent with the diploids being of the genotypes a/α and α/α, respectively, and suggests that the presence of the *tup1* allele in an α strain may alter its mating specificity, rather than changing the α allele to *a*. In crosses of the type *a tup1* $\times$ α *tup* $+$, the coincidence of α and *tup1* consistently produced this pattern of abnormal mating. Both *a* and α strains of other *tup* loci mated normally and only with the opposite mating type.

Diploids homozygous for *tup1* failed to sporulate, although all were *rho* +. Heterozygotes at this locus, or homozygotes at any of the other three *tup* loci, sporulated normally.

Many spontaneous *tup* mutants are also clumpy. It has not been determined whether it is in fact the *tup* mutation that is responsible for this characteristic.

Three of the *tup* genes have been located on the genetic map (Wickner, 1974a). The *tup1* gene is located on chromosome III, about 22 centimorgans to the right of *thr4*. the *tup3* gene is located on chromosome II between *lys2* and *tyr1*. The *tup4* gene is located within 3 centimorgans of the centromere of chromosome XV. The *mak1* gene (Wickner, 1974b) appears to be to the left of *tup4* and is also tightly centromere-linked, but whether the centromere is to the left or right of these genes, or is in between them, is not yet clear. The *tup2* gene has not yet been located.

Laskowski and Lehmann-Brauns (1973) isolated mutants in three *tup* genes from another strain of *S. cerevisiae* (they use the designation *typ*) which are linked. Presumably, at least two of these genes are different from *tup1*, *tup2*, *tup3*, and *tup4*, since none of the latter are linked.

VI.   Applications

Mutants such as those described here may permit the routine labeling of DNA in physiological studies. However, several problems remain:

1. Labeling of DNA is not necessarily uniform (Cryer *et al.*, 1973). Mitochondrial DNA appears to be labeled preferentially.

2. Different *tup* mutations may result in differences in labeling.

3. A large proportion of the dTMP in the medium is degraded. Multiply phosphatase-deficient strains may improve this situation.

4. In some genetic backgrounds the *tup* phenotype does not behave as a simple Mendelian trait. This complicates the construction of strains.

In spite of these problems, *tup* mutants promise to be useful in several areas:

1. Labeling of cellular DNA.

2. Selection of mutants in DNA synthesis.

3. Studies of mechanisms of DNA repair and recombination.

4. Studies of cellular permeability. It would be interesting to know, for example, to what other compounds *tup* mutants have become permeable.

5. Studies of the mechanisms of mating specificity.

The strains described herein are available on request from the author.

References

Brendel, M., and Haynes, R. H. (1972). *Mol. Gen. Genet.* **117**, 39–44.

Brendel, M., and Haynes, R. H. (1973). *Mol. Gen. Genet.* **126**, 337–348.

Cryer, D. R., Goldthwaite, C. D., Zinker, S., Lam, K.-B., Storm, E., Hirschberg, R., Blamire, J., Finkelstein, D. B., and Marmur, J. (1973). *Cold Spring Harbor Symp. Quant. Biol.* **38**, 17–29.

Grivell, A. R., and Jackson, J. F. (1968). *J. Gen. Microbiol.* **54**, 307–317.

Hartwell, L. H., Culotti, J., and Reid, B. (1970). *Proc. Nat. Acad. Sci. U.S.* **66**, 352–359.

Hawthorne, D. C., and Mortimer, R. K. (1960). *Genetics* **45**, 1085–1110.

Howard, F. B., Frazier, J., and Miles, H. T. (1969). *J. Biol. Chem.* **244**, 1291–1302.

Jannsen, S., Lochmann, E. R., and Laskowski, W. (1968). *Z. Naturforsch. B* **23**, 1500–1507.

Jannsen, S., Lochmann, E. R., and Megnet, R. (1970). *FEBS (Fed. Eur. Biochem. Soc.) Lett.* **8**, 113–115.

Jannsen, S., Witte, I., and Megnet, R. (1973). *Biochim. Biophys. Acta* **299**, 681–685.

Johnston, J. R., and Mortimer, R. K. (1959). *J. Bacteriol.* **78**, 292.

Laskowski, W., and Lehmann-Brauns, E. (1973). *Mol. Gen. Genet.* **125**, 275–277.

Lehman, I. R., Bessman, M. J., Simms, E. S., and Kornberg, A. (1958). *J. Biol. Chem.* **233**, 163–170.

Mortimer, R. K., and Hawthorne, D. C. (1966). *Genetics* **53**, 165–173.

Nickerson, W. J., and Webb, M. (1956). *J. Bacteriol.* **71**, 129–139.

Okada, T., Kanagisawa, K., and Ryan, F. J. (1961). *Z. Vererbungslehre* **92**, 403–412.

Perkins, D. D. (1949). *Genetics* **34**, 607–626.

Stacey, K. A., and Simson, E. (1965). *J. Bacteriol.* **90**, 554–555.

Toh-e, A., Ueda, Y., Kakimoto, S., and Oshima, Y. (1973). *J. Bacteriol.* **113**, 727–738.

Wickerham, L. J. (1946). *J. Bacteriol.* **52**, 293–301.

Wickner, R. B. (1974a). *J. Bacteriol.* **117**, 252–260.

Wickner, R. B. (1974b). *Genetics* **76**, 423–432.

Chapter 18

Mutants of Meiosis and Ascospore Formation

MICHAEL S. ESPOSITO AND ROCHELLE E. ESPOSITO

Erman Biology Center, Department of Biology
University of Chicago, Chicago, Illinois

I. Introduction

Formation of spores by both prokaryotic and eukaryotic microbes presents an opportunity to study the biochemical and genetic regulation of differentiation in single cells. Sporulation of yeasts is of particular interest, since it encompasses meiosis and ascospore development. It therefore allows inquiry into the nature of functions required for genetic events of major importance in sexual reproduction, as well as the mode of integration of the numerous cell functions that contribute to ascospore formation.

A principal approach to the study of sporulative development involves the isolation and characterization of mutants defective in the process of

sporulation. The basic premise in the analysis of variants is that their phenotypes can provide information regarding the coordinate control and
integration of landmark developmental events. Furthermore, it is expected
that the study of defective or abnormal sporulation will uncover new
functions normally required for sporulation in the wild type.

A comprehensive investigation of the genetic control of sporulation
requires procedures that permit the detection of both recessive and
dominant gene mutations. The detection of recessive mutations affecting
meiosis and ascospore formation is complicated by the fact that these
processes occur in diploid cells or cells of higher ploidy. Two systems have
been developed that allow the recovery of both recessive and dominant
mutations. One approach takes advantage of the life cycle of homothallic
strains to obtain mutants in the homozygous condition in diploid cells.
This procedure has been used with both *Saccharomyces cerevisiae* and
Schizosaccharomyces pombe. Another technique employs certain $n + 1$
disomic strains of *S. cerevisiae* that begin but do not complete sporulation.

In using these methods the following criteria have been applied to isolate
mutants: (1) absence of, or abnormal ascospore formation, and (2) absence
of intragenic recombination during meiosis. Mutations of the first type
affect events indispensable for ascospore formation which occur throughout
sporulation. Variants that perturb meiosis and spore development but form
ascospores are not detected by this criterion. Mutations of the second type
are selected for their specific effects on genetic recombination during the
early stages of sporulation.

The detailed methods of isolation and characterization of mutants with
defective sporulation are the main subject of this article. The properties
of the mutants and the information they have yielded regarding the control
of sporulation functions are discussed.

II. Isolation of Mutants

A. Procedures Employing Homothallic Strains

1. TEMPERATURE-SENSITIVE *spo* MUTANTS OF *S. CERVISIAE*

The life cycle of *S. cerevisiae* is summarized in Fig. 1. The species exists
in homothallic as well as heterothallic forms. In the heterothallic condition
haploids of *a* and *α* mating type copulate and give rise to diploids heterozygous at the mating-type locus. The diploids are stable and reproduce by
budding; when transferred to sporulation medium, they cease growth and

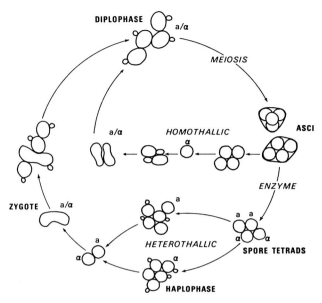

FIG. 1. Life cycles of heterothallic and homothallic forms of *Saccharomyces*. In homothallic strains the diplophase is the dominant state; the haplophase is transient. The heterothallic state is characterized by stable diploids and stable haploids. (Adapted from Mortimer and Hawthorne, 1969.)

undergo meiosis and ascospore formation. Individual ascospores give rise to haploid ascosporal clones. In the case of homothallic strains carrying the *D* (diploidization) gene (Winge and Roberts, 1949), directed mutation of the mating-type locus occurs during the mitotic divisions that follow ascospore germination (Hawthorne, 1963). Each ascospore forms both *a* and *α* progeny. Conjugation of *a* and *α* haploid cells occurs, resulting in ascosporal clones which consist primarily of diploid cells. These diploids are heterozygous at the mating-type locus but homozygous at all other loci, and reflect the genotype of the haploid spore.

Homothallic strains have been used to obtain mutants with defective sporulation in *S. cerevisiae* (Esposito and Esposito, 1968, 1969). Temperature-sensitive mutants were sought to facilitate genetic analysis and for their utility in physiological studies. The parental sporulation-proficient strain employed in mutant isolation exhibited 55% asci at 20°C, 75% asci at 30°C, and 45% asci at 34°C. The procedures used are described below.

A sporulated suspension of the homothallic strain consisting of approximately 80% asci was irradiated with ultraviolet light on the surface of solid growth medium to approximately 40% survival. The plates were incubated at 30 °C for 6 days. At the dose used about 75% of the surviving clones resulted from the growth of single ascospores. Survivors, capable of growth at the three temperatures, were tested for ascus production at 20°, 30°, and 34 °C. Percent sporulation was determined by hemocytometer counts. Among 896 survivors of mutagenic treatment, 75 mutants were found that demonstrated sporulation values at least 3 σ from the mean of the wild-type parental strain at a given temperature. Both heat-sensitive and cold-sensitive mutants were isolated.

2. Mutants of *Schiz. pombe*

The fission yeast *Schiz. pombe* also occurs in both homothallic and hetero-thallic forms (Fig. 2). In the heterothallic state there are two mating types, h^+ and h^-. Haploid cells of opposite mating type conjugate to form h^+/h^- zygotes. The zygotes spontaneously undergo meiosis and ascospore forma-tion, although some diploid cells divide by binary fission before sporulation occurs. Haploid cells carrying the mating-type allele h^{90} are homothallic. During the early stationary phase of growth, haploid cells conjugate to yield h^{90}/h^{90} zygotes which undergo meiosis and spore formation (Leupold, 1950, 1958, 1970).

Bresch *et al.* (1968) isolated mutants of *Schiz. pombe* utilizing a spore suspension from an h^{90} homothallic strain. Spontaneous, ultraviolet-induced, and nitrous acid–induced mutants which formed asporogenous clones were obtained. Failure to form spores was monitored by exposing colonies to iodine vapors. Colonies containing ascospores stain reddish brown, whereas colonies without asci remain unstained. Three-hundred

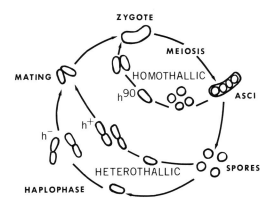

Fig. 2.   Life cycle of heterothallic and homothallic forms of *Schizosaccharomyces*. In both the homothallic and heterothallic cycles, the haplophase is dominant. Zygotes spontaneously undergo sporulation.

mutants isolated by this technique represented mutations affecting mating, meiotic division, and postmeiotic spore wall formation.

B.   The Use of a/α Disomic Strains

Haploid strains of baker's yeast disomic for chromosome III were isolated by Shaffer *et al.* (1971). Disomic strains heterozygous for the mating-type locus located on this chromosome do not form asci, but undergo premeiotic DNA synthesis and exhibit enhanced recombination when transferred to sporulation medium (Roth and Fogel, 1971). Genetic recombination in $n + 1$, a/α disomic strains can be monitored for markers present in heterozygous or heteroallelic condition on the homologous chromosome pair.

The increase in recombination exhibited by such strains was used to detect mutations affecting meiosis by Roth and Fogel (1971). An $n + 1$, a/α stock, heteroallelic at the *leu2* locus on chromosome III, was mutagenized with ethyl methanesulfonate (EMS). Survivors were transferred to sporulation medium and subsequently replica-plated to leucineless medium to detect prototrophic intragenic recombinants. The detection of both recessive and dominant mutations affecting recombination was anticipated, since all chromosomes except chromosome III were present in haploid condition in the parental strain. Among 940 survivors examined, 15 mutants were isolated, which yielded 10% or less of the wild-type level of prototrophs although they retained both mating-type alleles and were heteroallelic at the diagnostic locus.

C.   Mutations Affecting Mating-Type Control of Sporulation

In both *S. cerevisiae* and *Schiz. pombe*, diploid cells homozygous for mating-type alleles a/a, α/α, and h^+/h^+, h^-/h^-, respectively, do not sporulate (Roman *et al.*, 1955; Leupold, 1955). Diploids of *S. cerevisiae* of a/a and α/α genotype do not exhibit premeiotic DNA synthesis and the enhanced capacity for recombination observed when a/α diploids are transferred to sporulation medium (Roth and Lusnak, 1970; Roth, 1972).

Gene mutations affecting the control of sporulation by mating-type alleles have been obtained by Hall and Hopper (1973). Diploids of genotype a/a and α/α were treated with EMS, and surviving clones that had acquired sporulative ability were sought. Mutagenized diploids were heterozygous for both cycloheximide resistance and canavanine resistance, two recessive genetic markers located on different chromosomes. These markers were employed to aid in the detection of sporulation-proficient strains. Survivors were first replica-plated to sporulation medium, and the sporulation replicas then replica-plated to growth medium containing both cycloheximide and canavanine. Survivors that yielded segregants simultaneously displaying both recessive traits after exposure to sporulation medium were examined for their ability to form asci at 23° and 32 °C.

Mutants that had acquired sporulation ability at one or both of the temperatures represented mutations at the mating-type locus that restore heterozygosity and dominant mutations located elsewhere that permit sporulation of a/a and α/α diploids. The latter mutations allow the expression of gene functions usually not detectable in a/a and α/α diploids.

D. Detection of Dominant Mutations

Dominant mutations that prevent sporulation of *S. cerevisiae* have also been recovered among survivors of ultraviolet irradiation of haploid strains (Simchen *et al.*, 1972). Mutagenized haploids of *a* mating type were mated with an α strain, and the diploids tested for ascus production. A dominant mutation preventing sporulation was obtained by this procedure.

III.   Genetic Characterization of Mutants

Genetic analysis of sporulation mutants requires a system permitting hybridization of mutants with normal strains, and intercrossing of mutants with one another. In the case of sporulation mutants, hybridization studies are complicated by the fact that the mutants are generally defective in the production of gametes required for crosses. Procedures used in the genetic characterization of mutants are described below.

A.   Homothallic Systems

1. CONDITIONAL MUTANTS OF *S. cerevisiae*

Temperature-sensitive *spo* mutants isolated in homothallic strains of *S. cerevisiae* are employed in crosses using ascospores produced by mutants at permissive temperatures (Esposito *et al.*, 1972). In homothallic strains the haplophase is a transient state, since diploidization occurs shortly after spore germination. To construct hybrids ascospores from different strains are germinated together on nutrient medium (see Fig. 3). Copulation occurs between haploid cells of the two strains, and the hybrids are detected in mating mixtures by prototrophic selection (Pomper and Burkholder, 1949). The ability of the hybrid to sporulate at the restrictive temperature indicates whether the mutation is dominant or recessive.

Putative monohybrids for sporulation mutants are sporulated, and asci

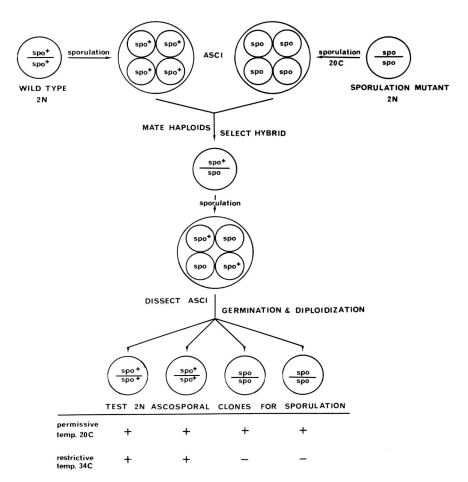

FIG. 3.   Procedure for construction and analysis of monohybrids of *spo* mutants isolated in *Saccharomyces*. The sporulation-proficient and mutant diploids are homothallic and homozygous for complementing auxotrophic markers. Haploid ascospores from strains are mixed and incubated on nutrient medium to allow mating. Hybrid diploids are selected on minimal medium. The diploids yield haploid ascospores when sporulated. Germination of ascospores is followed by diploidization, thus ascosporal clones consist of diploid cells which can be tested for their ability to sporulate. (Adapted from Esposito *et al.*, 1972.)

are dissected to determine whether the *spo* mutation segregates as a single gene mutation. The homothallic system is particularly useful for this purpose, since each ascospore gives rise to a diploid colony which can be tested for sporulation at permissive and restrictive temperatures. Asci of hybrids heterozygous for a single gene mutation affecting sporulation contain two spores that are temperature-sensitive with respect to sporulation and two that are wild type.

Recessive mutations of 11 loci (*spo 1* through *11*) and three dominant mutations (*SPO 20, 78,* and *98*) were recognized by these methods. The number of genes capable of mutating to the *spo* phenotype was estimated at 48 ± 27, based on the frequency of repeated isolation of mutants of the same locus.

2. ANALYSIS OF MUTANTS OF *Schiz. pombe*

Genetic studies of asporogenous mutants isolated in an h^{90} homothallic strain employ haploid isolates of the mutants in standard hybridization procedures (Bresch *et al.,* 1968). The analysis of mutants that copulate but fail to sporulate is facilitated by the fact that such strains can be maintained on medium that supresses conjugation (Leupold, 1970). Haploid isolates of mutants are thus always available for hybridization with wild-type strains and for complementation analysis.

Twenty-five loci involved in mating, meiosis, and spore formation have been identified from complementation studies involving 300 asporogenous mutants. One of the loci, *meiI-1,* required for completion of the first division of meiosis is identical to or part of the complex mating-type locus *h.*

B.   Genetic Analysis in Disomic Strains

Disomic *a/α* strains of *S. cerevisiae* do not mate. Genetic analysis of mutants isolated in these strains has been performed with *a* and *α* haploid derivatives that have lost one of the chromosomes in disomic condition by mitotic nondisjunction, and *a/a* and *α/α* mitotic recombinants that have retained the disome and *leu 2* heteroalleles (Roth, 1973).

When *a* or *α* haploid derivatives of mutants are employed in matings, hybrids involving intercrosses of mutants or crosses to the wild type are obtained by prototrophic selection; their phenotypes with respect to recombination and sporulation are assessed to test complementation and dominance, respectively.

In practice it is useful to perform matings between *a/a* and *α/α* disomic strains. Matings of heteroallelic *leu2, a/a,* or *α/α* mutants strains with heteroallelic *leu2, a/a,* or *α/α* nonmutant strains result in $2n + 2$ hybrids tetrasomic for chromosome III. When such strains are sporulated, they yield a certain proportion of ascospores which are $n + 1$, *a/α*, and heteroallelic at the *leu2* locus. Segregation of the mutation affecting recombination among spores of this genotype is used to determine whether a given mutant represents the alteration of a single gene as shown in Fig. 4. Roth (1973) reported the analysis of two mutant strains by this technique. One harbored a single recessive mutation (*mei1*), while the other was a double mutant (*mei2 mei3*).

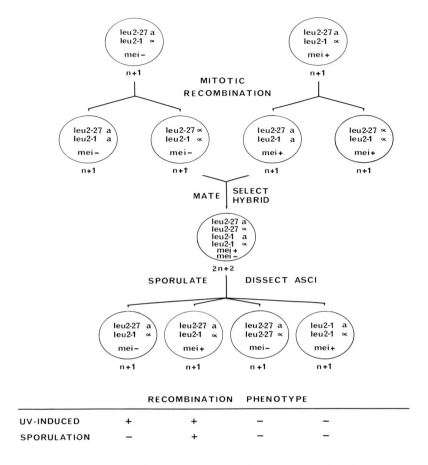

FIG. 4. Genetic analysis of *mei* mutants isolated in $n + 1$ a/α disomic strains. $n + 1$ *aa* and $\alpha\alpha$ mitotic recombinants are recognized by their ability to mate and retention of heteroalleles of the *leu2* locus. $2n + 2$ hybrids are obtained by prototrophic selection. On sporulation they yield tetrads which exhibit tetrasomic segregation for chromosome III. a/α spores heteroallelic at *leu2* fail to mate, and yield prototrophic recombinants after ultraviolet treatment. Spore clones of this type that do not yield prototrophic recombinants after sporulation are *mei*; those that yield recombinants are *mei +*. In the tetrad shown all four ascospores are a/α; two are homoallelic at *leu2*, and two are heteroallelic at *leu2*.

C. Mutations Affecting Control of Sporulation by Mating-Type Alleles

Gene mutations induced in diploids homozygous at the mating-type locus which permit sporulation can be divided into three phenotypic classes (Hall and Hopper, 1973). Mutants are observed that (1) lose mating ability and sporulate, (2) retain mating ability and sporulate at 23° and 32°C, and (3) retain mating ability and sporulate at 32° but not at 23 °C. Mutants in the last-mentioned two categories were tested for the presence of a dominant gene mutation allowing sporulation of a/a and α/α diploids. The mutant strains were mated with a/a or α/α wild-type strains to construct tetraploids. Segregation of the mutation among a/a and α/α diploid ascosporal clones obtained by sporulation of tetraploids illustrated that the mutations induced were dominant, heterozygous in the mutant diploid, acted in both a/a and α/α diploids, and were not linked to the mating-type locus.

D.   Additional Mutations Affecting Sporulation

1.   GENES WITH OTHER KNOWN EFFECTS

Gene mutations that disrupt or abolish sporulation of *S. cerevisiae* have been encountered in genetic studies unrelated to the detection of sporulation-defective mutants. Respiratory-deficient diploids do not sporulate. Respiratory deficiency is conferred by the cytoplasmic petite mutation, nuclear petite mutations, mutations of the tricarboxylic acid cycle, and lesions in fatty acid metabolism (Ephrussi and Hottinguer, 1951; Sherman and Slonimski, 1964; Ogur *et al.*, 1965; Keith *et al.*, 1969). Mutations affecting sporulation have also been detected among ultraviolet light and x-ray sensitive mutants (Cox and Parry, 1968; Mori and Nakai, 1969; Resnick, 1969), nonsense suppressors (R. K. Mortimer, personal communication; Rothstein, 1973), cell division cycle mutants (Simchen, 1974), and recessive mutants defective in x-ray-induced mitotic intragenic recombination. (Rodarte-Ramon, 1970, 1972; Rodarte-Ramon and Mortimer, 1972). Some of the latter mutants failed to sporulate or exhibited reduced meiotic intragenic recombination but did not affect intergenic recombination.

A cytoplasmic mutation that does not prevent spore formation, but apparently affects premeiotic DNA synthesis has been detected among survivors of haploid strains intensively labeled with ^{32}P (Hottinguer de Margerie, 1967; Moustacchi *et al.*, 1967). The mutation referred to as *AM* (abnormal meiosis) is dominant. Hybrids of crosses to *AM* strains are diploid with respect to DNA content during growth, but after meiosis yield many tetrads which exhibit tetraploid segregation patterns. Experimental data suggest that the mutation causes an extra round of premeiotic DNA synthesis.

Diploid strains of *Saccharomyces* that produce only two-spored asci containing diploid spores have been also reported (Grewal and Miller, 1972). The nuclear division occurring during sporulation resembles both mitosis and the first meiotic division of meiosis (Grewal and Miller, 1972; Moens,

1974). While the mode of chromosome movement is unknown, the behavior of these strains indicates that ascospore formation can occur in cells that do not undergo conventional meiosis involving both reductional and equational divisions.

2.   Genetic Background and Sporulation

Heterothallic strains of *S. cerevisiae* commonly employed in genetic studies have not been routinely subjected to selection for high sporulation. Accordingly, the genetic background for spore formation of certain strains and segregants from crosses is quite poor. The influence of polygenic modifiers on sporulative ability is shown in Fig. 5 which summarizes data

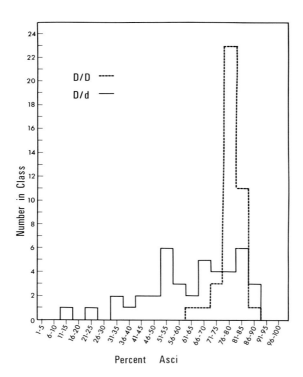

Fig. 5.   Evidence for modifiers reducing sporulation present in haploid heterothallic strains. The dashed-line histogram summarizes the sporulation values of ascosporal clones obtained from dissection of a *D/D* homothallic strain selected for high sporulation. The solid-line histogram represents the sporulation values of *D/D* ascosporal segregants from the cross of a haploid heterothallic (*d*) strain to the homothallic line selected for high sporulation. (Courtesy of R. Rothstein.)

communicated by R. Rothstein. A heterothallic haploid strain (i.e., α, d), was crossed with a homothallic haploid strain (i.e., a/α, D/D) selected for high sporulation of its meiotic products. The resultant heterozygote (a/α, D/d), was sporulated, and the two homothallic diploid ascosporal clones of each ascus were tested for sporulation. The range of sporulation values observed reflects the contribution of the haploid heterothallic parent to genotypes that depress sporulative ability.

This behavior suggests that caution be exercised in drawing the conclusion that a particular mutation affects sporulation when a diploid homozygous for the mutation fails to sporulate. A rigorous test of the association of the phenotypes involves the demonstration that (1) revertants of the mutation locus also acquire sporulative ability, or vice versa, and (2) the mutation and sporulation deficiency segregate together in crosses to sporulation-proficient strains.

Rothstein (1973) applied the latter criterion to test the association of sporulation deficiency and homozygosis for the ochre nonsence suppressors, *SUP3* and *SUP7*, noted by R. K. Mortimer (personal communication). In the case of *SUP7* poor sporulation and the *SUP7* mutation did not segregate together, whereas in the case of *SUP3* cosegregation was observed. The association of the *SUP3* mutation and sporulation deficiency was confirmed by the isolation of a mutation at the *SUP3* locus that resulted in the simultaneous loss of suppressor activity and the return of sporulative ability.

IV.   Analysis of Mutant Phenotypes

The phenotypes of mutants and their effects on the sporulation process are described in the following sections. Only those landmark events that have been studied in mutants are discussed. For more general reviews, see Fowell (1969), Haber and Halvorson (1972), and Tingle *et al.* (1973).

A.   Mutants of *S. cerevisiae*

Diploid cells of baker's yeast can be induced to sporulate by transferring cells from growth medium to a nitrogen-deficient sporulation medium containing acetate as a carbon source. Sporulation is maximal and most rapid when cells are adapted to respiration at the time of transfer and possess sufficient endogenous reserves to support macromolecular synthesis. Techniques for achieving rapid sporulation of *S. cerevisiae* are described elsewhere in this volume. Two regimens have been employed to establish sporulation cultures in the study of mutants. They differ with respect to whether glucose-grown cells in early stationary phase or acetate-grown cells in exponential phase are used to inoculate sporulation medium. The basic features of sporulation do not appear to depend on the origin of the inoculum, although sporulation is more synchronous and more extensive when acetate-grown cells are employed (Roth and Halvorson, 1969; Fast, 1973). The temporal order of landmark events relevant to the study of mutants is summarized in Fig. 6.

1. Completion of Mitotic Cell Division

When cells are transferred to sporulation medium at early stationary phase or from the logarithmic phase of growth, the population consists of budded and unbudded cells distributed over the mitotic cell cycle. In the wild type, budded cells transferred to sporulation medium complete cell division. Mitotic nuclear segregation occurs, and buds become separable from mother cells (M. S. Esposito and R. E. Esposito, 1974). Temperature-sensitive *spo* mutants were tested for their ability to complete the cell cycle in sporulation medium. Cells were grown in acetate nutrient medium at the restrictive temperature for sporulation (34°C), harvested during the

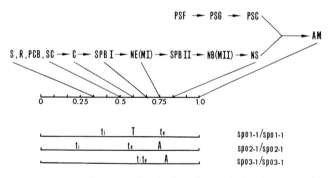

FIG. 6. Temporal order of events of meiosis and sporulation in yeasts. Arrows do not imply a causal relationship. S, Premeiotic DNA synthesis begins; R, commitment to recombination begins; PCB, polycomplex body formation; SC, synaptonemal complex formation; C, commitment to sporulation; SPBI, first spindle pole body duplication; NE(MI), nuclear elongation and meiosis I; SPBII, second spindle pole body duplication; NB(MII), nuclear budding and meiosis II; NS, nuclear separation; PSF, prospore wall formation; PSG, prospore wall growth; PSC, prospore wall closure; AM, ascospore maturation; t_i, point of irreversible inhibition of sporulation; t_e, point of escape from temperature sensitivity; T, stage of termination; A, stage at which development becomes abnormal. The approximate time of occurrence of individual events is shown on a 0 to 1 time scale, where 1 is the time of appearance of asci (at 30°C mature asci are first seen at 12 hours, i.e., 1 = 12 hours). (Adapted from Moens *et al.*, 1974.)

logarithmic phase, and transferred to sporulation medium prewarmed to 34°C. Completion of mitotic division by budded cells was monitored by measuring cell numbers in samples that had been sonicated for 5 seconds using a Biosonic II sonicator at minimum power. The sonication treatment separates mother cells and buds that have completed cell division. Three recessive *spo* mutants (*spo8*, *9*, and *10*) failed to complete cell division. In these strains neither cell division nor sporulation can proceed after the nutritional shift down to sporulation medium. The mutants include variants blocked in premeiotic DNA synthesis (*spo8* and *spo9*), as well as a mutant exhibiting premeiotic DNA synthesis (*spo10*). Their phenotypes may reflect defects in functions required only during nitrogen starvation, or in gene products required for growth and sporulation, which are defective specifically under conditions of sporulation. These alternatives can be evaluated by determining whether or not the mutations are allelic to conditional lethal mutations of genes required for vegetative growth.

2. PREMEIOTIC DNA SYNTHESIS

Premeiotic DNA replication begins approximately 4 hours after cells have been transferred to sporulation medium. Premeiotic DNA synthesis does not occur in *a/a* and *α/α* diploids (Roth and Lusnak, 1970), *mei1* and *mei2*, *mei3* (Roth, 1973), *spo7*, *8*, *9*, and *11*, and *SPO98* (M. S. Esposito and R. E. Esposito, 1973, 1974), and the dominant mutant 132 (Pinon *et al.*, 1974). These findings provide evidence for the existence of gene products required directly or indirectly for DNA replication during meiosis. These products are capable of function, or are not required in mitotic cells.

Pinon *et al.* (1974) characterized the DNA synthesized in *a/α*, *a/a*, and *α/α* diploids and in strain 132. No synthesis of nuclear and mitochondrial DNA was detected in *a/a* and *α/α* diploids during the period of premeiotic DNA synthesis in the *a/α* control. The diploid carrying the dominant sporulation mutation exhibited mitochondrial DNA synthesis and initiated but did not complete chromosomal DNA synthesis. The occurrence of mitochondrial DNA synthesis in the mutant suggests that some gene products required for premeiotic chromosomal replication are not needed for replication of the mitochondrial genome during sporulation.

3. SYNTHESIS AND TURNOVER OF RNA AND PROTEIN

When diploid cells are introduced into sporulation medium, cell division ceases but RNA and protein synthesis proceed. The RNA and protein content per cell at first rises and then declines, reflecting turnover of these molecules (Croes, 1967; Esposito *et al.*, 1969). Inhibitor studies with wild-type strains have shown that sporulation is dependent on the protein synthesis observed (Croes, 1967; Esposito *et al.*, 1969; Magee and Hopper, 1974).

Diploids of *spo* mutants (Esposito *et al.*, 1970; M. S. Esposito and R. E. Esposito, 1974), *mei* mutants (Roth, 1973), and *a/a* and *α/α* diploids (Pinon *et al.*, 1974; Magee and Hopper, 1974) have been examined for accumulation of RNA and protein during sporulation. None of the strains studied thus far fails to synthesize RNA and protein. Thus no evidence of sporulation-specific gene products required for the synthesis of these macromolecules has been obtained. Two sporulation-deficient strains have been reported by Chen and Miller (1968), which fail to develop proteinase-A activity during sporulation, at a time when this activity increases in sporulation-proficient strains. This observation suggests that protein turnover may be required for sporulation.

4. CARBOHYDRATE SYNTHESIS AND DEGRADATION

The accumulation of carbohydrates accounts for nearly two-thirds of the increase in dry weight exhibited by sporulating yeast (Roth, 1970). Kane and Roth (1974) monitored cellular carbohydrate synthesis by *a/α*, *a/a*, and *α/α* diploid cells in sporulation medium. Carbohydrates were fractionated into trehalose, glycogen, mannan, and an insoluble fraction. Synthesis of these carbohydrates occurred in both *a/a* and *α/α*, as well as *a/α*, diploids. Sporulating *a/α* diploids differ from *a/a* and *α/α* strains, however, in that the last-mentioned fail to exhibit glycogen degradation. In *a/α* diploids the decline in glycogen content accompanies ascospore formation, and may reflect the utilization of its turnover products in construction of the ascospore wall (Kane and Roth, 1974; Hopper *et al.*, 1974).

5. COMMITMENT TO GENETIC RECOMBINATION

When diploid cells of *S. cerevisiae* are exposed to sporulation medium, they acquire enhanced recombinational ability before they become committed to completion of sporulation (Sherman and Roman, 1963). The enhanced recombination exhibited by diploid cells that can return to mitosis has been referred to as commitment to recombination (R. E. Esposito and M. S. Esposito, 1974). This term is employed because the recombination detected may represent events of meiosis or events stimulated in mitotic cells by the synthesis of gene products required for recombination during meiosis.

Commitment to recombination is detected at the time of onset of pre-meiotic DNA synthesis (Roth, 1973) and involves both intragenic and intergenic recombination (R. E. Esposito and M. S. Esposito, 1974). Studies of asporogenous strains indicate that DNA synthesis and commitment to recombination are coordinately controlled. These events are simultaneously blocked in *a/a* and *α/α* strains and in diploids containing *mei1* and *mei2* (Roth, 1972, 1973).

Commitment to recombination has been detected in mutants that exhibit premeiotic DNA synthesis but do not form asci (Esposito *et al.*, 1970; R. E. Esposito and M. S. Esposito, 1974). Diploids homozygous for *spo1*, *2*, or *3* exhibit commitment to both intragenic and intergenic recombination. Intergenic recombination in certain genetic intervals in *spo* and control strains is at the full meiotic level in cells that return to mitotic division on plating. The results indicate that cells become capable of recombination at meiotic levels in certain intervals before they become committed to meiotic chromosome disjunction. Cytological studies described in Section IV,A,6 indicate the *spo1* diploids terminate cytological development before duplication of the spindle pole bodies for meiosis I (Moens *et al.*, 1974). Commitment to recombination thus occurs in cells that do not proceed beyond this cytological stage.

6. CYTOLOGY

Electron microscope studies of sporulation in *S. cerevisiae* have been performed by several investigators (Engles and Croes, 1968; Lynn and Magee, 1970; Moens and Rapport, 1971; Moens, 1971; Peterson *et al.*, 1972; Guth *et al.*, 1972; Beckett *et al.*, 1973; Illingworth *et al.*, 1973). Nine cytological stages can be distinguished. These are described below (see Fig. 7).

Stage 1. During the first stage the nucleus of the cell contains a single spindle pole body associated with the nuclear membrane. Microtubules, found in association with the spindle pole body radiate into the nucleoplasm. Cells remain in this condition for approximately 6 hours.

Stage 2. A structure, which has been termed the polycomplex body, appears in the nucleus at approximately 6 hours of sporulation. The polycomplex body contains multiple synaptonemal complex–like elements. When this structure appears, synaptonemal complex elements are also observed in the nucleoplasm. This stage lasts for approximately 2 hours.

Stage 3. At 8 hours the spindle pole body duplicates. The newly formed spindle pole bodies initially remain side by side and are connected to one another by an electron-dense bridge. Between 8 and 10 hours ascogenous cells pass from stage 3 to stage 8.

Stage 4. A spindle consisting of microtubules is formed between the spindle pole bodies which appear opposite one another on the nuclear membrane. When the spindle is formed, the synaptonemal complex elements of the polycomplex body are no longer visible. The spindle elongates, and the nucleus begins to assume an hourglass configuration.

Stage 5. The fifth stage is characterized by duplication of the spindle pole bodies at each end of the meiosis I spindle apparatus. At this stage the first division of meiosis is completed.

Stage 6. The duplicated spindle pole bodies face one another, and spindles

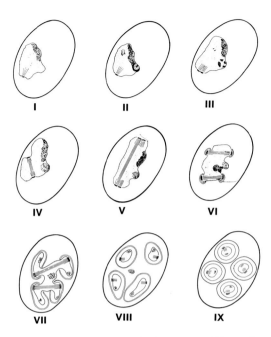

FIG. 7.   Meiosis and ascospore development in *S. cerevisiae*. The nucleus is lightly dotted, the nucleolus is hatched, and the polycomplex body is densely dotted; the spindle pole bodies are indicated by two short parallel lines at the nuclear envelope, the spindle microtubules by groups of straight parallel lines, the synaptonemal complex by the dotted tripartite structure in the nucleus, and the prospore wall membranes by the closed double lines on the cytoplasmic side of the spindle pole bodies. Both meiotic divisions occur without breakdown of the nuclear envelope. The stages are described in detail in the text. (Adapted from Moens and Rapport, 1971.)

for the second meiotic division are formed at each end of the nucleus. A prospore wall consisting of a double membrane is formed on the cytosol side of each spindle pole body, and the nucleolus begins to occupy a region of the nucleoplasm that will not be included in any of the four haploid ascospores.

Stage 7. The spindles of the second meiotic division increase in length, and the nuclear membrane buds at each of the four poles. The prospore wall membranes grow and surround each nuclear bud. The polycomplex body is no longer apparent, and a nucleolus is present in each of the nuclear buds.

Stage 8. At approximately 10 hours the prospore walls close, and the nucleus of the ascogenous cell is separated into four individual haploid nucleii. The nucleolus of what was previously the diploid cell is not included in any of the four haploid products and remains in a portion of the nucleoplasm not incorporated into any of the spores. At prospore wall closure the spindle pole body detaches from the prospore wall.

Stage 9. During the last stage, which requires approximately 2 hours, ascospore wall material is inserted between the unit membranes comprising the prospore wall. The inner prospore wall membrane becomes the plasmalemma of the spore, and the outer membrane apparently contributes to the proteinaceous coat of the mature ascospore. Immature prospores become spherical in structure, and ascospores visible by light microscopy are formed.

The fine structure of three *spo* mutants has been described (Moens *et al.*, 1974). Diploids homozygous for the mutation *spo1-1* terminate cytological development prior to stage 3. The spindle pole bodies of diploids incubated in sporulation medium at a restrictive temperature remain unduplicated.

Diploids homozygous for the mutations *spo2-1* and *spo3-1* exhibit abnormal cytological development. Diploids of *spo2-1* become abnormal at stage 5; the nucleus separates into two separate structures at the first division of meiosis. At the second division of meiosis, this aberrant behavior is repeated and ascogenous cells contain four separate nucleii. Prospore walls form at each spindle pole body. The prospore walls grow and close but rarely include any of the nuclei. Thus ascogenous cells containing four anucleate, closed prospore walls, and four separate, presumably haploid, nuclei are formed. Anucleate, closed prospores do not form mature ascospore walls (see Fig. 8).

Diploids homozygous for the mutation *spo3-1* exhibit abnormal cytological behavior at stage 7. Nuclear budding lags with respect to prospore wall growth and closure. Ascogenous cells containing anucleate and partially anucleate, immature, closed prospore walls and a nucleus arrested in early nuclear budding are observed.

The behavior of *spo2* and *spo3* diploids indicates that prospore wall formation, growth, and closure can occur in the absence of normal nuclear budding and do not depend on the integrity of the nuclear membrane. The failure of anucleate prospore walls to develop further with respect to insertion of ascospore wall material between the unit membranes suggests that this developmental stage is directed by the haploid nucleus usually present within the prospore. Ascogenous cells containing both immature anucleate as well as mature nucleated spores have been observed in single cells of *spo3*. Ascospore wall formation is thus an autonomous function of nucleated prospores.

7. TEMPERATURE-SHIFT STUDIES

The conditional nature of temperature-sensitive mutations can be exploited to detect the time of involvement (synthesis and function) of their

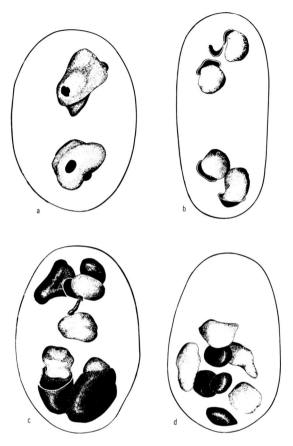

FIG. 8.   Fine-structural defects in 79-1B (*spo2-1*) sporulated at 34°C. (a) The division products of meiosis I have separated, and each half has a meiosis II spindle capped by spindle pole bodies (black in drawing); in normal sporulation the two halves stay in open connection until late in ascospore formation. (b) At meiosis II the division products also separate precociously, so that there are four presumptive ascospore nuclei in the ascus. Each nucleus has a developing prospore wall attached to it at the site of the spindle pole body (close together are nonsister nuclei). (c) The coordination between nuclear movement and ascospore wall formation appears to be disturbed to the extent that normal capsulation does not take place. (d) At the completion of sporulation, there are four free nuclei (dotted) and four anucleate spores. Sometimes somewhat better ascospores are formed, occasionally some normal-looking spores. (Moens *et al.*, 1974.)

gene products in sporulation (Esposito *et al.*, 1970). To obtain this information two parameters of *spo* mutants have been measured: t_i, the time during sporulation at the restrictive temperature at which a mutant becomes irreversibly affected and cannot sporulate when transferred to the permissive temperature; and t_e, the time at which a mutant incubated at the permissive temperature can sporulate when shifted to the restrictive temperature. When t_i precedes t_e, the time interval they span is a minimum estimate of the time from synthesis to function of the gene product. If t_i follows t_e, this indicates that mutant cells become irreversibly affected by the presence of the mutant gene product after the product functions in the wild type.

Temperature-shift experiments with three *spo* mutants (*spo1*, *spo2*, and *spo3*) indicate that the functions defective in the mutants are not continuously required during sporulation. The t_i and t_e points of *spo1-1* diploids straddle the cytological stage at which development terminates, suggesting that the function may be required beyond the point of termination. The t_i and t_e points of *spo2-1* and *spo3-1* diploids precede the cytological stages at which development becomes abnormal. This indicates that the gene products that cause abnormal development function before these stages (see Fig. 6).

8. PHENOTYPIC CHARACTERIZATION UNDER SEMIPERMISSIVE CONDITIONS

The properties of ascospores produced by temperature-sensitive mutants at semipermissive temperatures can be employed to further characterize the functions defective in mutants. Moens *et al.* (1974) noted that diploids of *spo3-1* yield many two-spored asci at a permissive temperature and that these ascogenous cells contain nuclear material which was not incorporated into the two spores formed. The question thus arose whether the spores formed contained a complete haploid genome and represented the products of normal meiosis. Ascospores present in two-spored asci formed at a semipermissive temperature (30°C) by *spo3-1* diploids were examined for viability, genetic recombination, and segregation of genetic markers present in the heterozygous condition (Esposito *et al.*, 1974). Viability, recombination, and segregation were normal in these spores, suggesting that they represent the products of normal meiosis. Segregation of a genetic marker linked to a centromere indicated that the two spores represent near-random inclusion of the four haploid genomes present in the ascogenous cell in prospores that mature. These results, together with the properties of *spo3-1* diploids at the restrictive temperature, indicate that the function defective in *spo3* mutants is specifically required for the integration of prospore wall development and nuclear budding.

B. Mutants of *Schiz. pombe*

1. CYTOLOGY

Mutants isolated in *Schiz. pombe* (Bresch *et al.*, 1968) have been examined at the level of light microscopy following Giemsa staining to determine the stage at which sporulation is disrupted. The results are summarized in Fig. 9.

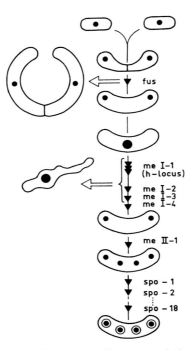

FIG. 9.   Cytological properties of mutations affecting sporulation detected in *Schiz. pombe*. (Bresch *et al.*, 1968.)

Mutants of the *fus1* locus are asporogenous, because fusion of haploid cells does not proceed to completion. Copulating pairs remain in end-to-end association, but cell wall dissolution fails to occur. The *fus*− cells are arrested in zygote formation but are not yet committed to the developmental pathway leading to spore formation. They can resume vegetative growth when transferred to fresh medium and give rise to haploid colonies (Egel, 1973).

Giemsa staining of the nuclei of sporulation-deficient mutants distinguishes three additional cytological stages of arrest. Mutants of four loci, *meiI-1*, *-2*, *-3*, and *-4*, terminate meiosis as mononucleate zygotic cells. These mutants have not been examined to determine whether termination of meiosis precedes or follows premeiotic DNA synthesis. Mutants of a single locus, *meiII-1*, undergo gametic nuclear fusion, complete the first meiotic division, and do not proceed further. They are recognized as binucleate cells following Giemsa staining. Mutants of 18 loci (designated *spo1* through *18*) terminate sporulative development at completion of the meiotic divisions. Mutants of these loci give rise to asporogenous clones containing tetranucleate ascogenous cells.

2. COMMITMENT TO SPORULATION

Mutants of the four loci that terminate sporulation after fusion of gametic nuclei and before completion of the first meiotic division have been examined to determine whether the zygotes formed can undergo diploid mitotic division if returned to fresh growth medium (Egel, 1973). Mutants of three loci, *meiI-1*, *-2*, and *-3*, are capable of diploid mitotic cell division. Zygotes homozygous for the *meiI-4* mutation are incapable of mitosis. A double mutant, *meiI-2*, *meiI-4*, is capable of diploid mitotic cell division. Since the double mutant displays the phenotype of *meiI-2* rather than that of *meiI-4*, we may conclude that *meiI-4* is defective in a function not required for diploid mitotic cell division. This observation is consistent with the conclusion that *meiI-4* zygotes do not divide mitotically because they have become committed to sporulation.

V. Summary

In the preceding discussion methods for isolation and properties of three classes of yeast mutants defective in sporulation were described: (1) asporogenous mutants capable of vegetative growth, (2) mutants defective in commitment to intragenic recombination during sporulation but capable of mitotic intragenic recombination, and (3) mutants produced by mutations affecting the control of sporulative ability by mating-type alleles.

Mutants of the first type reveal which functions are indispensable for completion of sporulation. Their study has provided information regarding the stages at which sporulation terminates or becomes aberrant when an event required for completion of development fails to occur normally. The pleiotropic effects of these mutations indicate which developmental events are coordinately controlled and those that are independently regulated. Mutants of the second type are selected for failure to exhibit an early sporulation event. Mutants in this class analyzed thus far do not undergo premeiotic DNA synthesis. These results suggest that the capacity for premeiotic DNA replication and expression of functions that enhance recombination are jointly controlled. Mutants in the third category have been most recently described and are produced by the action of dominant mutations which allow expression of functions required for sporulation in cells that do not normally sporulate.

While several features of the genetic control of sporulation can be studied with the above types of mutants, it seems appropriate in a discussion of the genetic control of yeast meiosis to indicate that systems selective for mutations specifically affecting classic reciprocal recombination and meiotic chromosome disjunction have not been reported. These events may be dispensable for spore formation, but their genetic control is of obvious importance to an understanding of meiosis. Hopefully, the versatility of yeast genetic systems will be exploited to recover such gene mutations. Mutants of this type and variants already isolated should allow a thorough characterization of the genes controlling sporulation events.

ACKNOWLEDGMENT

Preparation of this article and research in this laboratory was supported by NSF Grant GB27688, Cancer Center Grant CCRC-III-B-3, and the Wallace C. and Clara Abbott Memorial Fund from the University of Chicago.

REFERENCES

Beckett, A., Illingworth, R. F., and Rose, A. H. (1973). *J. Bacteriol.* **113**, 1054.
Bresch, C., Muller, G., and Egel, R. (1968). *Mol. Gen. Genet.* **102**, 301.
Chen, A. W., and Miller, J. J. (1968). *Can. J. Microbiol.* **14**, 957.
Cox, B. S., and Parry, J. M. (1968). *Mutat. Res.* **6**, 37.
Croes, A. F. (1967). *Planta* **76**, 209.
Egel, R. (1973). *Mol. Gen. Genet.* **121**, 277.
Engels, F., and Croes, A. (1968). *Chromosoma* **25**, 104.
Ephrussi, B., and Hottinguer, H. (1951). *Cold Spring Harbor Symp. Quant. Biol.* **16**, 75.
Esposito, M. S., and Esposito, R. E. (1968). *Genetics* **60**, 176.
Esposito, M. S., and Esposito, R. E. (1969). *Genetics* **61**, 79.
Esposito, M. S., and Esposito, R. E. (1973). *Colloq. Int., Cent. Nat. Rech. Sci.* **227**, 135–139.
Esposito, M. S., and Esposito, R. E. (1974). *Genetics* **78**, 215.
Esposito, M. S., Esposito, R. E., Arnaud, M., and Halvorson, H. O. (1969). *J. Bacteriol.* **100**, 180.
Esposito, M. S., Esposito, R. E., Arnaud, M., and Halvorson, H. O. (1970). *J. Bacteriol.* **104**, 202.
Esposito, M. S., Esposito, R. E., and Moens, P. B. (1974). *Mol. Gen. Genet.* (in press).
Esposito, R. E., and Esposito, M. S. (1974). *Proc. Nat. Acad. Sci. U.S.* **71**, 3172.
Esposito, R. E., Frink, N., Bernstein, P., and Esposito, M. S. (1972). *Mol. Gen. Genet.* **114**, 241.
Fast, D. (1973). *J. Bacteriol.* **116**, 925.
Fowell, R. R. (1969). *In* "The Yeasts" (A. H. Rose and J. S. Harrison, eds.), Vol. 1, pp. 303–383. Academic Press, New York.
Grewal, N. S., and Miller, J. J. (1972). *Can. J. Microbiol.* **18**, 1897.
Guth, E., Hashimoto, T., and Conti, S. (1972). *J. Bacteriol.* **109**, 869.
Haber, J. E., and Halvorson, H. O. (1972). *Curr. Top. Develop. Biol.* **7**, 61–82.
Hall, B., and Hopper, A. (1973). *Colloq. Int. Cent. Nat. Rech. Sci.* **227**, 141–144.
Hawthorne, D. C. (1963). *Proc. Int. Cong. Genet., 11th, 1963* Vol. 1, p. 34.
Hopper, A. K., Magee, P. T., Welch, S. K., Friedman, M., and Hall, B. D. (1974). *J. Bacteriol.* **119**, 619.
Hottinguer de Margerie, H. (1967). Ph.D. Thesis, University of Paris.
Illingworth, R. F., Rose, A. H., and Beckett, A. J. (1973). *J. Bacteriol.* **113**, 373.
Kane, S. M., and Roth, R. (1974). *J. Bacteriol.* **118**, 8.

Keith, A. D., Resnick, M., and Haley, A. B. (1969). *J. Bacteriol.* **98**, 415.
Leupold, U. (1950). *C.R. Trav. Lab. Carlsberg, Ser. Physiol.* **24**, 381.
Leupold, U. (1955). *Schweiz. Z. Allg. Pathol. Bakteriol.* **18**, 1141.
Leupold, U. (1958). *Cold Spring Harbor Symp. Quant. Biol.* **23**, 161.
Leupold, U. (1970). *In* "Methods in Cell Physiology" (D. M. Prescott, ed.), Vol. 4, pp. 169–175. Academic Press, New York.
Lynn, R. R., and Magee, P. T. (1970). *J. Cell Biol.* **44**, 688.
Magee, P. T., and Hopper, A. (1974). *J. Bacteriol* **119**, 952.
Moens, P. B. (1971). *Can. J. Microbiol.* **17**, 507.
Moens, P. B. (1974). *J. Cell Sci.* (in press).
Moens, P. B., and Rapport, E. (1971). *J. Cell Biol.* **50**, 344.
Moens, P. B., Esposito, R. E., and Esposito, M. S. (1974). *Exp. Cell. Res.* **83**, 166.
Mori, S., and Nakai, S. (1969). *Proc. Int. Congr. Genet., 12th, 1968* Vol. 1, p. 80.
Mortimer, R. K., and Hawthorne, D. C. (1969). *In* "The Yeasts" (A. H. Rose and J. S. Harrison, eds.), Vol. 1, pp. 386–460. Academic Press, New York.
Moustacchi, E., Hottinguer-de-Margerie, H., and Fabre, F. (1967). *Genetics* **57**, 909.
Ogur, M., Roshanmanesh, A., and Ogur, S. (1965). *Science* **147**, 1590.
Peterson, J., Gray, R., and Ris, H. (1972). *J. Cell Biol.* **53**, 837.
Pinon, R., Salts, Y., and Simchen, G. (1974). *Exp. Cell Res.* **83**, 231.
Pomper, S., and Burkholder, P. R. (1949). *Proc. Nat. Acad. Sci. U.S.* **35**, 456.
Resnick, M. (1969). *Genetics* **63**, 519.
Rodarte-Ramon, U. S. (1970). Ph.D. Thesis, University of California, Berkeley.
Rodarte-Ramon, U. S. (1972). *Radiat. Res.* **49**, 148.
Rodarte-Ramon, U. S., and Mortimer, R. K. (1972). *Radiat. Res.* **49**, 133.
Roman, H., Phillips, M., and Sands, S. (1955). *Genetics* **50**, 546.
Roth, R. (1970). *J. Bacteriol.* **101**, 53.
Roth, R. (1972). *Abstr., Annu. Meet., Amer. Soc. Microbiol.* p. 73.
Roth, R. (1973). *Proc. Nat. Acad. Sci. U.S.* **70**, 3087.
Roth, R., and Fogel, S. (1971). *Mol. Gen. Genet.* **112**, 295.
Roth, R., and Halvorson, H. O. (1969). *J. Bacteriol.* **98**, 831.
Roth, R., and Lusnak, K. (1970). *Science* **168**, 493.
Rothstein, R. (1973). *Genetics* **74**, Suppl., s234.
Shaffer, B., Brearby, I., Littlewood, R., and Fink, G. (1971). *Genetics* **67**, 483.
Sherman, F., and Roman, H. (1963). *Genetics* **48**, 255.
Sherman, F., and Slonimski, P. (1964). *Biochim. Biophys. Acta* **90**, 1.
Simchen, G. (1974). *Genetics* **76**, 745.
Simchen, G., Pinon, R., and Salts, Y. (1972). *Exp. Cell Res.* **75**, 207.
Tingle, M., Klar Singh, A. J., Henry, S. A., and Halvorson, H. O. (1973). *Symp. Soc. Gen. Microbiol.* **23**, 209–243.
Winge, O., and Roberts, C. (1949). *C.R. Trav. Lab. Carlsberg, Ser. Physiol.* **24**, 341.

Subject Index

Glusulase, 147, 171, 198
Glusulase method, 148–149
Glusulase treatment, 150–151
Growth of yeast cultures, methods for
  monitoring, 131
Growth rate, determination of, 122
Gurr's improved Giemsa solution, 18

H

Hansenula capsulata, 173
Hansenula wingei, 73, 79, 82
  agglutination of, 90
Haupt's adhesive, 13
Hawthorne's tetraploid, 5
Heat shocks of yeast, 217
Helicase, 147, 171
Helix pomatia, 60, 169
Helly's fixative, 7
  composition of, 9
Hemocytometer counts of yeast, 153
Hexadecane media, 107
Histidine biosynthesis in yeast, 267
Homothallic strains of yeast, 58
  procedures for, 304
Hughes Hilo flow pump, 75

I

Invertase, 175, 177–178

K

Kluyveromyces fragilis, 206
Kluyveromyces lactis, 221

L

Lallemand baking yeast, 5, 9
Light microscopy of yeasts, 1
Linkage groups of yeast, 222
Lipase, 174
Lipid, staining of in yeast, 7
Lipid synthesis in yeast, 58
Lipomyces, 19
Liquid media, yeasts growing in, 13
Lomofungin, 57
Ludox, 208
Lytic enzymes, 174

M

Macromolecular synthesis, measurement of,
  51

Mannan, 37
Mannanase, 174
Mapping studies of yeast, 222
Mapping techniques in yeast, 223
Mating cultures of yeast, delay of mitosis
  in, 84
Mating in yeasts
  procedures of synchronous, 80
  sequence of events during, 77
  synchronous, 71
Mating medium, 82
Mating methods for yeast, synchronous,
  82
Mating pheromone, 216
Mating procedures for yeast, 73
Mating process in yeast, 75
Mating reaction in yeast, kinetics of,
  75–76
Mating type in yeast, 89
Meiosis sporulation in yeasts, 315
Meiotic chromosomes, counting, 14
Metabolic pathways in yeast, 239
Methionine biosynthesis in yeast, 264
  regulation of, 265
Microburner, 196
Microculture, 6
Microforge, 196
Micromanipulator(s)
  Cailloux, 193
  Fonbrune Series A, 192
  in yeast studies, 189
  Lawrence Precision Machine, 194
  Leitz lever-activated, 193
  list of commercially available, 191
  Singer MKIII, 193
  types of, 190
  Zeiss sliding, 193
Micromonospora RA, 174
Micropolyspora, 174
Mitochondria, staining of, 7
Mitotic segregation in yeast, 229
mRNA of yeast, 56
Mushroom extract, 198
Mutant phenotypes in yeast, analysis of,
  314
Mutants incorporating dTMP in yeast,
  isolation of, 297
Mutants of *Saccharomyces cerevisiae*, 314
  characterization of, 287
  conditional, 308
  isolation of, 287

pH characteristic of, 55
RNA synthesized during, 56
study of, 50
total RNA during, 55
Sporulation media for yeast, 48, 52
Sporulation-specific events in yeast, 59
Staining of unfixed yeasts, 7
Staining of yeast with silver, 35
Streptomyces albidoflavus, 174
Streptomyces GM, 174
Streptomyces griseus, protease fraction of, 60
Streptomyces WL-6, 174
Synchronization of yeast, 90
Synchronous culture(s) of yeast, 148, 205
Synthetic glucose medium, 120
Synthetic medium for yeast, 115

T

Technicon counter, 158
Tetracycline, 61
Tetrad analysis of yeast, 223
Thermomyces lanuginosus, 174
Thymidylate biosynthesis in yeast, 289
Thymidylate mutants in yeast, isolation procedure for, 289
Thiolutin, 57
Torulopsis, 173
Tritium labeling of yeast, 280, 282
Tritium suicide, 274, 280

Tritium suicide selection, 279
efficiency of, 283
Turbidity of yeast cultures
continuous monitoring of, 163
measuring, 159

U

Urograffin, 208
Ustilago violacea, 3

V

Velocity separation of yeast, 206
in zonal rotors, 208
Vogel's medium, 173, 179

W

Wiskerhamia fluorescens, 6–7, 9

Y

Yeast genetics, laboratory manual for, 274
Yeast Genetics Stock Center, 49, 274

Z

Zygote formation in yeast
medium for, 92
synchronous, 89
timing of, 83
Zygote isolation in yeast, 78, 80
Zygote purification in yeast, 83
Zygotes in yeast, synchronous mass production of, 91
Zymolase, 173

CONTENTS OF PREVIOUS VOLUMES

Volume I

Volume II

Volume III

Volume IV

Volume V

Volume VI

Volume VII

Volume VIII

Volume IX

Volume X

A 5
B 6
C 7
D 8
E 9
F 0
G 1
H 2
I 3
J 4